Notes on Numerical Fluid Mechanics and Multidisciplinary Design

Volume 152

Notes on Numerical Fluid Mechanics and Multidisciplinary Design publishes state-of-art methods (including high performance methods) for numerical fluid mechanics, numerical simulation and multidisciplinary design optimization. The series includes proceedings of specialized conferences and workshops, as well as relevant project reports and monographs.

Indexed by SCOPUS, zbMATH, SCImago.

All books published in the series are submitted for consideration in Web of Science.

More information about this series at http://www.springer.com/series/4629

Rudibert King · Dieter Peitsch
Editors

Active Flow and Combustion Control 2021

Papers Contributed to the Conference "Active Flow and Combustion Control 2021", September 28–29, 2021, Berlin, Germany

Springer

Editors
Rudibert King
Department of Measurement and Control
Institute of Chemical and Process Engineering
Technische Universität Berlin
Berlin, Germany

Dieter Peitsch
Department of Aero Engines
Institute of Aeronautics and Astronautics
Technische Universität Berlin
Berlin, Germany

ISSN 1612-2909 ISSN 1860-0824 (electronic)
Notes on Numerical Fluid Mechanics and Multidisciplinary Design
ISBN 978-3-030-90726-6 ISBN 978-3-030-90727-3 (eBook)
https://doi.org/10.1007/978-3-030-90727-3

This Springer imprint is published by the registered company Springer Nature Switzerland AG
The registered company address is: Gewerbestrasse 11, 6330 Cham, Switzerland

Preface

The efficient conversion of energy to serve societal needs for mobility and improved lifestyle enjoys continuous attention since decades and especially in recent years by initiatives like 'Fridays for Future' and others. However, striving for more efficiency poses challenges for conventional energy conversion systems such as gas turbines. Thermodynamically, conventional configurations are limited in terms of further efficiency improvements by maximum temperature allowances to avoid unwanted emissions and ensure sufficient lifetime.

Since many years, energy research in Berlin has focused on the improvement of both the efficiency and effectivity of technical systems. Active flow control (AFC) demonstrated a strong potential to save costs for land, sea, and air vehicles already by reducing drag and increasing lift. The successful collaborative research center CRC 557 Control of turbulent shear flows at Technische Universität Berlin, funded by the Deutsche Forschungsgemeinschaft (DFG), led to a strong recognition of this approach and resulted in a series of conferences, where the findings were discussed with experts in the field. In parallel, pressure gaining combustion (PGC) promised a significant chance to increase thermal efficiency of gas turbines. However, these approaches of more efficient thermodynamic cycles introduce the challenge of unavoidable unsteadiness for adjacent components. In order to increase the operability range of combustors, compressors, and turbines, the effect of highly dynamic processes has thus moved into the focus of research activities.

In order to wide up the view onto the potential of PGC, the CRC 1029 substantial efficiency increase in gas turbines through direct use of coupled unsteady combustion and flow dynamics was proposed, which was granted by DFG in 2012 for a first 4-year period and continued for a second funding period in 2016. The objective is to achieve a higher thermodynamic efficiency of gas turbines, while keeping the additional control and aerodynamic challenges within bounds. Multidisciplinarity has been proven to be required definitely, which is expressed in this volume by the combined authorships from various disciplines. Thermodynamic understanding of energy conversion together with experimental and numerical fluid mechanics needs to be combined with mathematics and control theory in order to create the overall picture. The reduction in amplitude of pressure waves traveling within the system was achieved by the introduction of concepts like the shockless explosion combustion (SEC). Included in this volume are as well results for rotating detonation combustion (RDC), which in this context appears to be a promising concept for the future as well.

A series of successful conferences was set up in the past to convey the results of this detailed research, bringing together recognized experts in these various fields. Starting in 2006 and 2010 with 'Active Flow Control I and II', an interdisciplinary discussion was initiated in the community. This approach was followed later by adding the specific challenges of controlling unsteady combustion in the conferences 'Active Flow and Combustion Control' in 2014 and 2018. The present volume contains most of the presentations given at 'Active Flow and Combustion Control 2021'. The successful format of the preceding conferences was unchanged with invited lectures and single-track sessions only. Not all presenters could prepare a manuscript for this volume, but it still presents a well-balanced combination of theoretical, numerical, and experimental state-of-the-art results of active flow and combustion control.

The papers presented within this volume deal with all different aspects of flow and combustion control. They show the high potential of active measures within the flow and combustion regime, which is partially already demonstrated and partially predicted, when it comes to an improved operation of gas turbines overall and their individual components. While the practical demonstration of the overall concept in an interacting environment of compressor, combustor, and turbine is still pending, the basic understanding of the relevant parameters and influence coefficients is available by now.

All papers in this volume have been subjected to an international review process, supported by an International Program Committee. We would like to express our sincere gratitude to all involved reviewers: B. Atakan, H.-J. Bauer, A. Bauknecht, M. Bellenoue, M. Bohon, B. Boust, J. Braun, T. Breiten, F. di Mare, N. Djordjevic, M. Eck, D. Greenblatt, T. Grönstedt, F. Haucke, M. Heinkenschloss, E. Kaiser, R. Klein, M. Lemke, R. Liebich, R. Mailach, J. Moeck, S. Müller, R. Niehuis, N. Nikiforakis, B. Noack, K. Oberleithner, C.O. Paschereit, H. Pitsch, R. Radespiel, J. Reiss, M. Samimy, H.-P. Schiffer, P. Stathopoulos, J. Wild, D. Williams.

The research activities within CRC 1029 as well as the performed conferences were funded by the German Research Foundation (DFG). This funding as well as the continuous support by Dr. Michael Lentze and Dr. Matthias Beilein and their teams from DFG is gratefully acknowledged.

The members of CRC 1029 are indebted to their respective hosting organizations, TU Berlin and FU Berlin, for the continuous support, and to Springer and the editor of the series Notes on Numerical Fluid Mechanics and Multidisciplinary Design, W. Schröder, for handling this volume.

Last but not least, we are indebted to Mario Eck and especially Steffi Stehr for their irreplaceable support in organizing and administrating CRC 1029, organizing the conference and compiling this volume.

September 2021

Dieter Peitsch
Rudibert King
(Chairmen of AFCC 2021 and CRC 1029)

Contents

Part I: Constant Volume Combustion and Combustion Control

Pressure Gain and Specific Impulse Measurements in a Constant-Volume Combustor Coupled to an Exhaust Plenum

Bastien Boust[1](✉), Marc Bellenoue[1], and Quentin Michalski[2]

[1] PPRIME Institute, CNRS – ISAE-ENSMA – Poitiers University, 1 Avenue Clément Ader, 86961 Chasseneuil, France
{bastien.boust,marc.bellenoue}@ensma.fr
[2] RMIT University, Melbourne, Australia
quentin.michalski@rmit.edu.au

Abstract. In the framework of air-breathing propulsion, the thermodynamic cycle of turbomachines is likely to be adapted to pressure-gain combustion (PGC) technology, which raises integration challenges. In this study, a constant-volume combustion chamber, fed with air and isooctane, was experimentally coupled to a nozzle assembly featuring a plenum with variable volume and throat diameter. The spark-ignited, direct-fueled combustion chamber allowed for time-resolved measurements of combustion pressure and axial thrust, as well as direct imaging of the reacting flow. The main parameters that drive the cyclic operation of the facility – regarding combustion dynamics and thrust generation – have been recorded as a function of the exhaust geometry, including the air stagnation pressure, the air/fuel equivalence ratio, and the cycle frequency. The pressure gain measured in the exhaust plenum proves to be a relevant parameter of the PGC facility, by representing the increase in stagnation pressure upstream of a turbine in turbomachine applications. The cycle frequency directly drives the pressure gain that reaches up to 31% in our constant-volume combustion facility, which yields an outstanding increase in specific impulse by up to 23% compared to conventional constant-pressure combustion.

Keywords: Pressure-gain combustion · Engine performance · Airbreathing propulsion

1 Introduction

The thermodynamic cycles of current turbomachines are based on a constant-pressure heat release (Joule-Brayton cycle). Thermodynamic cycles in which the pressure rises through the combustion phase (Humphrey-Atkinson) hold a theoretical potential for substantial efficiency gain. Many solutions on how to implement Pressure-Gain Combustion (PGC) exist in the literature. These include rotating or pulsed detonation [1], the auto-ignition of a stratified mixture [2], a confined deflagration in a piston engine [3] or piston-less chambers [4–9].

R. King and D. Peitsch (Eds.): AFCC 2021, NNFM 152, pp. 3–15, 2022.
https://doi.org/10.1007/978-3-030-90727-3_1

The latter was investigated in recent works both experimentally [5] and numerically [6, 7] by LES on a combustion chamber in which the inlet and exhaust systems are rotary valves. The combustion chamber is operated with indirect injection where liquid isooctane is injected upstream of the intake valves in a carburation chamber. The authors addressed the cycle stability including the analysis of both the ignition phenomena and the cycle-to-cycle variation of the flame propagation. Studies on this prototype did not address the effect of the exhaust plenum pressure as it remained equal to the atmospheric pressure. However, downstream pressure is relevant for further integration in a turbomachine, as the turbine operates downstream of the combustion chamber and extracts work from exhaust gases while restraining the flow. The turbine presence could therefore induce the buildup of higher backpressure in the exhaust plenum. Moreover, the unsteadiness of the pulsating exhaust flow can be detrimental to turbine efficiency [10]. To overcome this, the flow unsteadiness should be levelled by introducing either an exhaust plenum with a volume adapted to the combustion chamber or a pulse converter manifold. In a previous study conducted on another CVC device [11], the authors investigated the influence of fixed back pressure on the overall combustion process. Higher back pressure, without or with limited scavenging of the chamber, led to a higher residual burnt gas dilution, which strongly slowed down the combustion and reduced the combustion peak pressure.

Thrust measurement has been previously suggested to measure the performance of such devices [12]. Thus, this article aims at investigating the effect of an exhaust plenum of adjustable volume and throat on the combustion and propulsive properties of a CVC combustor operated in a cyclic, pulsed regime. Particularly, the objectives of this study are to characterize the pressure gain of the facility and its specific impulse in the corresponding conditions. The influence of the main operating parameters will be sought, such as plenum volume, throat diameter, cycle frequency, equivalence ratio and air stagnation pressure.

2 Experimental Setup and Diagnostics

The device investigated in this study is a piston-less constant-volume combustion chamber, as thoroughly described in a previous study [5]. Its basic principle is the opening and closing of intake and exhaust ports by rotating valves that generate two cycles per revolution, so that a single cycle lasts 180° of the valve angle θ. The combustor (see Fig. 1a) is fed with hot, pressurized air generated by an electrical heater (100 kW, up to 200 °C), a dome pressure regulator (up to 0.5 MPa) and a Coriolis mass flow meter (Endress-Hauser 80F40, 0.5% uncertainty). The diameter (80 mm) and volume (65 L) of the air feedline are such that the intake pressure (P_{in}) measured in the intake plenum remains independent of the combustion chamber pressure fluctuations. The pressure sensors used in the intake plenum (P_{in}), combustion chamber (P_{cc}) and exhaust plenum (P_{ex}) are absolute water-cooled piezoresistive sensors, respectively Kistler 4007, 4011 and 4049 (0.2% uncertainty).

The nozzle diameter is selected using two nozzles based on the ISO9300 design with respective throat diameters of 10 and 20 mm, a toroidal convergent and a conical divergent of expansion ratio 1.4 (see Fig. 1b). Meanwhile the combustion chamber volume is fixed (V_{cc} = 0.65 L), the exhaust plenum volume V_{ex} can be chosen by adding spacers upstream of the nozzle. This study involves an exhaust plenum which volume V_{ex} is either: equivalent to the combustion chamber volume V_{cc}, equal to half of it (0.3 L), or higher (1 L). The device assembly (chamber, intake and exhaust plenums) is heated by maintaining a steady airflow at T_{in} = 180 °C. The chamber wall temperature is measured by a K thermocouple, T_w = 100–130 °C. Each test features the cyclic operation of the combustor during up to 4 s, to obtain more than 50 combustion cycles per test.

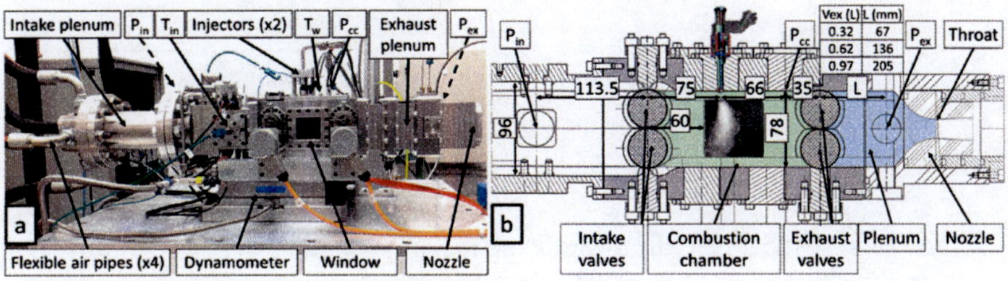

Fig. 1. a) Constant-volume combustion facility, b) Cross-section and dimensions.

The combustion chamber is fueled by direct liquid isooctane injection, with a stoichiometric air/fuel dilution of D_{st} = 15.0. The injection system consists of a 3.6 L vessel filled with isooctane, pressurized by 10 MPa nitrogen, and connected to two GDI injectors (Bosch HDEV 5.2) flush-mounted to the chamber upper wall (see Fig. 1b). The two injectors' total fuel mass flowrate, obtained from the previous measurement [13], is 20 g/s. The overall equivalence ratio in the combustion chamber, denoted OER, is derived from the mass of fuel injected during the cycle, m_f, and from the density of hot compressed air, ρ_{air}:

$$OER = D_{st} \frac{m_f}{\rho_{air} V_{cc}} \tag{1}$$

During firing tests, the spontaneous emission of the flame is recorded by a high-speed color camera (Phantom V310), while the fuel plume is observed by Mie scattering using a high-power halogen lamp (see Fig. 2b). In the visualization window, the fuel plume is roughly 26° wide with a 12° tilt angle (see Fig. 2a). In the orthogonal direction, the plume spans over 42°.

The spray plume penetrates up to 110 mm in atmospheric conditions, so part of the fuel is likely to reach the bottom of the chamber. However, during firing tests the wall-wetting effect is certainly limited by the elevated chamber pressure (0.3 MPa) and hot chamber wall (100–130 °C).

Ignition is performed at the chamber's rear wall, in the visualization area (see Fig. 2b), with a standard non-resistive sparkplug (1 mm gap) 2.5 mm off the wall. A top-plug automobile ignitor placed directly on the sparkplug is used to generate the spark (100 mJ deposit during approximately 2 ms).

The axial thrust of the device, F_x, is measured by a piezoelectric dynamometer (Kistler 9255C, bandwidth 2.2 kHz) that was carefully scaled in-situ. The efforts exerted

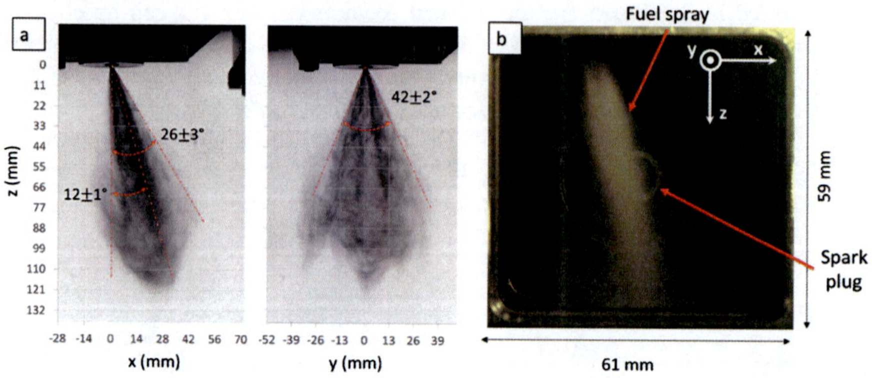

Fig. 2. a) Backlight imaging of a single spray in quiescent air, b) Mie scattering of the two sprays in real flow conditions.

by the flexible hoses that feed air towards the intake plenum have been carefully quantified: their contribution (–3.8 ± 0.25 daN/MPa) is fluctuating cyclically, and remains small compared to the magnitude of thrust obtained in this study. As the working fluid always exerts a thrust on the device, even when inert air flows through the nozzle, removing the thrust contribution of inert cycles recorded before a reacting test is necessary. Thus, in this study, the axial thrust measured during a combustion test is corrected as follows:

$$F_x(t) = F_x(t)|_{combustion} - F_x(t)|_{inert,ensemble-average} \tag{2}$$

A 0D analysis developed, deployed, and validated on another device [13] is used to extract further information, such as the residual burnt gas dilution or the initial temperature, from the pressure traces measured in the chamber.

In summary, this facility offers a fully instrumented, cyclic CVC chamber, with a versatile exhaust assembly featuring variable plenum volume and nozzle diameter. It allows physical analyses of such a PGC device's combustion and propulsion behavior in its operation range, based on time-resolved pressure and thrust measurements, making it possible to derive the pressure gain and specific impulse. In the following, the facility will be implemented using three plenum volumes (0.32, 0.62 and 0.97 L), two nozzle diameters (10 and 20 mm), and its operating conditions will be controlled in terms of air pressure (0.25–4.0 MPa), overall equivalence ratio (0.5–1.5), and cycle frequency (16–50 Hz).

3 Results and Discussion

3.1 Effect of a Nozzled Plenum on the Unsteady Combustor Dynamics

The presence of a nozzle downstream of the combustor is expected to have multiple effects on its operation: it may choke the exhaust flow and subsequently increase the amount of residual burned gas (RBG) in the combustion chamber. In this section, three exhaust nozzle configurations are explored: a 10 mm diameter nozzle, a 20 mm diameter nozzle, and without a nozzle. In this section, the facility is operated at 0.3 MPa, stoichiometry, with a cycle frequency of 16 Hz (hence a 62.5 ms period) to favor air-fuel mixing and ensure a 100% ignition rate. For simplicity, cyclic measurements are plotted versus time t.

Reference Case with an Open Plenum. Without a nozzle, the flow is choked in most phases due to the elevated initial pressure (0.3 MPa); therefore, the boundary conditions are very repeatable, namely the intake plenum pressure P_{in} and the opened exhaust plenum pressure P_{ex}. The intake plenum exhibits a strong but stable pressure fluctuation of ±0.028 MPa (See Fig. 3A). The ignition timing is set to 44.9 ms, which is 12.7 ms after the end of injection. It is adjusted so that most of the combustion pressure peaks occur near the opening of the exhaust valves. The overall equivalence ratio computed with the 0D model is 1.0, and the associated dilution of 1.6%. These values are obtained on the assumption of perfect and continuous gas mixing, especially during the scavenging phase. The dilution is computed based on the CO_2 mass fraction ratio between that at the end of combustion and that at the ignition timing of the following cycle. On the flame pictures (see Fig. 3C) corresponding to the events of a firing cycle (see Fig. 3B), we observe a few rich and sooty areas in the flame, indicated by yellow spots. This implies that the equivalence ratio is locally fuel-lean or stoichiometric, and the mixture is relatively homogeneous. The maximum combustion pressure is of 1.01 ± 0.06 MPa in average and the combustion duration is of $t_{10\text{–}90} = 2.8 \pm 0.3$ ms (from 10% to 90% of the pressure increase). This duration represents well the free propagation of the flame, starting from an ignition kernel to the wall. Both peak pressure and combustion duration are similar to those found previously by the same device operating with a port-fuel injection [5].

Comparison to the Case of A Plenum with Nozzle. In the following, the three conditions investigated refer to the open plenum (without a nozzle), 10 mm and 20 mm diameter exhaust nozzles. The last two configurations are set up with an exhaust plenum of 0.62 L equivalent to the combustion chamber volume (see Fig. 4).

Operating the device with an exhaust nozzle certainly influences the composition of the fresh charge in the combustion chamber. For clarity, pressure traces are normalized by the stagnation pressure recorded in the 65 L reservoir feeding the device with hot air, P_{res}. The average maximum pressure ratio decreases from 3.3 in the open plenum case to 2.8 with a 20 mm nozzle, and 2.3 with a 10 mm nozzle. Meanwhile, the combustion duration $t_{10\text{–}90}$ increases from 2.8 ± 0.3 ms in the open plenum, to 4.0 ± 0.3 ms with a 20 mm nozzle, and 4.9 ± 0.3 ms with a 10 mm nozzle. In the case of the 10 mm nozzle, the nozzle choking prevents the complete scavenging of the plenum, since P_{ex} no longer reaches ambient pressure between two exhaust phases, t = 25–50 ms (see Fig. 4); in turn, this phenomenon also prevents the complete scavenging of the combustion chamber.

Therefore, it is likely that the open plenum corresponds to a well-scavenged, fresh air/fuel charge, whereas the 20 mm and 10 mm nozzles generate an increasing dilution of the fresh gases by RBG, thus decreasing the combustion peak ratio and increasing the combustion duration, as already evidenced in former CVC applications [13]. The

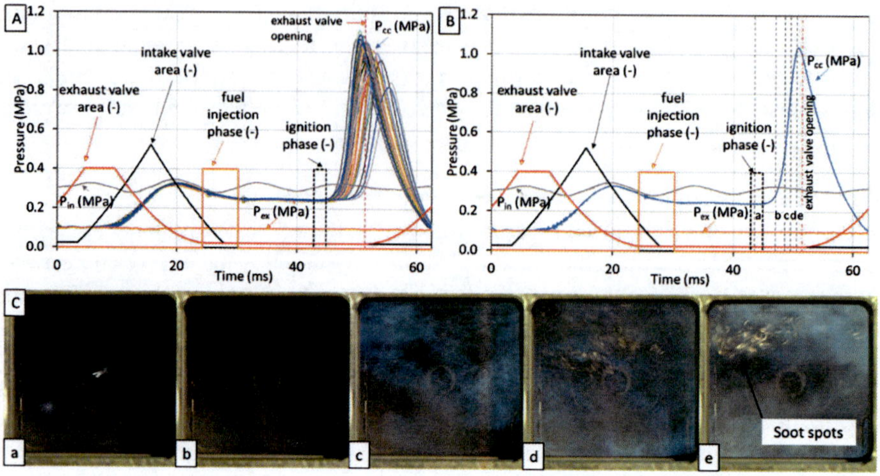

Fig. 3. A) Instantaneous P_{cc} of consecutive cycles, ensemble-averaged P_{in} and P_{ex} from a 3 s sequence without a nozzle. B) Instantaneous pressure traces during a single cycle from that sequence. C) Flame imaging during the combustion phase of that cycle.

air mass flowrate confirms this trend through the facility, which is lower for the 10 mm nozzle (21 g/s versus 58 g/s for the open plenum), and by the RBG dilution computed with the 0D analysis that is higher for the 10 mm nozzle (11.5% versus 1.6% for the open plenum). Consequently, the 20 mm nozzle is selected in the following study to allow for moderate RBG dilution, as the device will be operated at a cycle frequency higher than 16 Hz.

In summary, a nozzled exhaust plenum profoundly affects the behavior of the combustion chamber. The absence of a nozzle favors the complete scavenging of the combustion chamber, whereas the presence of a nozzle allows RBG to dilute the fresh air/fuel charge. Thus, as the nozzle diameter decreases, the combustion pressure is lower, and the combustion time is longer. The exhaust plenum undergoes incomplete scavenging, and its pressure no longer returns to ambient, which may be beneficial to damp the pressure fluctuations upstream of a turbine. Finally, the management of the plenum pressure is a key issue not only for integration purposes but also for the chemical control of the combustion process.

Effect of the Plenum Volume on the Exhaust Dynamics. In this section, three configurations of plenum volume V_{ex} are implemented, namely 0.32 L, 0.62 L and 0.97 L, all using the same 20 mm diameter exhaust nozzle selected previously. As the cycle duration is no longer constant, the signals are plotted as a function of the phase angle θ that ranges 0–180° during each cycle.

For this study, the facility is operated at stoichiometric OER with the same cycle frequency (25 Hz), the same stagnation pressure (0.3 MPa) and temperature (180 °C), the same phasing for injection ($\theta = 105°$) and ignition ($\theta = 140°$). The tests corresponding to the three volumes V_{ex} (see Fig. 5) have a significant success rate of 98% on average. Thrust is delivered over a period corresponding to the opening duration of the exhaust

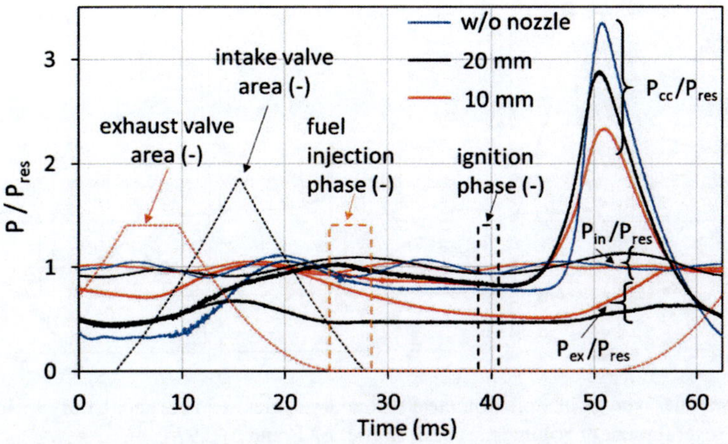

Fig. 4. Normalized instantaneous pressure traces with a 10 mm nozzle (red curves), 20 mm nozzle (black curves) and without a nozzle (blue curves). Plenum volume 0.62 L.

valve; the intake and exhaust sections are respectively denoted by S_{intake} and $S_{exhaust}$. As V_{ex} increases, thrust magnitude increases noticeably. The evolution of thrust over time seems to follow a pseudo-periodic, damped oscillation of apparent frequency 37 Hz, that occurs at exhaust valve opening and vanishes during the constant-volume phase. This oscillation may result from the mechanical excitation of the device structure by the unsteady thrust impulse generated periodically by the nozzle.

Obviously, the plenum volume V_{ex} plays a role in damping the pressure fluctuations generated by the exhaust valve opening: Fig. 5 shows that the fluctuation in P_{ex} decreases with increasing plenum volume. More generally, this effect has been recorded for cycle frequency in the range 25–50 Hz (see Fig. 7b); the time-average magnitude of P_{ex} increases with increasing plenum volume. As a result, the plenum volume drives the instantaneous dynamics of the exhaust pressure and its time-averaged value, which in turn drives the time-averaged thrust.

In summary, the increase in plenum volume yields an increase in the magnitude of the thrust pulse produced during the exhaust valve opening for a given nozzle diameter. For the three plenums implemented in this study, the plenum volume seems to have a marginal effect on the combustion chamber; however, it still drives the dynamics of the exhaust pressure and, subsequently, the thrust of the device.

3.2 Effect of the Main Operating Parameters on the Time-Averaged Combustor Performance

In the following, the effect of cycle frequency, equivalence ratio, and stagnation pressure are investigated and their effect on the performance of the PGC device. The experimental conditions are set to obtain as many secondary cycles as possible, i.e. cycles that take place after a successfully burned cycle because these cycles represent the cyclic operation of a CVC chamber.

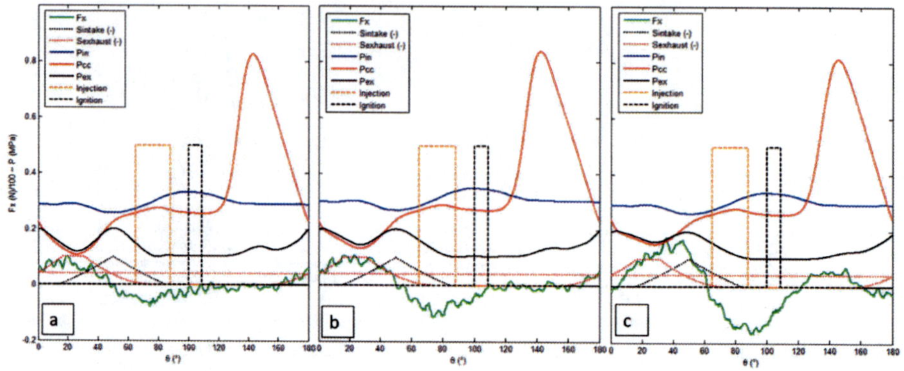

Fig. 5. Ensemble-average of stoichiometric secondary cycles at 25 Hz and 0.3 MPa with a 20 mm nozzle and several plenum volumes: a) 0.32 L, b) 0.62 L and c) 0.97 L.

Effect of Inlet Pressure and Equivalence Ratio. The conditions presented so far involved the same reacting mixture, namely stoichiometric air/isooctane at stagnation conditions 180 °C and 0.3 MPa; in other words, the previous results have been obtained with a constant amount of fuel per cycle. In contrast, this section is dedicated to the respective effects of air supply pressure and equivalence ratio on the average combustor outputs.

The effect of mixture properties is investigated firstly in terms of the air/fuel equivalence ratio (see Fig. 6a), considering the 0.62 L plenum equivalent to the combustion chamber. As opposed to the previous tests, the amount of fuel per cycle is varied accordingly. As usual in such CVC chambers, the maximum pressure ($P_{cc\ max}$) is optimum for near-stoichiometric OER, which corresponds to the minimum value of the combustion time ($t_{10\text{–}90}$). As the OER increases, both maximum exhaust pressure ($P_{ex\ max}$) and the time-averaged thrust ($F_{x\ avg}$) increase and reach a plateau for OER = 1.2–1.5. Thus, there is a shift between the optimum OER for combustion (that is 1.1 for $P_{cc\ max}$) and the optimum OER for thrust generation, which may be due to the presence of extra air brought by the scavenging phase. Indeed, the OER has been computed for the chamber volume, whereas the scavenged air contained by the exhaust plenum also contributes to the heat release and thrust generation.

Secondly, the effect of air stagnation pressure P_{res} is observed in the range 0.25–0.40 MPa (see Fig. 6b), as it is known to strongly affect both CVC processes and nozzle flow dynamics. Regarding the combustion phase, the increase in P_{res} yields an increase in $P_{cc\ max}$ and a decrease in $t_{10\text{–}90}$. Thus, there is a benefit in operating the combustion chamber at high pressure because it brings a shorter combustion time, which is favorable for operating at high cycle frequency. Moreover, as P_{res} increases, both the exhaust plenum pressure $P_{ex\ max}$ and the time-averaged thrust $F_{x\ avg}$ increase subsequently.

In summary, the combustion process is mainly sensitive to equivalence ratio, as combustion pressure and time are optimal near stoichiometry, and stagnation pressure to a lesser extent. This recalls the usual behavior of premixed flames, in which fundamental flame speed strongly depends on equivalence ratio and temperature, but less on pressure. Thereby, the faster the combustion propagates, the lower the heat loss, and the higher the pressure peak. Regarding propulsion, the air stagnation pressure has a quasi-linear effect on both the time-averaged thrust and the maximum plenum pressure, which is a comparable feature between our unsteady CVC device and classical steady choked nozzles.

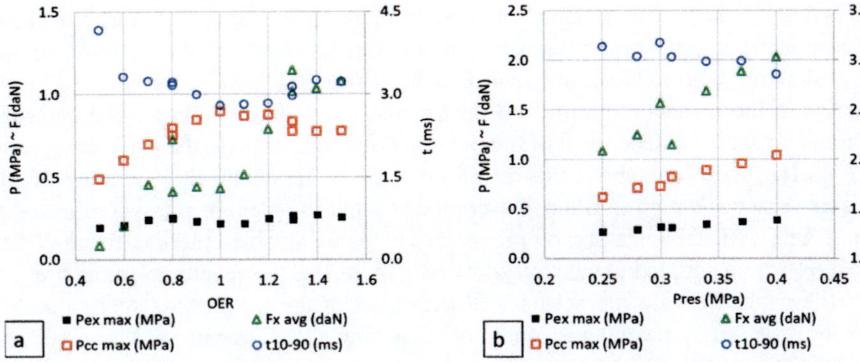

Fig. 6. Combustion and propulsion properties for secondary cycles: a) versus equivalence ratio at 0.3 MPa, 25 Hz and plenum 0.62 L, b) versus air stagnation pressure at stoichiometry, 30 Hz and plenum 0.97 L.

Effect of Cycle Frequency on the Pressure Gain. Overall, the analysis conducted so far evidenced the main parameters regarding the operational performance of the device, regarding either the combustor – that drives the heat release–, or the choked exhaust plenum – that converts it in thrust. Among the operational parameters of the facility, it is clear that the repetition rate of combustion cycles is the main lever to produce thrust. However, one can expect that a higher cycle frequency limits the time allocated to the scavenging phase, which leads to a higher dilution by RBG in the fresh charge. This effect makes cycle frequency a critical parameter regarding the dynamics of CVC systems.

In the following, the intrinsic efficiency of the combustion chamber (resp. of the exhaust plenum) is quantified via the so-called "pressure gain" as usually done in PGC devices. The pressure gain of the combustion chamber is defined as:

$$P_{cc,gain}(\%) = 100 \bullet \left(\frac{P_{cc,avg}}{P_{res}} - 1 \right) \quad (3)$$

The pressure gain related to the combustion chamber is similarly affected by the cycle frequency, whatever the plenum volume (see Fig. 7a). It firstly decreases between 25 and 35 Hz, under the effect of increasing dilution by RBG, as observed earlier (see Fig. 5a–c). Then, the pressure gain increases with increasing frequency. As the cycle duration decreases, P_{cc} gradually reaches equilibrium with P_{ex} during the scavenging phase, and thus its evolution is choked by the throat. This behavior results in converging values of the pressure gain for the combustor and the plenum above 35 Hz (see Fig. 7b). Above this frequency, the cycle duration becomes too short for the fresh charge to burn completely in the combustion chamber. As a result, one can state that increasing the cycle frequency increases the maximum pressure in the exhaust plenum, thus increasing the stagnation pressure upstream of the nozzle, which in turn should contribute to generating more thrust.

The absolute value of the pressure gain, which reaches 31% in the present study at 50 Hz, is quite close to that obtained in another device featuring a PGC turbomachine demonstrator [9], whereby the pressure gain amounts to 28% for a cycle frequency of 13.7 Hz, although thermodynamic conditions differ from the present study.

Finally, in the present experimental device and operating conditions, the pressure gain parameter seems independent of the plenum volume as soon as the cycle frequency exceeds 40 Hz. Therefore, above this threshold of cycle frequency, the operation of our PGC device reaches a regime where the combustor and the plenum act overall as a gas generator with adjustable stagnation pressure. This is a valuable outcome for the PGC system likely to be coupled to a turbine downstream of the gas generator. In the present facility, the choked nozzle represents well the effect of the turbine. As such, a single turbine could be fed by several gas generators featuring a PGC operation. Moreover, the volume of the exhaust plenum was not observed to have a critical influence so that it may be reduced for integration purposes in turbomachines.

In summary, the pressure gain computed in the plenum is a parameter of interest to characterize the efficiency of our PGC device, which reaches a maximum value of 31% in this study. It is more relevant than the pressure gain computed in the chamber, especially at high frequency when the combustion time competes with the cycle duration.

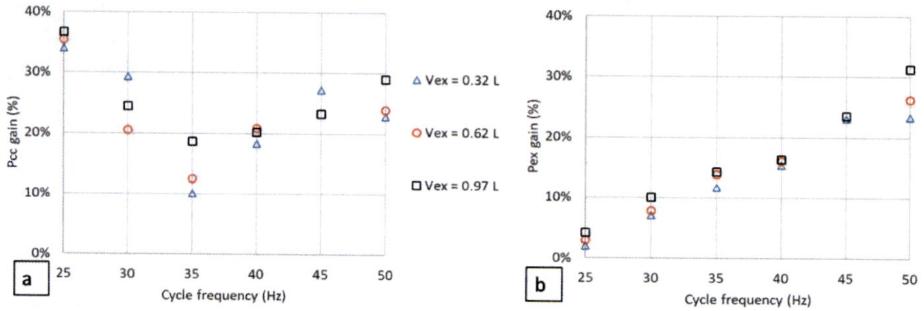

Fig. 7. Effect of cycle frequency on the pressure gain recorded: a) in the combustion chamber, b) in the exhaust plenum. Stoichiometric combustion cycles, 0.30 MPa, 20 mm throat.

Effect of Cycle Frequency on Propulsion Performance. The time-averaged value of thrust $F_{x\ avg}$ is computed in the same operating conditions, as it is a resulting output of the overall system performance. The specific impulse of the device I_{sp} is also derived from thrust, considering the actual fuel flowrate $\dot{m}_f$ and the standard gravity g_0:

$$I_{sp} = \frac{F_{xavg}}{\dot{m}_f \bullet g_0} \tag{4}$$

The time-averaged thrust $F_{x\ avg}$ gradually increases (see Fig. 8a) as the cycle frequency increases, because more cycles are performed during a given time lapse. Noticeably, thrust exhibits the same behavior for the three plenum volumes V_{ex} used in this study; hence these volumes are large enough to allow for successful coupling with the combustion chamber.

To assess the gain in efficiency brought by the pressure-gain technology, the specific impulse resulting from our CVC device is compared to the specific impulse that would be obtained if the same plenum and nozzle were fed by a Constant-Pressure Combustion (CPC) chamber, i.e. the same process as would be implemented in a simple turbojet operating with the same compression ratio (3:1). For this purpose, the thermodynamic properties (composition, temperature, specific heat ratio) of the burned gas resulting from the CPC of air and isooctane are computed with the same initial conditions (pressure, temperature, equivalence ratio) as in the CVC facility, based on usual assumptions: isobaric combustion, chemical equilibrium and adiabaticity. Then, the thrust of the CPC process is obtained by expanding this pressurized flow of burned gas through the same nozzle as in the CVC facility (nozzle pressure ratio and geometry). This method is essential to assess the difference in efficiency between the CVC and CPC processes, despite minor drawbacks. Firstly, current turbomachines cannot sustain the elevated temperatures (up to 2000 °C) obtained in this near-stoichiometric CPC computation, which clearly makes the comparison theoretical. Secondly, our CVC facility is subjected to wall heat losses, whereas the CPC computation is adiabatic, which is rather realistic owing to the thermal management of optimized turbomachine chambers; this comparison might therefore be detrimental to the CVC process due to experimental heat losses.

As a result, the CVC process offers an improvement in specific impulse compared to the CPC process as the pressure gain reaches approximately 25% (see Fig. 8b), above a cycle frequency of 45 Hz. Indeed, in such conditions, the specific impulse of the CVC cycle exceeds that of the CPC cycle by 4%, 5% and 23% for the plenum volumes 0.32 L, 0.62 L and 0.97 L, respectively. The largest plenum volume (0.97 L) may favor the gain in performance by allowing better scavenging of the combustion chamber, while reducing the influence of the exhaust plenum on the combustion chamber (backpressure and dilution by RBG). As suggested by Fig. 8b, cycle frequency plays a critical role in the increase in pressure gain that, in turn, drives the propulsion performance of the PGC device. However, a physical limit may be encountered when increasing the cycle frequency: the time necessary for combustion (several ms, see Fig. 6), as well as for other phases of fluid transfer, is likely to compete with the cycle duration.

In summary, the cycle frequency directly affects the thrust delivered by the PGC device. Compared to a CPC chamber operating in the same conditions with the same nozzle, our CVC facility yields an increase in specific impulse (up to 23%) that seems to be driven by the pressure gain. This is all the more noticeable as the CPC computation is adiabatic, while the CVC data proceed from real experiments with heat losses.

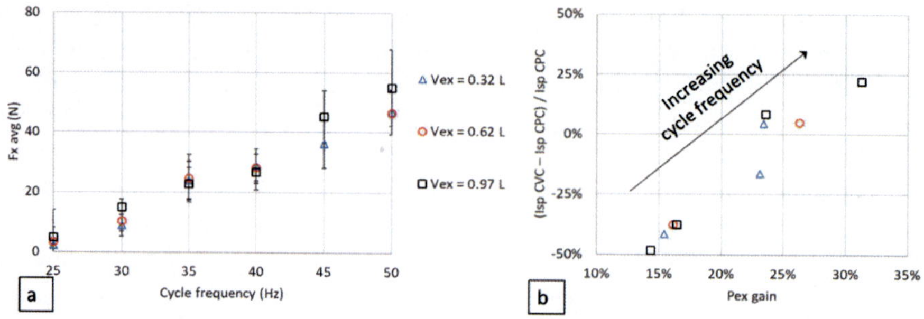

Fig. 8. Effect of cycle frequency: a) on time-averaged thrust, b) on the improvement of specific impulse by the PGC device. Stoichiometric combustion cycles, 0.30 MPa, 20 mm throat.

4 Conclusion

This article reports the experimental study of an air-breathing pressure-gain combustion (PGC) facility, featuring a constant-volume combustor (CVC) coupled to an adjustable exhaust plenum. The facility was operated with compressed air (0.25–0.40 MPa, 180 °C) and direct-fueled isooctane. The operation of this facility showed the interactions between the CVC chamber and the exhaust plenum, based on the physical outputs of the device, such as the pressure gain and the axial thrust.

As the nozzle diameter decreases, the amount of residual burned gas in the combustion chamber increases, which affects the heat release process: the diluted combustion regime exhibits a longer combustion time and lower combustion pressure. The effect of the air/fuel equivalence ratio, which was set in the range 0.5–1.5, showed the usual optima of combustion pressure and duration at stoichiometric conditions. Investigating the cycle frequency in the range 25–50 Hz revealed its key effects on the system dynamics: as the cycle frequency increases up to 35 Hz, the residual burned gas dilution increases, which in turn decreases the pressure gain in the combustion chamber. At cycle frequency above 35 Hz, this effect is alleviated as the flow exiting from the combustor gets choked by the nozzle throat, so that the pressure gain is the same in the chamber and in the plenum.

It is noticeable that – whatever the plenum volume – the pressure in the exhaust plenum increases almost linearly with increasing cycle frequency, and so does the time-averaged thrust and the specific impulse. For comparison purposes, the specific impulse of a constant-pressure combustion (CPC) chamber coupled to the same nozzle is computed with the same compression ratio. Although the present CVC facility is not adiabatic, its specific impulse noticeably exceeds the ideal CPC process by up to 23%.

Finally, the pressure gain in our CVC facility reaches around 31% at 50 Hz, which means a 31% increase in the stagnation pressure of the gas generator. The pressure gain in the exhaust plenum remains a critical parameter for such PGC applications, as it drives the inlet pressure of a turbine located downstream. Thus, designing PGC-based turbomachines will require optimizing the plenum feeding by single or multiple CVC chambers.

References

1. Kailasanath, K.: Review of propulsion applications of detonation waves. AIAA J. **38**(9), 1698–1708 (2000). https://doi.org/10.2514/2.1156
2. Reichel, T.G., Schäpel, J.S., Bobusch, B.C., Klein, R., King, R., Paschereit, C.O.: Shockless explosion combustion: experimental investigation of a new approximate constant volume combustion process. J. Eng. Gas Turbines Power **139**(2), 021504 (2016). https://doi.org/10.1115/1.4034214
3. Kaiser, S., Seitz, A., Donnerhack, S., Lundbladh, A.: Composite cycle engine concept with hectopressure ratio. J. Prop. Power **32**(6), 1413–1421 (2016). https://doi.org/10.2514/1.B35976
4. Tarnawski, P., Ostapski, W.: Pulse-powered turbine engine concept implementing rotating valve timing system: numerical CFD analysis. J. Aerosp. Eng. **32**(3), 04019017 (2019). https://doi.org/10.1061/(ASCE)AS.1943-5525.0001001
5. Boust, B., Michalski, Q., Bellenoue, M.: Experimental investigation of ignition and combustion processes in a constant-volume combustion chamber for air-breathing propulsion. In: AIAA–Joint Propulsion Conference (2016). https://doi.org/10.2514/6.2016-4699
6. Labarrere, L., Poinsot, T., Dauptain, A., Duchaine, F., Bellenoue, M., Boust, B.: Experimental and numerical study of cyclic variations in a constant volume combustion chamber. Combust. Flame **172**, 49–61 (2016). https://doi.org/10.1016/j.combustflame.2016.06.027
7. Exilard, G.: Large Eddy Simulation of Constant Volume Combustion in a Ground-Breaking New Aeronautical Engine. Paris Saclay University, Paris (2018)
8. Michalski, Q., Boust, B., Bellenoue, M.: Experimental investigation of ignition stability in a cyclic constant-volume combustion chamber featuring relevant conditions for air-breathing propulsion. Flow Turbul. Combust. **102**(2), 279–298 (2019). https://doi.org/10.1007/s10494-019-00015-1
9. Salminen, H.J., Ciccateri, F., Ijäs, T., Ulmala, O., Mikulski, M.: Experimental demonstration of a novel deflagration-based pressure gain combustion technology. In: AIAA–Joint Propulsion Conference (2020). https://doi.org/10.2514/6.2020-3871
10. Watson, N., Janota, M.S.: Turbocharging the Internal Combustion Engine. Macmillan Education, London (1982)
11. Michalski, Q., Boust, B., Bellenoue, M.: Toward a cyclic self-ignited constant-volume combustion for air-breathing propulsion applications. In: AIAA–Joint Propulsion Conference (2018). https://doi.org/10.2514/6.2018-4478
12. Kaemming, T.A., Paxson, D.E.: Determining the pressure gain of pressure gain combustion. In: AIAA–Joint Propulsion Conference (2018). https://doi.org/10.2514/6.2018-4567
13. Michalski, Q., Boust, B., Bellenoue, M.: Influence of operating conditions and residual burned gas properties on cyclic operation of constant-volume combustion. In: King R. (ed.), Active Flow and Combustion Control 2018, vol. 127, pp. 215–233. Springer (2019). https://doi.org/10.1007/978-3-319-98177-2_14

Control of Auto-ignitive Wave Propagation Modes from Hot Spots by Mixture Tailoring in Shockless Explosion Combustion

Lisa Zander(✉), Johann Vinkeloe(✉), and Neda Djordjevic

Fachgebiet für Verbrenungskinetik, Technische Universität Berlin,
Müller-Breslau-Str. 8, 10623 Berlin, Germany
{lisa.zander,johann.vinkeloe}@tu-berlin.de

Abstract. Shockless Explosion Combustion is a novel combustion concept that achieves pressure gain combustion by quasi-homogeneous auto-ignition of the fuel/air mixture. Shockless Explosion Combustion is, like other combustion concepts based on auto-ignition, prone to premature ignition and detonation formation in the presence of reactivity gradients, so called hot spots. Two measures to inhibit detonation formation and to achieve quasi-homogeneous auto-ignition, dilution and fuel blending, are investigated by means of zero-dimensional simulations of generic hot spots. Experimental ignition delay times measured in a high pressure shock tube are used to select suitable chemical-kinetic models for the numerical investigation and the calculation of temperature sensitivities of ignition delay times. The main focus of this investigation are the two non-dimensional regime parameters ξ and ε, as they enable characterization of the mode of auto-ignitive wave propagation from hot spots. ξ is the ratio between the speed of sound and the auto-ignitive wave propagation velocity and ε describes the ratio between the time a pressure wave travels through the hot spot and the excitation time. Dilution of the combustion mixture with steam and CO_2 aims at extending excitation times and therefore decreasing the parameter ε. Fuel blending of Dimethyl ether with hydrogen or methane aims at reducing the temperature sensitivity of ignition delay time and low values of ξ. It is demonstrated that both measures are effective at mitigating detonation development while maintaining quasi-homogeneous auto-ignition in presence of hot spots.

Keywords: Ignition delay time · Excitation time · Knock · Auto-ignition · Detonation peninsula · Dilution · Fuel blending

1 Introduction

The novel combustion concept Shockless Explosion Combustion (SEC) is a way to implement pressure gain combustion in a gas turbine [1]. It promises a substantial increase in efficiency over a conventional gas turbine cycle [2,3], while it

L. Zander and J. Vinkeloe—These authors share first-authorship.

R. King and D. Peitsch (Eds.): AFCC 2021, NNFM 152, pp. 16–34, 2022.
https://doi.org/10.1007/978-3-030-90727-3_2

circumvents the disadvantages associated with using detonation waves to achieve pressure gain combustion like high pressure peaks or deflagration-to-detonation and exergy losses [1].

In SEC the pressure rise is achieved by quasi-homogeneous auto-ignition of the fuel-air mixture. The ensuing pressure wave generated by the volumetric heat release is reflected at the end of the combustion chamber as a suction wave. It is used to refill the combustion tube with a fresh fuel-air mixture to initiate the next combustion cycle. The quasi-homogeneous auto-ignition that is required for the process is achieved by matching the residence time of the fuel-air mixture in the combustion chamber with its ignition delay time through fuel stratification. Realizing quasi-homogeneous auto-ignition in a real combustion environment demands on-the-fly control of fuel stratification and is still a topic of on-going research [4]. Nevertheless, it is experimentally proven that the homogeneity of ignition correlates with the pressure increase in SEC [5,6].

Combustion concepts that are based on auto-ignition are prone to detonation formation and inhomogenoues ignition. Premature ignition in a local spot with increased reactivity can lead to the development of undesired detonation waves. These local spots with increased reactivity are here referred to as hot spots and can be caused by a temperature or concentration gradient. This is not only a challenge for SEC, but also for different auto-ignition based combustion concepts such as HCCI (homogeneous charge compression ignition) [7]. Detonation development from hot spots can be explained with the SWACER (Shock Wave Amplification by Coherent Energy Release) mechanism [8]. Premature ignition leads to localized heat release. This local heat release causes a pressure rise, which propagates as an acoustic pressure wave. If this pressure wave is in phase with heat release resulting from the auto-ignition of subsequent discrete mixture volumes along the reactivity gradient, the heat release can reinforce the pressure wave and a detonation wave may form.

Zeldovich et al. [9] demonstrated the importance of the temperature gradient of the hot spot for detonation formation. Different modes of auto-ignitive wave propagation can be initiated depending on the value of the temperature gradient [10]. Premature ignition along the reactivity gradient of the hot spot generates an auto-ignitive wave. In the following a hot spot with a temperature gradient is considered as in the analysis in [11]. The velocity of the ensuing auto-ignitive wave u_{ai} that results from the gradient in ignition delay time $\partial\tau_i/\partial r$ can be expressed by the temperature gradient of the hot spot $\frac{\partial T}{\partial r}$ and the temperature sensitivity of ignition delay time $\frac{\partial \tau_i}{\partial T}$ [11],

$$u_{ai} = \left(\frac{\partial \tau_i}{\partial r}\right)^{-1} = \left(\frac{\partial \tau_i}{\partial T}\frac{\partial T}{\partial r}\right)^{-1}. \tag{1}$$

When the auto-ignitive wave velocity is close to the speed of sound, the pressure wave is reinforced by the heat release and a detonation wave may form. However, when the auto-ignitive wave velocity exceeds the speed of sound characterizing the pressure wave propagation, reinforcement of the pressure wave by the heat release is not possible and the auto-ignitive wave will propagate without being affected by pressure waves as a supersonic auto-ignitive wave. In the limiting case of infinite auto-ignitive wave velocity the gas auto-ignites homogeneously as a thermal explosion. When the auto-ignitive wave velocity is below the speed of sound, wave propagation occurs with two different propagation mechanisms depending on the velocity: subsonic auto-ignitive propagation, which is driven by the reactivity gradient in the hot spot and flame propagation, which is driven by diffusive processes.

The aim to classify the wave propagation led to the development of non-dimensional parameters ξ and ε [11]. The non-dimensional parameter ξ normalizes the speed of sound of the gas a to the velocity of auto-ignitive wave,

$$\xi = \frac{a}{u_{ai}}, \tag{2}$$

and expresses the influence of the auto-ignitive wave velocity on the propagation mode based on the work of Zeldovich [10] which is discussed above. In theory, the heat release of the auto-ignition can reinforce the pressure wave when ξ equals one, i.e. the auto-ignitive wave propagates with a velocity at the speed of sound, and detonations may occur. In practice, lower and upper bounds in ξ are defined for detonation formation depending on the conditions and fuel/oxidizer mixture, as the initial temperature gradient of the hot spot may change during the induction period [11]. This is attributed to heat conduction, mass diffusion and gas expansion [11]. For small values of ξ the regime of supersonic auto-ignitive wave propagation is observed, while subsonic auto-ignitive wave propagation appears at large values of ξ. A second non-dimensional parameter ε is proposed, to account for the rapidness of heat release rate, which is expressed by the excitation time [11]. It describes the ratio of the transit time of an acoustic wave though the hot spot relative to the excitation time of the gas τ_e, which is the characteristic time scale of the heat release,

$$\varepsilon = \frac{r_{hs}/a}{\tau_e}, \tag{3}$$

where r_{hs} describes the radius of the hot spot. Small values of ε indicate that heat release is much slower than acoustic waves traveling through the hot spot. Therefore, reinforcement of the pressure wave becomes less likely.

To classify modes of wave propagation from a hot spot a regime diagram as $\xi - \varepsilon$ diagram has been proposed [11]. The area in regime diagram where detonation formation is observed is usually referred to as detonation peninsula.

Furthermore, a criterion is proposed to discern whether detonation formation is possible [12–14]. If the excitation time τ_e is much longer than the difference in ignition delay time between the hot spot and the surrounding gas $\Delta\tau_i$, the surrounding gas ignites before the formation a of detonation wave is possible. Only for negative $\Delta\tau_i$ premature ignition appears in the hot spot that can possibly lead to the formation of a detonation wave. Hence, in the case of premature ignition the following condition

$$\left|\frac{\Delta\tau_i}{\tau_e}\right| < 1 \tag{4}$$

is fulfilled when the surrounding gas ignites before a detonation wave can form. The criterion in Eq. (4) has also been used to classify experimentally observed detonation transition [15].

In the following it is demonstrated, that the condition in Eq. (4) can be reformulated in terms of the non-dimensional parameters ξ and ε. $\Delta\tau_i$ can be expressed by the mean ignition delay time gradient $\overline{\frac{\partial\tau_i}{\partial r}}$ and the radius r_{hs} of the hot spot,

$$\Delta\tau_i = -r_{hs}\overline{\frac{\partial\tau_i}{\partial r}} = -r_{hs}\bar{u}_{ai}^{-1} \tag{5}$$

and can thus be related to the mean auto-ignitive wave velocity $\bar{u}_{ai}$. In other words, the difference in ignition delay time $\Delta\tau_i$ equals the time that the auto-ignitive wave that originates at the maximum reactivity in the hot spot needs to reach the surrounding gas. As the temperature sensitivity of ignition delay time may vary with temperature, also the ignition delay time gradient and auto-ignition velocity may vary within the hot spot. Therefore, the mean values $\overline{\frac{\partial\tau_i}{\partial r}}, \bar{u}_{ai}$ are chosen such that relation (5) holds. The ratio in criterion (4) can be rearranged using Eq. (5) for the case of premature ignition in the hot spot to

$$\left|\frac{\Delta\tau_i}{\tau_e}\right| = \frac{r_{hs}\bar{u}_{ai}^{-1}}{\tau_e}\frac{a}{a} = \frac{a}{\bar{u}_{ai}}\frac{r_{hs}/a}{\tau_e} = \xi\varepsilon, \text{ for } \overline{\frac{\partial\tau_i}{\partial r}} > 0. \tag{6}$$

Thus, when premature ignition appears in the hot spot the formation of a detonation wave is suppressed due to the auto-ignition of the surrounding when the following condition holds,

$$\xi\varepsilon < 1. \tag{7}$$

While it is shown above that the product of ξ and ε can be used interchangeably to the ratio $\frac{\Delta\tau_i}{\tau_e}$ to discern suppression of detonation development for the case of auto-ignition of the surrounding gas, it has also been used in the literature to distinguish between auto-ignitive wave propagation and deflagration [16,17]. It is proposed that values of $\xi\varepsilon$ over 1500 are associated with deflagration [16]. However, the border between both propagation modes is not very sharp [16].

Temperature or concentration inhomogeneities that may cause premature ignition are always present in technical systems, which also pose a challenge for SEC. Both detonations, as well as subsonic auto-ignitive wave propagation are to be avoided in SEC, as it is not designed for the former and the latter results in burned gas expansion with insufficient pressure gain. SEC requires supersonic auto-ignitive wave propagation or thermal explosion to a achieve (quasi-)homogenous ignition even in the presence of hot spots, i.e. low values of ξ and/or ε.

The parameters that influence the wave propagation mode are highly dependent on the physicochemical properties of the fuel-air mixtures and the thermodynamic conditions. Especially, the ignition and heat release characteristics of a mixture, namely the temperature sensitivity of ignition delay time and the excitation time, play an important role for the mode of auto-ignitive wave propagation. This enables tailoring of the mixture with focus on these properties. In this work two approaches are investigated: the extension of excitation time by dilution with CO_2 and steam [14,18], which is targeted at decreasing ε; and decreasing temperature sensitivity of ignition delay time by fuel blending [19,20], which is targeted at decreasing ξ.

Dimethyl ether (DME) is used as a fuel, respectively fuel component, in both investigations. The reason is that it is suited for auto-ignition based concepts due to its relatively high reactivity [21]. DME's negative temperature coefficient (NTC) behavior, i.e. an increase in ignition delay time with increasing temperature within a certain temperature range, makes it suitable as a fuel blend component. Due to the application of SEC in gas turbines, usually similar conditions as in conventional gas turbines are of interest [19]. Rähse et al. [2] demonstrated, that a gas turbine with SEC achieves a significantly increased efficiency for pressures between 24 and 50 bar and temperatures between 823 to 1039 K. For the investigations in this study a thermodynamic condition of 35 atm and 887 K, is chosen which resembles the condition at the inlet of a gas turbine combustor. It is determined by compression from ambient conditions with a pressure ratio of 35 and a compression efficiency of 90%.

The overall goal of this study is to assess the impact of dilution and fuel blending on the non-dimensional detonation parameters in the presence of hot spots.

2 Experimental and Numerical Methods

In order to determine the temperature sensitivity of ignition delay time and choose suited chemical kinetic models to investigate the effects of the proposed mixture tailoring, ignition delay times are measured behind reflected shock waves in a high-pressure shock tube. Detailed information about the facility and the measurement procedure can be found in [20,22,23].

Zero-dimensional homogeneous reactor simulations are performed for both, the choice of the chemical-kinetic model and the determination of the relevant gas properties for the calculation of ξ and ε as explained below. The reactor models are implemented in *Python* with the software package *Cantera* [24]. Excitation time is defined as the time that elapses between 5% of the maximum heat release rate and the maximum heat release rate. This definition is very common and used in e.g. [11,18,25]. Furthermore, it is chosen because it is also used in the simulations in [25–27], which the obtained results will be compared to.

Generic hot spots of a defined radius and fixed temperature gradient are used to investigate the effect for the two investigated measures, i.e. dilution and fuel blending. Equations (2) and (3) are then used to calculate the respective ξ and ε parameters, utilizing excitation times and temperature sensitivity of ignition delay time determined by reactor simulations.

Firstly, the influence of temperature on the non-dimensional parameters is studied for a fixed hot spot. A single generic hot spot with a radius of 5 mm and a temperature gradient of ±1 K/mm is chosen, as these are commonly used radii [16,18,25] and temperature gradients [20,28]. In the NTC region a hot spot with a temperature decrease is needed to excite an auto-ignitive wave. Therefore the sign of the temperature gradient is adjusted depending on whether a hot spot with a temperature increase or decrease is present. For this investigation, the non-dimensional parameters are evaluated at the surrounding gas temperature to enable comparison with figures that depict the temperature sensitivity of ignition delay time and excitation time over the temperature.

Secondly, the influence of the two investigated measures is studied for the thermodynamic condition that resembles a gas turbine combustor. By varying the hot spot radius and temperature gradient multiple points in the $\xi - \varepsilon$ diagram are created. The radius of the generic hot spots is varied between ±1 and ±10 mm and the temperature gradient is varied between 1 and 10 K/mm. These are common orders of magnitudes for similar kinds of investigation [16,25,28]. The effect of the two proposed measures is investigated on the points in the regime diagram by using the same hot spot properties, however the mixture is altered either by diluting or fuel blending. Note, that the values of ξ and ε depend on where in the hot spot they are defined, as they are temperature dependent. For the construction of regime diagrams they are evaluated in the middle of the hot spot at $0.5r_{hs}$ and a temperature of

$$T_{hs} = T_0 - 0.5r_{hs}\left(\partial T/\partial r\right)_{hs},$$

where T_0 defines the surrounding gas temperature, as this is a common definition [11,18].

3 Extension of Excitation Time by Dilution

The effect of dilution on the non-dimensional parameters is studied for a stoichiometric DME/air mixture and by adding steam or CO_2. The chemical-kinetic model AramcoMech 2.0 [29–35] is chosen for the investigation because it reproduces experimental ignition delay times of both undiluted DME/air mixtures [23] and CO_2-diluted DME/air mixtures [22] well. For comparison to simulation data from [25] the same chemical-kinetic model from Zhao et al. [36] is chosen, which also shows acceptable agreement with the experimental data [22,23].

3.1 Influence of Dilution on Excitation Time and Temperature Sensitivity of Ignition Delay Time

The modeling results indicate that dilution can significantly increase the excitation time (Fig. 1). The effect of CO_2-dilution is slightly stronger, which may be attributed to a physical effect (i.e. increased heat capacity) and/or a chemical effect. An increase in the duration of the heat release can mitigate the reinforcement of the pressure wave by the heat release and may prevent detonation formation.

Adding dilution also affects the temperature sensitivity of ignition delay time (Fig. 1). The temperature sensitivity of all investigated mixtures exhibits large negative values for low temperatures (Fig. 1), i.e. an increase in temperature will strongly decrease the ignition delay time. Positive values are attained in the NTC region. The absolute temperature sensitivity increases with dilution for most of the considered temperatures (Fig. 1). Due to the shift of the NTC region that is caused by dilution the absolute temperature sensitivity is reduced on the borders of the NTC region at around 800 and 930 K when dilution is added to the DME/air mixture (Fig. 1).

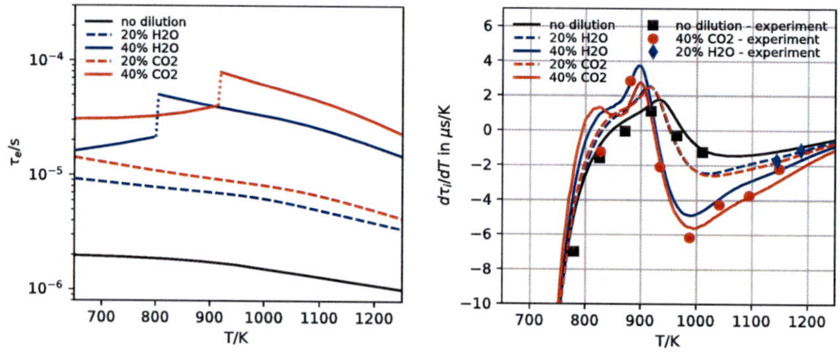

Fig. 1. Excitation time (left) and temperature sensitivity of ignition delay time (right) for undiluted and diluted stoichiometric DME/air mixture. Curves - prediction with AramcoMech 2.0, symbols - extraction from experimental ignition delay time.

3.2 Influence of Dilution on the Non-dimensional Regime Parameters

In the following the effect of dilution on the non-dimensional detonation parameters ξ and ε is studied over a range of temperatures by utilizing a generic hot spot with fixed properties as explained in Sect. 2 (Fig. 2). The non-dimensional parameter ε is strongly decreased by dilution over the whole investigated temperature range (Fig. 2). The effect on ε is slightly stronger for CO_2 dilution due to its longer excitation time (Fig. 1).

The temperature-dependent effect of dilution on the non-dimensional parameter ξ is illustrated in Fig. 2. The auto-ignitive wave velocity decreases strongly with temperature at low temperatures, which results in very high values of ξ (Fig. 2). The large values in ξ indicate that an auto-ignitive wave with subsonic propagation speed is expected at low temperatures. In the NTC region and at high temperatures, dilution decreases the auto-ignitive wave velocity below

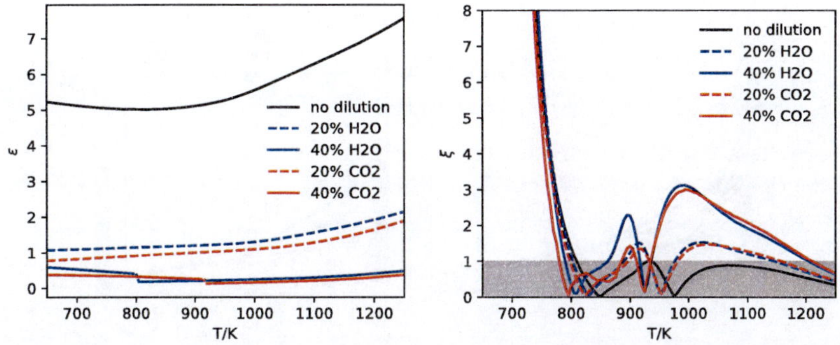

Fig. 2. Non-dimensional parameter ε for hot spot with radius of 5 mm (left), and non-dimensional parameter ξ for temperature gradient of ± 1 K/mm for undiluted and diluted DME/air mixture (calculated with AramcoMech 2.0).

the speed of sound, which also results in an increase in ξ above one (Fig. 2). Nonetheless, ξ is close to one in the intermediate and high temperature range and therefore in a range that is possibly prone to detonation formation. Therefore, the effect of dilution on the location in the $\xi - \varepsilon$ diagram will be studied in the following at 35 atm and 887 K.

The ξ and ε parameters of different hot spots with radii of 2, 5 and 8 mm and temperature gradients of 1, 4, 7 and 10 K/mm are used to create multiple points in the $\xi - \varepsilon$ diagram. Keeping the hot spot properties constant, the DME air mixture is further diluted and the effect on the location of the points is studied (Fig. 3). The arrows in Fig. 3 link the undiluted mixture to the diluted mixtures for hot spots with the same properties and hence demonstrate the effect of dilution. The excitation time, temperature sensitivity of ignition delay time and speed of sound are evaluated in the middle of the hot spot as explained in Sect. 2. The temperature at the location of evaluation lies within the NTC region for all considered mixtures and ensures that a hot spot with a decrease in temperature can be used for all considered points in the $\xi - \varepsilon$ plane. In order to link each hot spot in Fig. 3 to its temperature of evaluation, the arrows in Fig. 3 are color coded with respect to the evaluation temperature.

For all considered hot spots diluting the DME/air mixture significantly shifts towards smaller ε values. The non-dimensional parameter ξ is increased when adding dilution compared to the undiluted mixture (Fig. 3). When adding 20% dilution the effect on ξ is similar between steam and CO_2 dilution, due to their similar temperature sensitivity of ignition delay time (Fig. 1). However, dilution with 40% steam increases ξ stronger compared to CO_2 dilution. This is due to the fact that the applied chemical-kinetic model predicts the temperature sensitivity of mixtures with 40% steam to be significantly larger compared to the same amount of CO_2 dilution (Fig. 1). There is a small temperature range, where the temperature sensitivity of the mixture with 40% CO_2 is lower than that of the mixture with 20% CO_2 (Fig. 1). This results in ξ values of the mixture

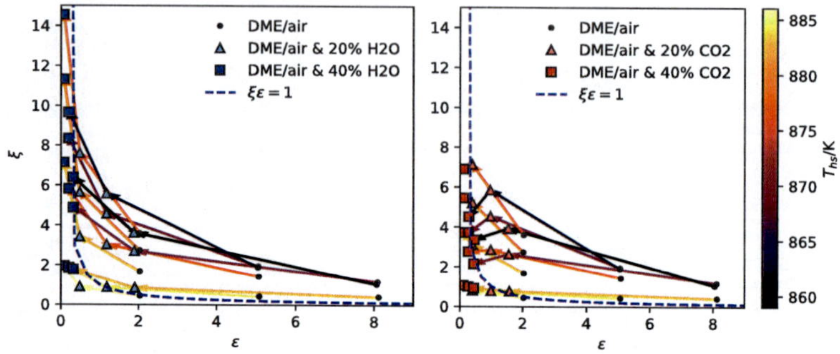

Fig. 3. Regime diagram at of stoichiometric DME/air mixture with and without dilution. Surrounding gas is at 35 atm and 887 K (calculated with AramcoMech 2.0).

with 40% CO_2 being lower compared to the mixture with 20% CO_2 dilution for hot spots with a larger temperature diminishment. For all points below the curve $\xi\varepsilon = 1$ detonation formation is prohibited according to the criterion in Eq. (7). When diluting the DME/air mixture with 40% steam or CO_2 the location in the $\xi - \varepsilon$ diagram is below or very close to the curve $\xi\varepsilon = 1$ for all considered hot spots in Fig. 3, which indicates that dilution decreases the propensity to detonation development.

To relate the effect of dilution on the location in the $\xi - \varepsilon$ diagram to the actual wave propagation mode, a second set of zero-dimensional simulations is conducted at 40 atm and 982 K and compared to results obtained at the same conditions of a spherical one-dimensional hot spot from [25] (Fig. 4). The pressure is comparable to the condition considered in this study and the temperature is also within the NTC region. Note, that the depicted detonation peninsula from [25] is only determined for the undiluted DME/air mixture and that it may differ for the diluted mixtures. Therefore, additionally the curve $\xi\varepsilon = 1$ is shown in Fig. 4, as it demarcates the region where detonation formation can be excluded, according to the criterion in Eq. (7). Compared to the detonation peninsula by Dai et al. [25] the criterion is rather conservative. This is because it considers the strictest case where detonation formation is not possible due to the ignition of the gas surrounding the hot spot. Other cases, where e.g. insufficient reinforcement of the pressure wave by the heat release leads to the prevention of detonation formation are not considered by criterion (7). However, these cases are considered in the regime diagram in Fig. 4 that is obtained through one-dimensional simulations [25]. The results demonstrate again, that dilution significantly reduces the ε parameter (Fig. 4). With the simulation results from [25] it is possible to evaluate the effect of the change in ε on the propagation mode and possible mitigation of detonation formation. For the undiluted mixture there are four hot spots that are within or on the borders of the detonation peninsula. Dilution with 20% steam or CO_2 shifts the points out of the detonation peninsula. Dilution with 40% steam or CO_2 shifts the points even into the region below the curve $\xi\varepsilon = 1$, indicating that detonation formation is mitigated according

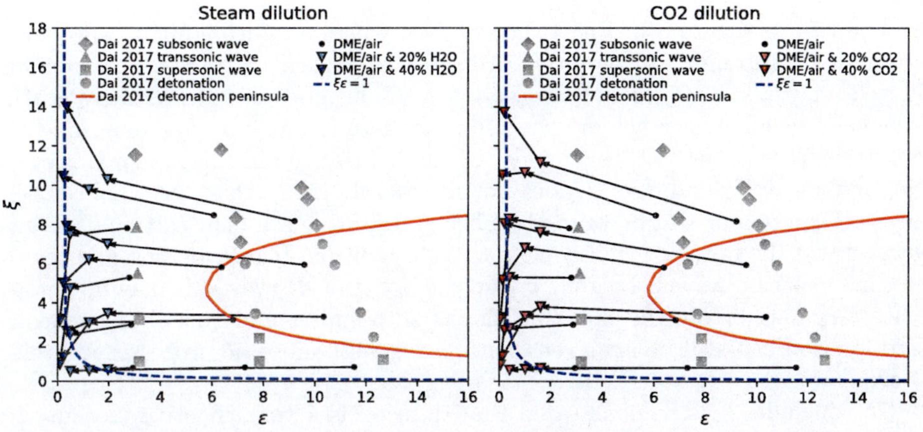

Fig. 4. Regime diagram of stoichiometric DME/air mixture with and without dilution compared to data from Dai et al. [25]. Surrounding gas is at 40 atm and 982 K (calculated with the model from [36].)

to the stricter criterion in Eq. (7). Hence, hot spots that would result in the formation of a detonation wave will not result in a detonation when dilution with CO_2 or steam is added. Instead, the ξ and ε values are shifted into the region where Dai et al. [25] observe transonic or supersonic auto-ignitive waves. The shift from the detonation regime to supersonic auto-ignitive deflagration is desired for stable SEC operation. The resulting very rapid ignition of the fuel-air mixture can achieve the aerodynamic confinement that is necessary for pressure gain in SEC without the formation of a detonation wave. In the case of transonic auto-ignitive wave propagation no mutual reinforcement appears between the pressure and reaction wave even though the speed of the auto-ignitive wave is close to the speed of sound [25]. Transonic auto-ignitive wave propagation is therefore closely related to small ε values. The appearance of transonic wave propagation is attributed to the small hot spot radius in [25], while it is achieved by an increase in excitation time in this study. Note, that the boundaries of the detonation peninsula and the region of transonic wave propagation are observed at ξ values larger than one in the one-dimensional simulations [25]. This can be attributed to the transient evolution of the temperature gradient and mass diffusion [11,37], which alter the ignition delay time gradient in the hot spot. For hot spots with a higher temperature difference to the surrounding gas the location in the regime diagram is within the subsonic propagation regime for the undiluted mixture. For these points, the ξ value increases with dilution compared to the undiluted mixture, while the ε value is strongly reduced.

The regime diagram in Fig. 4 shows cases, where an auto-ignitive wave with transonic or subsonic auto-ignitive propagation will ensue from a hot spot for diluted mixtures. This raises the question whether ignition in these cases is still homogeneous enough to achieve the necessary aerodynamic confinement in SEC. With the applied methods this cannot be finally assessed in this study. However, even if the ξ value is large, it does not necessarily mean that the whole gas in the SEC combustor will be burned ineffectively through a deflagration. The gas surrounding the hot spot is very close to auto-ignition. It may be possible that a subsonic reaction wave is created within the hot spot. However, it may not propagate very long before the surrounding gas auto-ignites and quasi-homogeneous auto-ignition can still be achieved, considering that subsonic processes are relatively slow.

To conclude, it is demonstrated that dilution is a very effective measure to prevent detonation formation in SEC and facilitate quasi-homogeneous auto-ignition in presence of reactivity gradients.

4 Decreasing the Temperature Sensitivity of Ignition Delay Time by Fuel Blending

Fuel candidates like hydrogen and methane can be produced from renewable resources and are therefore interesting for SEC application, but both fuels have generally long ignition delay times and very high temperature sensitivities of ignition delay times at the relevant conditions.

To decrease the non-dimensional parameter ξ by reducing the temperature sensitivity of ignition delay time, fuel blends with a fuel component with NTC behavior can be used. A ternary fuel blend of DME, methane and hydrogen, which was proposed by Cai and Pitsch [19] and experimentally investigated in [20], is further studied here. The intention in this case is to show that DME is capable to reduce the high temperature sensitivity of ignition delay time of H_2 and CH_4, while at the same time reducing the relatively large ignition delay time characteristic for those fuels to a level relevant for SEC.

Suited mechanisms and a validation for the ternary fuel blend containing DME, H_2 and CH_4 can be found in our previous study [20]. It was shown, that the chemical-kinetic model AramcoMech 3.0 [38] and the mechanism from Cai and Pitsch [19] are in good agreement with the conducted ignition delay time measurements. For the further analysis the AramcoMech 3.0 is used, based on the findings in the mentioned study.

4.1 Effect of Fuel Blending on Temperature Sensitivities of Ignition Delay Times

The temperature sensitivities of ignition delay times of the pure components are compared to those of fuel blends to show the effectiveness of the fuel blending. Using an Arrhenius-fitting and derivation approach the available experimental data is used to extract temperature sensitivities of ignition delay times for comparison with mechanism predictions, as shown in Fig. 5. Predicted temperature sensitivities of ignition delay times show good agreement to the temperature sensitivities derived from the experimental ignition delay time data.

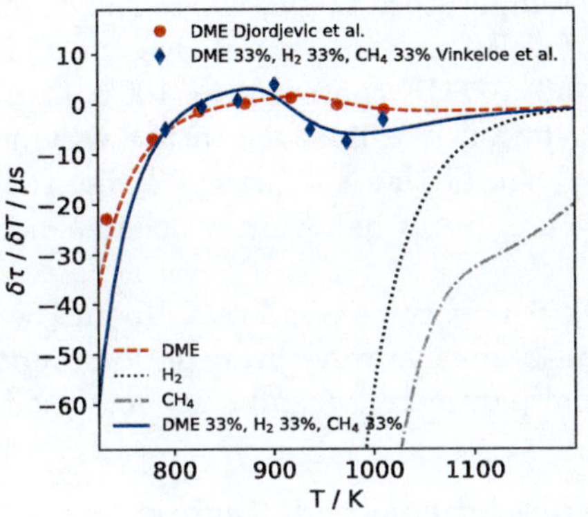

Fig. 5. Temperature sensitivities of ignition delay times derived from experimental data (data for DME from Djordjevic et al. [22] and data for ternary equimolar fuel blend from Vinkeloe et al. [20]) and numerical simulations (AramcoMech 3.0 [38]) at stoichiometric conditions and a nominal pressure of 35 bar.

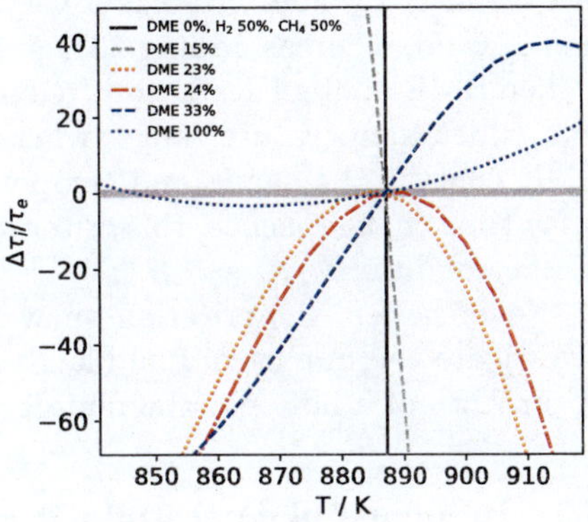

Fig. 6. Ratio of difference in ignition delay time between hot spot and surrounding gas to excitation time over varying hot spot peak temperature; 35 atm, 887 K and stoichiometric conditions and the ternary fuel blend DME/H_2/CH_4 with constant ratio between H_2 and CH_4

Compared to pure hydrogen and methane, DME and the equimolar fuel blend exhibit significantly reduced temperature sensitivity of ignition delay time in a temperature range between 700 and 1100 K. In between 800 and 900 K temperature sensitivities of ignition delay times of DME and the ternary fuel blend exhibit the same order of magnitude (Fig. 5).

Due to the low temperature sensitivity of ignition delay time the ternary fuel blend is interesting for the SEC application. Nevertheless, the probability of developing detonations in hot spots depends also on the hot spot's temperature gradient, size, fuel/oxidizer mixture properties and transient coupling of the pressure wave and the auto-ignitive wave. The criterion in Eq. 4 is applied to investigate the impact of the fuel blends. For absolute values of this criterion lower than unity, detonation formation is improbable. In that case, the gas surrounding the hot spot has enough time to ignite, before the pressure is rising due to premature ignition in a hot spot. Hence, the formation of a detonation is inhibited.

Figure 6 shows this criterion for the ternary fuel blend and varying hot spot temperatures. The considered fuel blends contain an equimolar hydrogen/methane fuel blend, blended with varying percentages of DME. Different blends lead to varying temperature ranges where the criterion is lower than 1, which is associated with conditions, where no detonation wave can form. The fuel blend of H_2 and CH_4 has a very small temperature range, where the criterion is valid, whereas adding 23% and 24% DME has a strong positive effect. A higher DME content leads to a reduction and a DME content near 100% leads to a wider temperature range, where the criterion is valid. Due to the varying DME content the shape and temperature range of the NTC range of ignition delay time change, hence, this criterion has a non-linear behavior, which depends on the conditions.

Nevertheless, the criterion shown in Fig. 6 is very conservative. To further investigate the impact of fuel blending on auto-ignitive wave propagation mode in presence of a hot spot the non-dimensional parameters ξ and ε are analyzed.

4.2 Influence of Fuel Tailoring on Non-dimensional Regime Parameters

The introduced detonation regime parameters ξ and ε give more information about the possible mode of auto-ignitive wave propagation in hot spots. As shown above fuel blends have a significant impact on the temperature sensitivities of ignition delay times, and therefore also on ξ (compare Eq. (2)). Due to the different fuel components, properties of the combustion mixtures like the speed of sound and the transient process of heat release vary. Hence, in addition to ξ also ε is affected by fuel blends.

Figure 7 shows values of ξ over the temperature of the surrounding gas for various blending ratios, assuming a temperature gradient of $-1\,\text{K/mm}$ (or $1\,\text{K/mm}$ in case of a positive temperature sensitivity) inside a generic hot spot. DME is the component in the considered ternary blend, that has the greatest impact on the overall ignition delay time and its temperature sensitivity. Hence, the percentage of DME has an impact on the overall magnitude of ξ as well as the temperature range of low values of ξ. Without DME it is not possible to get near the target of $\xi < 1$. For high percentages of DME in the fuel blend the temperature range where $\xi < 1$ increases. The relative hydrogen to methane ratio in the ternary fuel blend has only minor impact on ξ [20] and is thus not further studied here.

Due to the varying fuel composition the excitation time differs, which has an impact on ε. Figure 8 shows the impact of different fuels and fuel blend ratios on the non-dimensional parameter ε. For small percentages of DME in the ternary fuel blend ε remains at the same level. Higher percentages of DME decrease ε significantly.

When fuel blends are used to reduce ξ for SEC application, the impact on ε should also be taken into account. Compared to dilution the impact of the fuel on ε is lower for low percentages of DME.

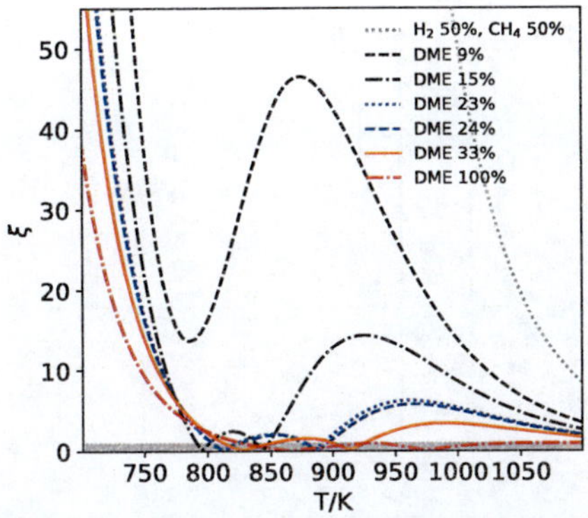

Fig. 7. Impact of fuel blends with varying blending ratios on the non-dimensional parameter ξ over temperature (AramcoMech 3.0, temperature gradient ± 1 K/mm, 35 atm, 887 K and stoichiometric conditions)

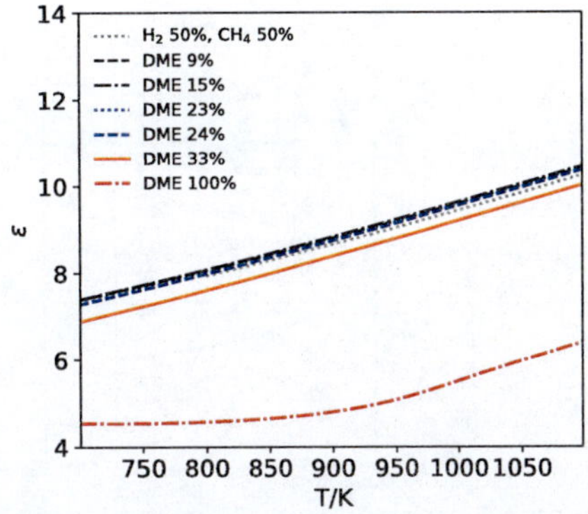

Fig. 8. Impact of fuel blends with varying blending ratios on the non-dimensional parameter ε over temperature (AramcoMech 3.0, hot spot radius 5 mm, 35 atm, 887 K and stoichiometric conditions)

At the condition of 35 atm and 887 K Fig. 9 shows the $\xi - \varepsilon$ diagram for the considered ternary fuel blends. For the numerical simulations, hot spot temperature gradients range from ± 1 K/mm to ± 10 K/mm and hot spot radii range from 1 mm to 10 mm. The conditions of the surrounding gas are kept constant, while varying fuel blend ratios. Simulations with the same hot spot parameters are connected with grey lines to illustrate the effect of DME addition.

Reference points (black asterisks) are hot spots with negative temperature gradients, because the fuel blend of CH_4/H_2 exhibits negative temperature sensitivities of ignition delay times. It can be shown that small percentages of DME in the ternary fuel blend with H_2 and CH_4 reduce the very high values of ξ for the H_2/CH_4 blend very effectively, whereas the impact on ε is rather weak (Fig. 9). DME contents higher than 33% have only a reduced impact on ξ, but an increasing impact on ε. Depending on the hot spot temperature gradient and radius, DME addition reduces ξ values in some cases to the necessary order of magnitude to fulfill the conservative criterion $\xi\varepsilon = 1$. But for most of the considered cases this criterion is not fulfilled, even for high DME contents.

Nevertheless, to reach values for ξ and ε under the $\xi\varepsilon = 1$ curve is a very strict requirement for detonation prevention. Usually a detonation peninsula is used to identify points where detonation formation occurs. In contrast to the $\xi\varepsilon = 1$ curve, which has a fixed position in the $\xi\varepsilon$ diagram, a peninsula is dependent on many parameters like temperature, pressure and fuel. Hence, fuel blending has an impact on the shape and position of the detonation peninsula.

Detonation peninsulas from literature are used here to compare the auto-ignitive wave propagation modes of pure fuels to the effect of fuel blending on ξ and ε. For this purpose numerical simulations at the same conditions and varying hot spot parameter (temperature gradients range from -1 K/mm to -10 K/mm

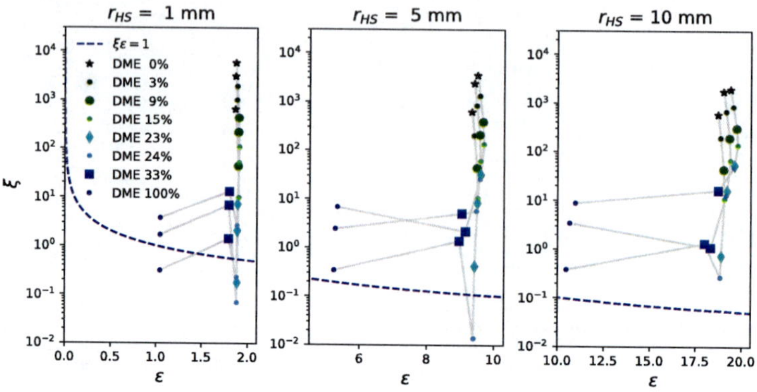

Fig. 9. $\xi - \varepsilon$ diagram at 35 atm, 887 K and stoichiometric conditions using AramcoMech 3.0

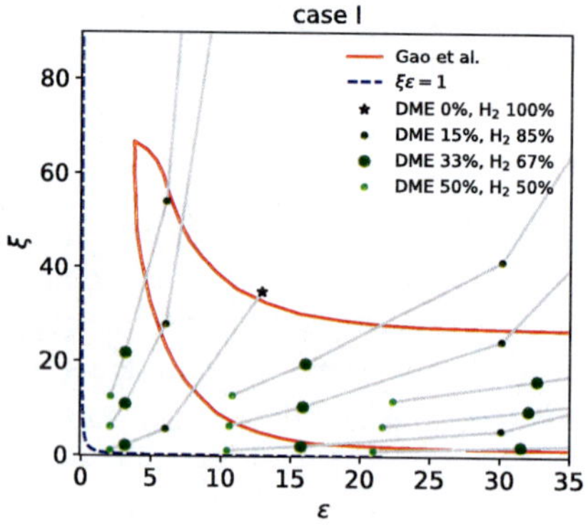

Fig. 10. Impact of blending DME and hydrogen on ξ and ε compared to regime diagram from Gao et al. [27] at 40 atm, 1000 K and stoichiometric conditions simulated with AramcoMech 3.0

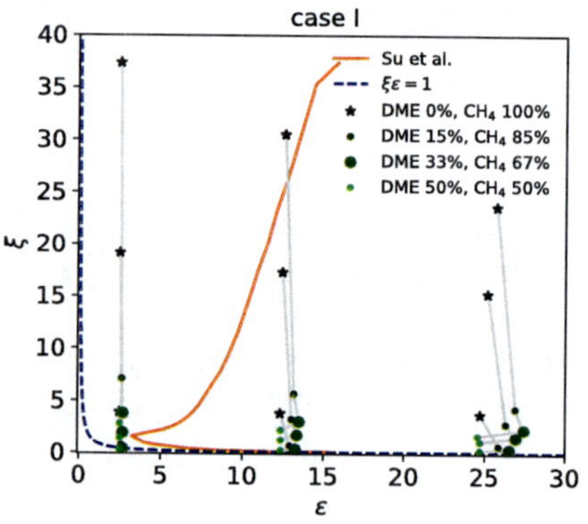

Fig. 11. Impact of blending DME and methane on ξ and ε compared to regime diagram from Su et al. [26] at 40 atm, 1300 K and stoichiometric conditions simulated with AramcoMech 3.0

and hot spot radii range from 1 mm to 8 mm) are conducted. Two dual fuel blend cases are investigated, because there are no detonation peninsulas available for fuel blends of hydrogen and methane. A detonation peninsula for hydrogen/air mixtures was investigated by Gao et al. [27] at 40 atm and 1000 K. Su et al. [26] conducted simulations to create a detonation peninsula for methane/air mixtures at 40 atm and 1300 K. Figure 10 shows the impact of blending DME and hydrogen, simulated with the AramcoMech 3.0. Values of both, ξ and ε decrease with DME addition. Depending on the hot spot parameters, DME addition can move the points out of the detonation peninsula for pure hydrogen/air mixtures and into the targeted area of supersonic auto-ignitive wave propagation. Figure 11 depicts the corresponding comparison for blends containing DME and methane. Similarly, a reducing effect on ξ can be observed, while the impact on ε is much lower than for hydrogen DME blends. Not all hot spot trajectories can be moved out of the detonation peninsula because the blending ratio in both dual fuel cases are not optimized and DME addition over 50% is not considered, because the detonation peninsula is only valid for pure H_2 or CH_4. Nevertheless a clear trend of the reducing effect of DME addition on ξ can be observed.

To conclude, it is demonstrated that adding a fuel exhibiting NTC behavior, such as DME, can be an effective measure to prevent detonation formation in SEC in presence of hot spots, when fuels with a strong temperature dependency of ignition delay time, such as H_2 and CH_4, are used. Additionally, the high reactivity of the blended DME reduces ignition delay times of those fuels at the relevant conditions to a level to make their utilization in SEC feasible.

5 Conclusions

The novel combustion concept SEC requires a quasi-homogeneous auto-ignition to achieve pressure gain combustion, which is associated with an advantage in efficiency. At the same time, processes based on auto-ignition are prone to the formation of undesired detonation waves, which needs to be avoided to evade damage to the machinery. To achieve that goal, the ignition and heat release characteristics of the fuel air mixture are modified by dilution aiming at extending the excitation time; and by fuel blending aiming at reducing the temperature sensitivity of ignition delay time. The effect on the non-dimensional regime parameters is investigated for a range of generic hot spots by means of 0-dimensional constant volume reactor simulations.

Increasing the excitation time by dilution with steam or CO_2 leads to a strong reduction in the non-dimensional parameter ε over the whole considered range of temperatures, indicating that the rapidness of heat release is decreased such that the reinforcement of the pressure wave through the heat release is mitigated. In comparison to findings from Dai et al. [25], dilution moves points in the regime diagram in the desired direction (low ε), out of the detonation peninsula into the area of super sonic auto-ignitive wave propagation leading to quasi-homogeneous auto-ignition.

It is shown that criterion $\xi\varepsilon < 1$ for regions where detonation formation is improbable is more conservative compared to detonation peninsula from [25]. Depending on the temperature and utilized diluent, dilution can lead to an increase in temperature sensitivity of ignition delay time and therefore also the ξ value. This unwanted effect can be eliminated by tailoring the fuel blend. Nevertheless, it was shown that tailored dilution of the combustion mixture with CO_2 or steam is a promising measure to mitigate detonation development and thus engine knock in auto-ignition based combustion concepts.

Reduction of the temperature sensitivity of ignition delay time and hence reduction of ξ is the main target of fuel blend tailoring in this study. In a ternary fuel blend with DME, H_2 and CH_4, DME is capable of increasing the overall reactivity and simultaneously reducing the high temperature sensitivities of ignition delay times of methane and hydrogen, which could not be used for SEC application as pure fuels. Numerical results show a strong impact of fuel blend ratio on ξ and a wider temperature range, where $\xi < 1$ or even $\xi\varepsilon < 1$, which is associated with inhibited detonation formation.

To achieve the goal of quasi-homogenous auto-ignition and inhibited detonation formation (low values for ξ and ε), fuel blending and dilution can be applied at the same time. Another advantage when using both methods at the same time is, that a larger parameter space is possible to adjust the ignition delay time, its temperature sensitivity and the excitation time simultaneously. Beside the SEC the presented findings are also interesting for other technical applications based on auto-ignition like the HCCI-engine, which are also prone to engine-knock or detonation formation.

Acknowledgements. The authors gratefully acknowledge support by the Deutsche Forschungsgemeinschaft (DFG) as part of collaborative research center SFB 1029 "Substantial efficiency increase in gas turbines through direct use of coupled unsteady combustion and flow dynamics" in project A08. Thanks for support with the measurements goes to Claudia Lugo Mayor, Adam Michael Altenbuchner, Olaniyi Oyeniyi and Yannick Kather.

References

1. Bobusch, B.C., Berndt, P., Paschereit, C.O., Klein, R.: Shockless explosion combustion: an innovative way of efficient constant volume combustion in gas turbines. Combust. Sci. Technol. **186**(10–11), 1680–1689 (2014)
2. Rähse, T.S., Paschereit, C.O., Stathopoulos, P., Berndt, P., Klein, R.: Gas dynamic simulation of shockless explosion combustion for gas turbine power cycles. In: Volume 3: Coal, Biomass and Alternative Fuels; Cycle Innovations; Electric Power; Industrial and Cogeneration Applications; Organic Rankine Cycle Power Systems of Turbo Expo: Power for Land, Sea, and Air, June 2017. V003T06A005
3. Rähse, T.S., Stathopoulos, P., Schäpel, J.S., Arnold, F., King, R.: On the influence of fuel stratification and its control on the efficiency of the shockless explosion combustion cycle. J. Eng. Gas Turbines Power **141**(1) (2019)
4. Arnold, F., Tornow, G., King, R.: Part load control for a shockless explosion combustion cycle. In: King, R. (ed.) Active Flow and Combustion Control 2018. Notes on Numerical Fluid Mechanics and Multidisciplinary Design, vol. 141, pp. 135–150. Springer, Cham (2019). https://doi.org/10.1007/978-3-319-98177-2_9
5. Yücel, F.C., Habicht, F., Arnold, F., King, R., Bohon, M., Paschereit, C.O.: Controlled autoignition in stratified mixtures. Combust. Flame **232**, 111533 (2021)
6. Yücel, F.C., Habicht, F., Bohon, M., Paschereit, C.O.: Autoignition in stratified mixtures for pressure gain combustion. In: Proceedings of the Combustion Institute, October 2020
7. Sheppard, C.G.W., Tolegano, S., Woolley, R.: On the nature of autoignition leading to knock in HCCI engines. SAE Trans. **111**, 1828–1840 (2002)
8. Lee, J.H., Knystautas, R., Yoshikawa, N.: Photochemical initiation of gaseous detonations. Acta Astronaut. **5**, 971–982 (1978)
9. Zeldovich, Y.B., Librovich, V.B., Makhviladze, G.M., Sivashinsky, G.I.: On the development of detonation in a nonuniformly heated gas. Astronaut. Acta **15**, 313–321 (1970)

10. Zeldovich, Y.B.: Regime classification of an exothermic reaction with nonuniform initial conditions. Combust. Flame **39**(2), 211–214 (1980)
11. Gu, X.J., Emerson, D.R., Bradley, D.: Modes of reaction front propagation from hot spots. Combust. Flame **133**(1–2), 63–74 (2003)
12. Berndt, P.: Mathematical modeling of the shockless explosion combustion. Ph.D. thesis, Freie Universität Berlin, Berlin (2016)
13. Berndt, P., Klein, R.: Modeling the kinetics of the shockless explosion combustion. Combust. Flame **175**, 16–26 (2017)
14. Zander, L., Tornow, G., Klein, R., Djordjevic, N.: Knock control in shockless explosion combustion by extension of excitation time. In: King, R. (ed.) Active Flow and Combustion Control 2018, vol. 141, pp. 151–166. Springer, Cham (2019). https://doi.org/10.1007/978-3-319-98177-2_10
15. Quintens, H., Strozzi, C., Zitoun, R., Bellenoue, M.: Experimental investigation of end-gas autoignition-to-detonation transition for an N-Decane/O2/Ar mixture. Shock Waves **30**(3), 287–303 (2020)
16. Bates, L., Bradley, D., Paczko, G., Peters, N.: Engine hot spots: modes of autoignition and reaction propagation. Combust. Flame **166**, 80–85 (2016)
17. Gorbatenko, I., Bradley, D., Tomlin, A.S.: Auto-ignition and detonation of n-butanol and toluene reference fuel blends (TRF). Combust. Flame **229**, 111378 (2021)
18. Dai, P., Chen, Z., Gan, X.: Autoignition and detonation development induced by a hot spot in fuel-lean and CO2 diluted n-heptane/air mixtures. Combust. Flame **201**, 208–214 (2019)
19. Cai, L., Pitsch, H.: Tailoring fuels for a shockless explosion combustor. In: King, R. (ed.) Active Flow and Combustion Control 2014. Notes on Numerical Fluid Mechanics and Multidisciplinary Design, pp. 299–315. Springer, Cham (2015). https://doi.org/10.1007/978-3-319-11967-0_19
20. Vinkeloe, J., Zander, L., Szeponik, M., Djordjevic, N.: Tailoring the temperature sensitivity of ignition delay times in hot spots using fuel blends of dimethyl ether, Methane, and Hydrogen. Energy Fuels **34**(2), 2246–2259 (2020)
21. Arcoumanis, C., Bae, C., Crookes, R., Kinoshita, E.: The potential of di-methyl ether (DME) as an alternative fuel for compression-ignition engines: a review. Fuel **87**(7), 1014–1030 (2008)
22. Djordjevic, N., Rekus, M., Vinkeloe, J., Zander, L.: Shock tube and kinetic study on the effects of CO_2 on dimethyl ether autoignition at high pressures. Energy Fuels **33**(10), 10197–10208 (2019)
23. Zander, L., Vinkeloe, J., Djordjevic, N.: Ignition delay and chemical- kinetic modeling of undiluted mixtures in a high-pressure shock tube: nonideal effects and comparative uncertainty analysis. Int. J. Chem. Kinet. **53**(5), 611–637 (2021)
24. Goodwin, D.G., Speth, R.L., Moffat, H.K., Weber, B.W.: Cantera: an object-oriented software toolkit for chemical kinetics, thermodynamics, and transport processes (2021). Version 2.5.1
25. Dai, P., Qi, C., Chen, Z.: Effects of initial temperature on autoignition and detonation development in dimethyl ether/air mixtures with temperature gradient. Proc. Combust. Inst. **36**(3), 3643–3650 (2017)
26. Jingyi, S., Dai, P., Chen, Z.: Detonation development from a hot spot in methane/air mixtures: effects of kinetic models. Int. J. Engine Res. **22**, 2597–2606 (2020)
27. Gao, Y., Dai, P., Chen, Z.: Numerical studies on autoignition and detonation development from a hot spot in hydrogen/air mixtures. Combust. Theor. Model. **24**(2), 245–261 (2020)

28. Bradley, D., Kalghatgi, G.T.: Influence of autoignition delay time characteristics of different fuels on pressure waves and knock in reciprocating engines. Combust. Flame **156**(12), 2307–2318 (2009)
29. Burke, S.M., et al.: An experimental and modeling study of propene oxidation Part 1: speciation measurements in jet-stirred and flow reactors. Combust. Flame **161**(11), 2765–2784 (2014)
30. Burke, S.M., et al.: An experimental and modeling study of propene oxidation Part 2: ignition delay time and flame speed measurements. Combust. Flame **162**(2), 296–314 (2015)
31. Burke, U., Metcalfe, W.K., Burke, S.M., Heufer, K.A., Dagaut, P., Curran, H.J.: A detailed chemical kinetic modeling, ignition delay time and jet-stirred reactor study of methanol oxidation. Combust. Flame **165**, 125–136 (2016)
32. Kéromnès, A., et al.: An experimental and detailed chemical kinetic modeling study of hydrogen and syngas mixture oxidation at elevated pressures. Combust. Flame **160**(6), 995–1011 (2013)
33. Li, Y., Zhou, C.-W., Somers, K.P., Zhang, K., Curran, H.J.: The oxidation of 2-butene: a high pressure ignition delay, kinetic modeling study and reactivity comparison with isobutene and 1-butene. Proc. Combust. Inst. **36**(1), 403–411 (2017)
34. Metcalfe, W.K., Burke, S.M., Ahmed, S.S., Curran, H.J.: A hierarchical and comparative kinetic modeling study of C1–C2 hydrocarbon and oxygenated fuels. Int. J. Chem. Kinet. **45**(10), 638–675 (2013)
35. Zhou, C.-W., et al.: A comprehensive experimental and modeling study of isobutene oxidation. Combust. Flame **167**, 353–379 (2016)
36. Zhao, Z., Chaos, M., Kazakov, A., Dryer, F.L.: Thermal decomposition reaction and a comprehensive kinetic model of dimethyl ether. Int. J. Chem. Kinet. **40**(1), 1–18 (2008)
37. Dai, P., Chen, Z., Chen, S., Ju, Y.: Numerical experiments on reaction front propagation in n-heptane/air mixture with temperature gradient. Proc. Combust. Inst. **35**(3), 3045–3052 (2015)
38. Zhou, C.-W., et al.: An experimental and chemical kinetic modeling study of 1,3-butadiene combustion: ignition delay time and laminar flame speed measurements. Combust. Flame **197**, 423–438 (2018)

Autoignition Modes in a Shockless Explosion Combustor

Fatma Cansu Yücel[1(✉)], Fabian Habicht[1], Myles Bohon[2], and Christian Oliver Paschereit[1]

[1] Chair of Fluid Dynamics, Technische Universität Berlin, Straße des 17. Juni 135, 10623 Berlin, Germany
f.yuecel@tu-berlin.de
[2] Chair of Pressure Gain Combustion, Technische Universität Berlin, Straße des 17. Juni 135, 10623 Berlin, Germany

Abstract. Pressure gain combustion as an alternative to isobaric combustion has been in the focus of research for the past decades as it potentially allows for increasing the thermal efficiency of conventional gas turbines significantly. Beside the most known concepts, such as pulse detonation and rotation detonation combustors, the shockless explosion combustor (SEC) has been proposed. In contrast to the previously mentioned detonation-based concepts the SEC process is based on a thermal explosion, hence avoiding entropy generation caused by propagating detonation waves. Conceptually, this is achieved through a homogeneous autoignition of the fuel–oxidizer mixture, which is realized by the proper stratification of the fuel concentration throughout the combustor, leading to a gradual rise in pressure. Since the process of autoignition is highly sensitive to perturbations, local deviations in the initial state of the mixture lead to a variety of autoignition modes. In this work, an SEC test rig is used to investigate the impact of different fuel injection profiles on the formation of autoignition modes. Pressure transducers are used to measure the pressure rise subsequent to the autoignition event. k-means clustering is applied to a set of pressure data to classify the measured pressure profiles. The same method was used to cluster the respective injection profiles. The results reveal that the gradient in reactivity is of major importance and can be used for increasing the pressure rise through ignition.

Keywords: Pressure gain combustion · Modes of autoignition · k-means clustering

1 Introduction

Implementing pressure gain combustion into a conventional gas turbine is a promising concept for increasing the overall thermal efficiency of the thermodynamic cycle. One concept among others is the shockless explosion combustor (SEC), which has been studied numerically and experimentally in previous works [1,2,6,8]. The concept is based on a periodic combustion process (Fig. 1),

R. King and D. Peitsch (Eds.): AFCC 2021, NNFM 152, pp. 35–46, 2022.
https://doi.org/10.1007/978-3-030-90727-3_3

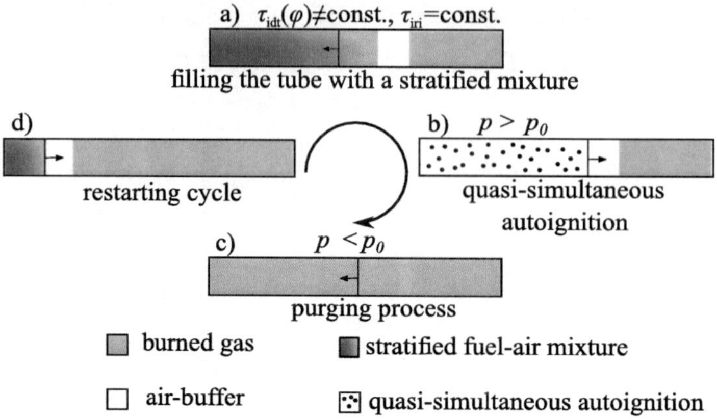

Fig. 1. Sketch of the SEC cycle.

including the quasi-instantaneous autoignition of a previously injected fuel–air package. The combustor is fed with a continuous air flow in which a well-defined fuel profile is added resulting in an axially stratified fuel–air mixture (Fig. 1a). The injected fuel profile has been tailored in order to compensate for the variation in residence time such that a simultaneous ignition of the entire combustor volume is achieved (Fig. 1b). Subsequent to the combustion event a pressure wave is induced that travels downstream. At the acoustically open outlet of the combustor the pressure wave is reflected as an expansion wave (Fig. 1c). The expansion wave then travels upstream and induces a pressure drop at the combustor inlet, which supports the refilling process (Fig. 1d).

The amplitude of the pressure rise induced by the ignition depends on the ignition homogeneity [8]. In particular, a perfectly homogeneous autoignition, is associated with a maximum rise in pressure. Since the ignition delay time τ_{idt} is a function of the local temperature T, pressure p and equivalence ratio φ the ignition time relative to the start of the injection τ_{iri} can be well controlled by the proper adjustment of the local equivalence ratio when assuming T and p to remain constant. However, a perfectly homogeneous autoignition is not feasible in applications and thus regarded as more of a theoretical consideration. In experiments, small but unavoidable perturbations in the injection process (e.g. pressure, temperature, and mixture composition) lead to deviations in the local ignition time τ_{iri}. The resulting spatial gradient in τ_{iri} causes an autoignition front, propagating with the velocity u_{ai} away from an exothermic center (region with increased reactivity). The propagating reaction front can be classified in four regimes that were first identified by Zel'dovich [11]. Based on his observations Zel'dovich introduced a dimensionless parameter:

$$\xi = \frac{a}{u_{\mathrm{ai}}} \tag{1}$$

with the speed of sound a. Gu et al. defined lower and upper limits of ξ, ξ_l and ξ_u, respectively. Based on these limits the occurrence of different modes can be summarized as follows [4,5]:

$\xi > \xi_u$:	subsonic autoignitive flame propagation or deflagration,
$\xi_l < \xi < \xi_u$:	coupling of a reaction front with a pressure wave forming a detonation,
$0 < \xi < \xi_l$:	supersonic autoignitive flame propagation,
$\xi = 0$:	thermal explosion (homogeneous autoignition).

Despite the unavoidable cycle-to-cycle variation in the local ignition time in the application, an overall increase in the pressure amplitude was achieved in a previous study [10] by applying a closed-loop control of the fuel injection based on the cycle-averaged pressure rise. However, a significant cycle-to-cycle variation in the recorded pressure rise remained. Measurements with an optically accessible combustor were conducted to examine single ignition events to further analyze the underlying effects. The results revealed the formation of different autoignition modes to be responsible for the cycle-to-cycle variation.

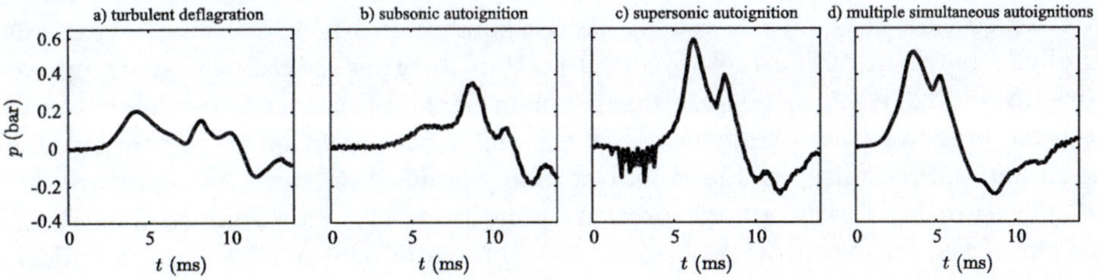

Fig. 2. Pressure histories of different autoignition modes observed by optical measurements in combination with pressure measurements in [10].

Four different autoignition modes were identified in [10] based on optical measurements in combination with pressure records: a) turbulent deflagration, b) subsonic autoignition, c) supersonic autoignition and d) aerodynamic confinement by multiple simultaneous autoignition fronts. Figure 2 shows example pressure histories for each case. It was observed that the simultaneous initiation of multiple autoignitions can lead to a similar rise in pressure compared with a supersonic autoignition. However, the differentiation between these two modes requires optical measurement and cannot be identified based on pressure data exclusively since the rise in pressure appears similar in both cases. Moreover, it was observed that the ignition characteristics of the applied fuel impacted the autoignition process. The used fuel dimethyl-ether (DME), as most hydrocarbon fuels, exhibits a negative temperature coefficient, meaning within a certain temperature range the ignition delay time decreases with an increasing temperature leading to a multi-stage ignition behaviour [7]. Single-stage ignitions are accompanied by a fast and quasi-instantaneous heat release, whereas multi-stage ignition processes are characterized by a temporally distributed heat release during each stage. An increasing heat release contributes to a faster propagating autoignition front, and thus, leading to a greater aerodynamic confinement. Consistently, a two-stage ignition was found in the experiments to correlate with a lower rise in pressure (Fig. 2b) while a single-stage ignition was associated with a higher pressure rise (Fig. 2c and d).

Taking these observations as a basis, in this work, the formation of autoignition modes as a consequence of the fuel concentration distribution is evaluated based on pressure measurements. Therefore, a set of pressure data is evaluated by applying a clustering algorithm allowing for the identification of characteristic pressure histories, which are subsequently correlated to the fuel concentration distribution inside the combustor. The probability of appearance of the different ignition modes is determined with respect to the applied fuel injection profile.

2 Experimental Setup and Measurement Procedure

A test rig was designed to allow for the investigation of the autoignition process in stratified mixtures under atmospheric pressure and high temperature conditions. A cross section of the used test facility is sketched in Fig. 3. An air preheater is applied upstream of the fuel injection station in order to rise the air temperature up to 1023 K. A restriction downstream of the preheater is implemented to prevent pressure waves that occur subsequent to the ignition from propagating upstream. DME is used as fuel due to its short ignition delay times at the applied temperatures and atmospheric pressure conditions [3]. This enables the examination of the autoignition process in the combustor section. The fuel is guided through a vaporizer in order to ensure gaseous injection and is subsequently injected downstream of the restriction into the air flow by ten circumferentially distributed ports, each individually controlled by a high-speed solenoid valve (Staiger VA 204-716). Adjusting the number of open valves allows for a precise injection of a pre-defined fuel trajectory within a given injection period t_{inj}. The feeding line pressure is controlled by applying a high-speed dome-loaded pressure regulator (Swagelok RD6) which prevents a pressure drop during the fuel injection duration. A convection tube with a length of 500 mm and an inner diameter of 40 mm is mounted between the injection station and the combustor to compensate for the ignition delay time, which ensures the ignition event to take place inside the combustor section. The combustor consists of a 500 mm stainless steel tube equipped with four high-speed, piezoresistive, pressure transducers P1 to P4 (Kulite EWCTV-312), as shown in Fig. 3. Two low-speed, thin-film metal, static pressure transducers FA and FF (Festo SPTW) are installed to monitor the air and fuel supply pressures, respectively. Thermocouples T1 and T2 record the wall temperature during each run.

Different injection trajectories are defined by the simultaneous operation of a number of valves. By this, a defined fuel profile is injected into a continuous air flow with a frequency of 5 Hz. The local equivalence ratio is varied between

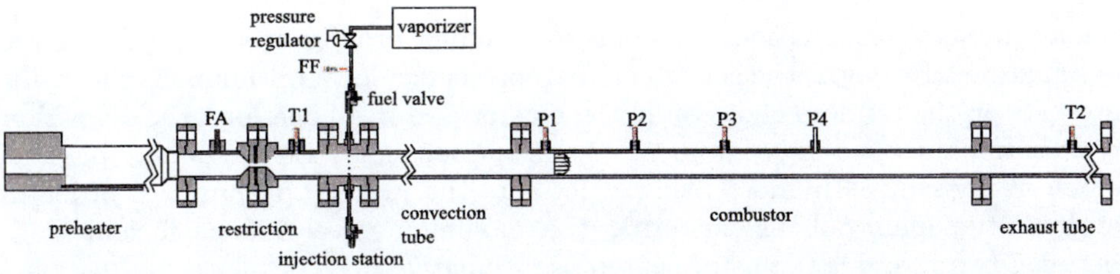

Fig. 3. Sketch of the atmospheric test rig equipped with high frequency pressure transducers (P1–P4), thermocouples (T1–T2) and low frequency pressure transducers (FF and FA).

$\varphi \in [1, 1.6]$ by adjusting the number of simultaneously open valves during the injection. The air mass flow rate was set to a steady-state value of 30 kg/h resulting in a mean air flow velocity of 18.5 m/s for the given boundary conditions. At the beginning of each cycle a defined mixture profile is added to the continuous air flow for a total injection duration of $t_{\mathrm{inj}} = 30\,\mathrm{ms}$. Each injection trajectory is defined by ten individual valve settings each operated for 3 ms. The average number of open valves is restricted to a range between 7 and 7.8, ensuring a comparable fuel mass flow rate in each cycle with a maximum deviation of 10% in the global equivalence ratio. Pressure data are recorded subsequent to ignition of 40 individual cycles with a sampling frequency of 10 kHz.

2.1 *k*-means Clustering

A total number of 1080 cycles from measurements with 27 arbitrary injection trajectories were examined. The pressure signal at sensor P2 (see Fig. 3) was evaluated for each individual cycle for $50\,\mathrm{ms} \leq t \leq 150\,\mathrm{ms}$. The recorded pressure signals were classified using *k*-means clustering. The iterative algorithm starts with a number of *k* cluster centers $C_{k,i}$, which are initiated from randomly chosen pressure traces. Each cycle data is then assigned to the nearest cluster center based on the euclidean distance

$$\mathrm{d}(p_j, C_{k,i}) = \|p_j - C_{k,i}\|_2, \tag{2}$$

where p_j denotes the pressure signal of cycle j. Subsequently, the cluster centers are recalculated from the cycle-averaged pressure signals from all cycles that are associated to the respective clusters. By iterating this procedure until the assignment of the pressure signals does not change anymore, an optimum classification is found. The results are characterized by the sum of distances to the cluster centers, according to

$$D_k = \sum_{i=1}^{k} \sum_{j=1}^{N_i} \mathrm{d}(p_j, C_{k,i}), \tag{3}$$

where N_i denotes the number of cycles that are assigned to the cluster i. For convenience in the discussion the individual clusters for pressure traces are given roman numeral designations and clusters for fuel profiles are given alphabetic designations. However, the solution only represents a local optimum as the result depends on the random choice of initial cluster centers. Therefore, the algorithm is executed independently for 50 times and the solution with the minimum value of D_k is chosen, as it most likely represents the global optimum. In order to evaluate the quality of the clustering, the maximum correlation coefficient R_{max} between two individual cluster centers is calculated:

$$R_{\text{max}} = \max\left(R\left(p_i, p_j\right)\right)\;,\quad i, j = 1...k. \tag{4}$$

In this work, the value of R_{max} is used to identify a reasonable k to ensure fundamentally different clusters with minimum resemblance, while minimizing the variation within each cluster.

2.2 Fuel Injection Modelling

To ensure a precise stratification of the injected fuel profile, previous investigations on the applied injection strategy have been conducted. The results clearly demonstrate the ability to inject a stratified mixture within the injection duration, which remains preserved during convection until the onset of ignition. Detailed results can be found in [9]. However, a summary is provided in the following for the sake of completeness. In general, the measurements revealed that by adjusting the number of open valves the fuel concentration distribution inside the combustor can be controlled. It was found that the preservation of the fuel profile is independent of the temperature and mainly dependent on the spatial width $\chi = \bar{u}_{\text{air}} t_{\text{inj}}$ of the injected profile, where $\bar{u}_{\text{air}}$ is the mean air flow velocity and t_{inj} is the injection duration.

Further investigations were conducted in [10] at varying temperatures of the air flow. The volumetric flow rate was kept constant allowing for a quantification of the impact of the Reynolds number on the diffusion processes. By this, the impact of turbulent fluctuations on the preservation of the injected fuel profile was analyzed. Based on the measurement data a simulation tool was developed solving the one-dimensional diffusion equation. Figure 4a illustrates an example injection trajectory (black line) with a duration of 30 ms assembled from ten consecutive valve settings. Since the injection in the experiments is affected by the inertia of the valves a polynomial Bézier curve was used to model the gradual change in the injected fuel flow rate. Moreover, the initial opening speed and the time of the valves are used as parameters for the adjustment of the valve behavior. The parameters were determined by matching the measured concentration profiles with the calculation results for a number of generic injection trajectories. The obtained model parameters were subsequently used to assess the fuel concentration at the injection station for arbitrary injection trajectories (red line).

Comparing the measurement data with the simulation results (Fig. 4b) revealed that the fuel distribution inside the combustor can be very well reproduced by the one-dimensional calculations. Hence, the fuel concentration distribution before the onset of ignition can be accurately predicted.

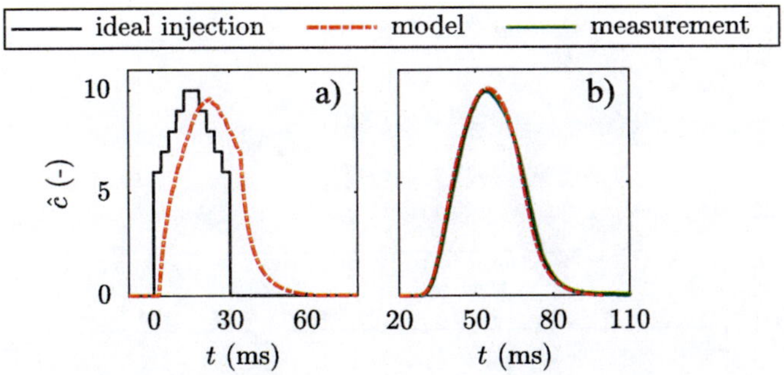

Fig. 4. Actual and modeled fuel injection trajectory (a), measured and modeled fuel concentration profile (b).

3 Results

In this section, first, all pressure signals will be clustered using the previously introduced k-means algorithm. Next, the one-dimensional simulation tool will be used to calculate the fuel distribution before the onset of ignition. The obtained fuel profiles are than clustered likewise and subsequently correlated to the obtained pressure clusters.

3.1 Clustering the Pressure Signals

Figure 5 shows the different cluster centers $C_{\mathrm{k,i}}$ (red lines) obtained for $k = 2$, 3 and 4. The clusters are sorted in ascending order based on the maximum pressure amplitude. This parameter is chosen since it is a driving parameter in the performance of the SEC. Further, it has been proven to be a suitable control parameter for an optimization approach of the SEC process [10]. The gray lines visualize the individual pressure traces of cycles that are assigned to the respective cluster. The number of cycles that are assigned to a certain clusters is noted on top of each figure. As expected, an increasing k leads to a higher congruence of the associated pressure traces. However, comparing the clusters obtained for $k = 3$ with the autoignition modes presented in Fig. 2, a great correspondence of the respective clusters and modes is visible. Cluster $C_{3,\mathrm{I}}$ shows features which have been observed for turbulent flames, while $C_{3,\mathrm{II}}$ show great correspondence with a two-stage ignition, inducing a faster propagation autoignition front. $C_{3,\mathrm{III}}$ exhibits similar features as a single-stage autoignition, which is associated with a high increase in pressure and a supersonic autoignition front.

When comparing the clusters obtained for each k, it is apparent that an increasing k results in a greater resemblance of some of the obtained clusters. While all clusters obtained for $k = 2$ and $k = 3$ seem fundamentally different, $C_{4,\mathrm{I}}$ and $C_{4,\mathrm{II}}$ show notable similarities. However, in order to objectively determine a reasonable k, a cross-correlation of the clusters obtained for each k is calculated

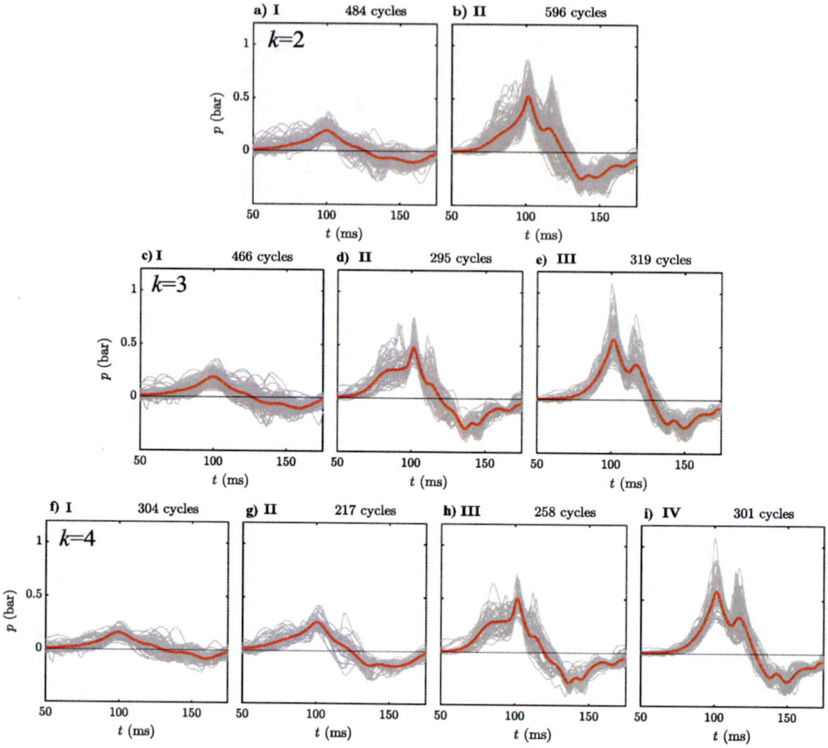

Fig. 5. Result of the k-means algorithm for $k = 2$, 3 and 4.

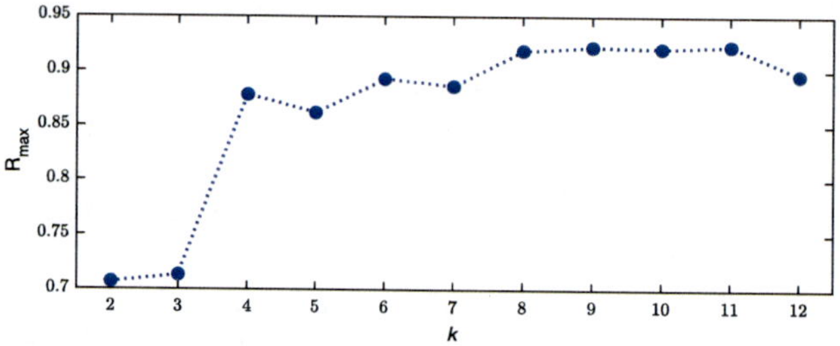

Fig. 6. Correlation coefficients for $k \in [1, 12]$

in the following. By this, it can be extracted in how far the individual clusters resemble each other. Figure 6 shows the maximum correlation coefficient $R_{\max}$ for two individual cluster centers with respect to the number of clusters. $R_{\max}$ slightly increases from $k = 2$ to $k = 3$, and increases significantly for $k = 4$ and reaches a local maximum of about 88 %. For larger values of k only a small increase is visible. Hence $k = 3$ is chosen for further examinations to ensure fundamentally different clusters with minimum resemblance.

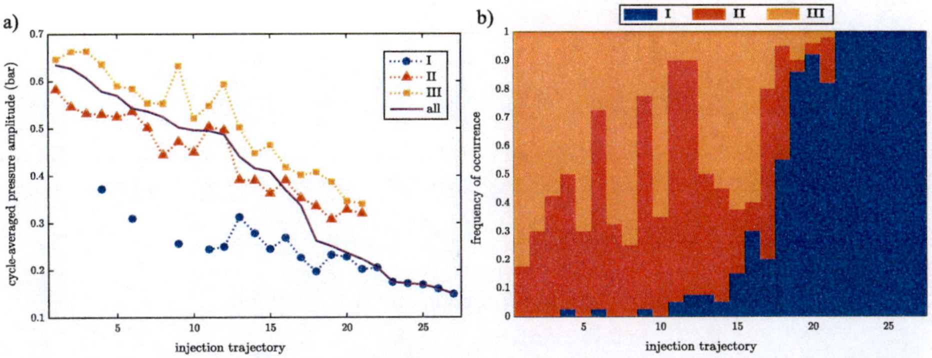

Fig. 7. Result of the k-means algorithm for $k = 3$ for all considered injection trajectories. The injection trajectories are sorted by the cycle-averaged pressure amplitude.

To evaluate the effect of the applied injection trajectory on the probability to observe a specific ignition mode, the frequency of occurrence of the ignition cluster is examined for each injected fuel profile. The applied injection trajectories are sorted with respect to the cycle-averaged maximum pressure amplitude, which is visualized in Fig. 7a (purple line). The amplitude varies from 0.65 bar down to 0.15 bar for the injection trajectories 1 to 27, respectively. It can clearly be seen that pressure signals from clusters **II** and **III** are generally correlated to larger pressure amplitudes. Cluster **I** evokes only small pressure amplitudes compared. At intermediate pressure levels cluster **I** occurs sporadically and is non-existent at high pressures. The opposite applies to clusters **II** and **III**. The frequency of occurrence of each respective cluster for the applied injection trajectory is shown in Fig. 7b. Clusters **II** and **III** appear frequently at injection trajectories that are associated with large pressure amplitudes. For trajectories that are linked to low pressure levels, cluster **III** is predominant. Overall, the data reveals that the frequency of occurrence of the identified clusters can be shifted by the applied injection trajectory, which ultimately causes a variation in the cycle-averaged pressure amplitude.

3.2 Clustering the Fuel Concentration Profiles

To systematically investigate the dependence between the injected fuel profile and the combustion process, the fuel distribution inside the combustor prior to the first detected pressure rise is assessed numerically by applying the previously introduced one-dimensional simulation tool. The obtained axial distributions in fuel concentration are then categorized by k-means clustering. Figure 8 shows the respective fuel clusters obtained for $k = 4$. The gray lines denote the fuel profile, that are assigned to the respective cluster, while the red line represent the cluster centers. Each cluster represents a unique shape of the fuel concentration distribution inside the combustor before the onset of ignition. The fuel clusters

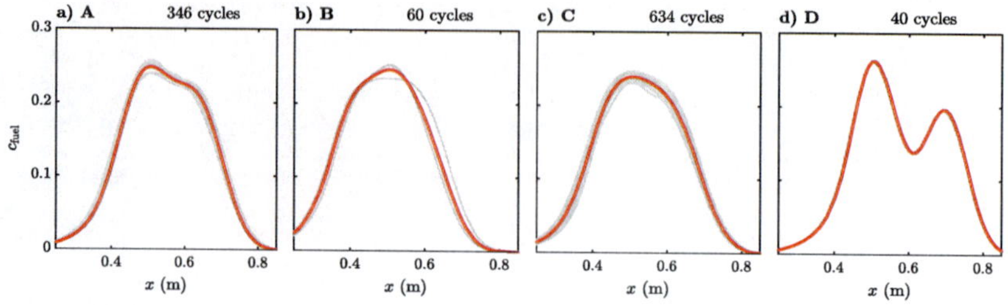

Fig. 8. Result of the k-means algorithm for $k = 4$.

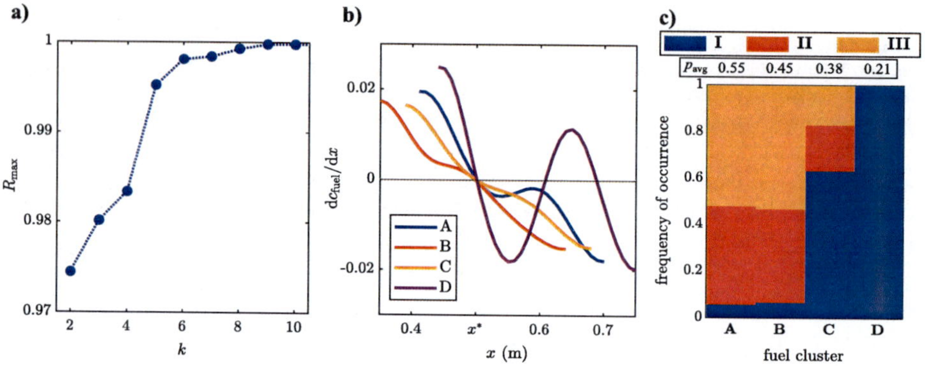

Fig. 9. Frequency of occurrence of ignition patterns with respect to the fuel clusters.

A–**D** are again sorted with regard to the average pressure rise in the associated combustion cycles.

The maximum correlation coefficient between two clusters R_{max} is determined and visualized in Fig. 9a to identify a reasonable number of k clusters. In general, an increasing k leads to a higher value of R_{max}. However, the largest increase is visible from $k = 4$ to $k = 5$. To ensure the identification of fundamentally different injection profiles with reasonably low resemblance, $k = 4$ is considered in the following. This number was chosen as it allows for distinguishing the main features of the fuel profiles, while ensuring a considerable difference between each pair of fuel profile clusters, as visible in Fig. 8. In Fig. 9b, the gradient in fuel distribution for each fuel cluster is visualized. Analogous to the fuel concentration profiles in Fig. 8, the maximum fuel concentration is aligned to $x^* = 0.5$ m, resulting in a zero value for the gradients of all graphs at this position in Fig. 9b. Assuming constant temperature, the ignition event is expected to occur at the axial position of the maximum fuel concentration. According to the observations from Zel'dovich [11], the resulting gradient in mixture reactivity close to this point is decisive for the formation of a certain ignition mode. For fuel cluster **A**, a nearly constant negative gradient is observed for $x > x^*$. In a reasonably large number of cycles, this presumably leads to the formation of a supersonic autoignition front or multiple simultaneous autoignition kernels, both resulting in a pressure signal associated to the ignition cluster **III**, as shown in Fig. 9c. Similar considerations can be formulated for fuel cluster **B**. Here, a nearly constant positive gradient is observed for $x < x^*$ over a reasonable length. For fuel cluster **C**, the absolute value of the gradient in fuel concentration exhibits a steeper increase in both axial direction from x^* when compared to the fuel clusters **A** and **B**. Thus, the formation of an ignition mode associated with cluster **III** is hindered and the ignition cluster **I** is observed more frequently. Fuel cluster **D** is associated with only minor rise in pressure. Here, two fuel concentration peaks are visible with steep gradients next to the associated axial positions. Clearly, this fuel distribution induces the lowest average pressure rise and the ignition cluster can be entirely assigned to cluster **I**. This goes along with previous investigations in [8] and [10], where optical measurements revealed that a fuel concentration distribution similar to cluster **D** leads to the formation of two turbulent flame fronts, which are initiating with a certain time delay. Hence, no aerodynamic confinement is achieved impeding a large rise in pressure.

4 Conclusion

In this work, the effect of fuel stratification on the formation of autoignition modes is analyzed based on pressure measurements. An SEC test rig is used to measure the pressure rise subsequent to the ignition event for various arbitrary fuel profiles. k-means clustering is used to define a number of ignition clusters based on pressure measurements inside the combustor. It was found that the clusters resemble different autoignition modes observed in previous investigations, namely: turbulent deflagration, subsonic autoignition and supersonic autoignition. In addition, a previously developed one-dimensional simulation tool is used to assess the axial distribution in fuel concentration prior to the ignition event based on the applied injection trajectory. The fuel profiles are subsequently clustered likewise and correlated with the respective ignition clusters. The results show that the gradient in the fuel distribution promotes the occurrence of different autoignition modes. In particular, a decrease in fuel concentration along the combustor length was associated with an increased homogeneity while a fuel distribution with two concentration peaks impedes an aerodynamic confinement causing a low rise in pressure. The cycle-averaged pressure amplitude for a given fuel cluster was found to be highly dependent on the frequency of occurrence of the individual ignition clusters. In conclusion, the cycle-to-cycle variation in the pressure amplitude, which has been observed in previous investigations, can be attributed to the formation of different autoignition modes. The results reveal that the gradient in fuel concentration close to the position of the first ignition greatly impacts the predominant autoignition mode. Based on the findings obtained with regard to the gradient in fuel concentration a more robust control algorithm can be implemented.

Acknowledgment. The authors gratefully acknowledge support by the Deutsche Forschungsgemeinschaft (DFG) as part of Collaborative Research Center SFB 1029

"Substantial efficiency increase in gas turbines through direct use of coupled unsteady combustion" on projects A03. Special thanks go out to Andy Göhrs and Thorsten Dessin for their technical support.

References

1. Berndt, P., Klein, R., Paschereit, C.O.: A kinetics model for the shockless explosion combustion. In: Turbo Expo: Power for Land, Sea, and Air, vol. 49767, p. V04BT04A034. American Society of Mechanical Engineers (2016)
2. Bobusch, B.C., Berndt, P., Paschereit, C.O., Klein, R.: Shockless explosion combustion: an innovative way of efficient constant volume combustion in gas turbines. Combust. Sci. Technol. **186**(10–11), 1680–1689 (2014)
3. Burke, U., et al.: An ignition delay and kinetic modeling study of methane, dimethyl ether, and their mixtures at high pressures. Combust. Flame **162**(2), 315–330 (2015)
4. Dai, P., Qi, C., Chen, Z.: Effects of initial temperature on autoignition and detonation development in dimethyl ether/air mixtures with temperature gradient. Proce. Combust. Inst. **36**(3), 3643–3650 (2017)
5. Gu, X.J., Emerson, D.R., Bradley, D.: Modes of reaction front propagation from hot spots. Combust. Flame **133**(1–2), 63–74 (2003)
6. Reichel, T., Schäpel, J.S., Bobusch, B.C., Klein, R., King, R., Paschereit, C.O.: Shockless explosion combustion: experimental investigation of a new approximate constant volume combustion process. J. Eng. Gas Turb. Power **139**(2), 021504 (2017)
7. Warnatz, J., Maas, U., Dibble, R.: Combustion. Springer, Berlin (2006). https://doi.org/10.1007/978-3-540-45363-5
8. Yücel, F.C., Habicht, F., Bohon, M., Paschereit, C.O.: Autoignition in stratified mixtures for pressure gain combustion. Proc. Combust. Inst. **38**(3), 3815–3823 (2021)
9. Yücel, F.C., Habicht, F., Jaeschke, A., Lückoff, F., Oberleithner, K., Paschereit, C.O.: Investigation of the fuel distribution in a shockless explosion combustor. In: Turbo Expo: Power for Land, Sea, and Air, vol. 84133, p. V04BT04A058. American Society of Mechanical Engineers (2020)
10. Yücel, F.C., Habicht, F., Arnold, F., King, R., Bohon, M., Paschereit, C.O.: Controlled autoignition in stratified mixtures. Combust. Flame **232**, 111533 (2021)
11. Zeldovich, Y.B.: Regime classification of an exothermic reaction with nonuniform initial conditions. Combust. Flame **39**(2), 211–214 (1980)

Fuel-Rich Natural Gas Conversion in HCCI Engines with Ozone and Dimethyl Ether as Ignition Promoters: A Kinetic and Exergetic Analysis

Dominik Freund, Christoph Horn, and Burak Atakan(✉)

Institute for Combustion and Gas Dynamics (IVG), Chair of Thermodynamics, University of Duisburg-Essen, Lotharstraße 1, D-47057 Duisburg, Germany
burak.atakan@uni-due.de

Abstract. Fuel-rich operated HCCI engines are suitable for the polygeneration of work, heat, and base chemicals like synthesis gas ($CO + H_2$). Under favorable conditions, these engines are exergetically more efficient than separate steam reformer and cogeneration gas engines. However, to achieve ignition, reactive fuel additives like dimethyl ether or ozone must be supplied, which have some, probably negative and not yet quantified, impacts on the exergetic efficiency.

Therefore, the aim of this work is to compute and evaluate the effect of DME and ozone on the exergy input and exergetic efficiency of fuel-rich operated HCCI engines, which convert natural gas at equivalence ratios of 1.5 to 2.5.

Results of a single-zone-model (SZM) and a multi-zone model (MZM) are compared to analyze the influence of inhomogeneities in the cylinder on the system's exergetic efficiency. Natural gas as fuel is compared with previous neat methane results.

The single-zone model results show that natural gas is much more reactive than methane. Ethane and propane convert partially in the compression stroke and lead to ethene, propene, and OH radicals. However, the ethane and propane conversions do not favor but slightly reduce the formation of methyl hydroperoxide, which is an important buffer molecule for fuel-rich methane ignition. But in addition, further buffer molecules like ethene or ethyl hydroperoxide are intermediately formed. The product selectivities are neither influenced by the natural gas composition, nor by the chosen additive.

Compared to ozone, the DME molar and mass fractions needed for ignition are up to 11 times higher, and its exergy contribution to the total mixture is even 95 times higher. Therefore, the system's exergetic efficiency is much higher when ozone is chosen as additive: reasonable values of up to 82.8% are possible, compared to 67.7% with DME. The multi-zone model results show that the efficiency is strongly dependent on the fuel conversion and thus unconverted fuel should be recycled within the polygeneration system to maintain high efficiencies. Comparing the total exergetic efficiency, ozone is a favorable additive for fuel-rich operated HCCI polygeneration.

Keywords: HCCI engines · Polygeneration · Ozone · Dimethyl ether · Partial oxidation

R. King and D. Peitsch (Eds.): AFCC 2021, NNFM 152, pp. 47–65, 2022.
https://doi.org/10.1007/978-3-030-90727-3_4

1 Introduction

Fossil resource usage can be reduced by more efficient and flexible energy and chemical compound conversion technologies. A promising approach is the polygeneration of work, heat, and synthesis gas in fuel-rich operated internal combustion (IC) engines which is a mid-term alternative to separate cogeneration gas engines and steam reforming processes.

Several works in the 2000s and 2010s showed that partial oxidation of methane or natural gas is feasible in spark-ignition (SI), compression ignition (CI), and homogeneous charge compression ignition (HCCI) engines. Each ignition type has its own challenges for stable operation, though.

CI engines typically need an additional fuel injection, e.g. diesel or n-heptane, leading to stratification within the cylinder. Karim et al. [1] investigated the partial oxidation of methane with highly oxygenated air in a dual-fuel CI engine experimentally in 2008 for equivalence ratios of 2.0 to 3.5 and found that this process is feasible and yields up to 80% synthesis gas in the dry exhaust gas. However, a diesel injection was still needed for ignition and combustion control.

Ignition in SI engines is dependent on the flame speed and thus limited to lower equivalence ratios. However, Lim et al. [2] modified a CI engine to perform spark ignition and achieved stable operation of methane/air-mixtures for equivalence ratios up to 2. For ignition, high intake temperatures of up to 450 °C were needed and the spark ignition timing was adjusted between 45 and 30° crank angle (°CA) before top dead center. This operation range could be extended by adding 5% hydrogen and 10% ethane to the fuel mixture – to represent natural gas. Then, a stable operation at an equivalence ratio (ϕ) of up to 2.8 was possible.

A very high flexibility can be achieved with HCCI engines, which are kinetically controlled and thus do not depend on fuel injection timing or flame speed. Our previous experimental work has shown that polygeneration with methane fueled HCCI engines is feasible and exergetic efficiencies of up to 81.5% are achieved [3, 4]. However, methane containing fuels (such as biogas or natural gas) represent a challenge for achieving HCCI due to their inertness leading to high octane-numbers and relatively high specific heat capacities. Ignition can be promoted with reactive additives such as ethers [3] or higher alkanes [5]. In our previous work we found that high additive mass fractions of up to 28% in the fuel mixture are needed when dimethyl ether (DME) [3, 6], diethyl ether (DEE), or n-heptane [4] are used.

The production of these additives leads to additional, yet partially unknown, exergy losses, and the added DME is typically produced from synthesis gas, and thus, is a product of the polygeneration system. Therefore, from a holistic viewpoint, alternatives are preferable. Ozone may be a much more suitable additive, since it is produced on demand via corona discharge [7] using surrounding air and much smaller amounts are needed to ignite methane (typically, less than 5000 ppm). Keum et al. [8] investigated the effect of ozone as an ignition promoter under stoichiometric conditions in 2018 and found that 100 ppm ozone provide a significant promotion effect in HCCI engines, whereas SI engines benefit from an increased flame speed only from an addition of 3000 to 6000 ppm. Sayyosouk et al. [9] modelled a zero-dimensional HCCI engine fueled with iso-octane at lean conditions ($\phi = 0.3$) and found that even 4 ppm ozone promote ignition effectively.

These results encouraged us to investigate the effect of ozone in fuel-rich operated HCCI engines [10]. There, the influence of ozone on the kinetics of methane in fuel-rich HCCI engines has been investigated in comparison to DME [10]. We have found that ozone is a much more effective additive than DME. Ozone decomposes very early in the compression stroke, does not noticeably increase the heat capacity of the mixture, and it leads to the formation of methyl hydroperoxide (CH_3OOH), which leads to OH radical formation near the end of the compression stroke via formaldehyde (CH_2O) and hydrogen peroxide (H_2O_2) formation. For a practical application, the HCCI engine is likely fueled with natural gas instead of neat methane. Therefore, in the present work the kinetic investigation is extended towards natural gas.

As discussed before, Lim et al. found that the addition of hydrogen and ethane extended the operation stability of fuel rich operated SI engines from $\phi = 2$ to $\phi = 2.8$. Duan et al. [11] compared neat methane with natural gas experimentally in a fuel lean operated HCCI engine. With 2.88% ethane and 0.41% propane in the natural gas mixture, the in-cylinder pressure increased by up to 4 bar, compared to neat methane. Although the combustion phasing (CA50) was shifted slightly towards earlier crank angles, this effect was not significant. Kaczmarek et al. [12] measured ignition delay times of natural gas/DME/air mixtures in a shock tube at fuel rich mixtures ($\phi = 2$, 10) and concluded that under those conditions, natural gas is much more reactive compared to methane due to its ethane and propane contents (which were set to 9% and 1%, respectively). As the natural gas composition may vary during the operation of a fuel-rich HCCI engine, the evaluation of the effect of ozone on the ignition is analyzed here and compared to the operation with neat methane.

Additionally, a comparison between ozone and DME, based on exergy methods, is carried out here. This comparison includes a comparative modelling study on the influence of DME and ozone on the engine's exergy input and their influences on the exergetic efficiency of the polygeneration system. The exergetic efficiency of the ozone generator is calculated with a Python model and the exergetic efficiency of the DME production is taken from the work of Zhang et al. [13]. Zhang et al. conducted a profound exergy analysis on the steam gasification of biomass and determined an exergetic efficiency of 43.5%.

This paper aims to answer the question. whether ozone is an exergetically favorable additive compared to DME, representative for other oxygenated hydrocarbons, when natural gas fueled HCCI engine are operated at equivalence ratios of 1.5 to 2.5. Our previous work [10] showed that trends are well reproduced with a single zone model, while the fuel conversion is overestimated for fuel-rich conditions. Thus, a comparison with a multi-zone model is included here, and the importance of the colder zones on the outcome is analyzed.

2 Methodology

The engine model is written in Python and the thermodynamics and reaction kinetics are computed by using the module Cantera [14]. A four-stroke single-zone model (SZM) and recently a two-stroke multi-zone model (MZM) were developed. The multi-zone model contains seven zones (numbered as subscript i) with inter-zonal heat and mass transfer and a non-reactive crevice zone of 1.5 vol-% of the dead volume. The heat transfer between the zones $\dot{q}_{i-1\rightarrow i}$ Eq. (1) is modelled as heat conduction according to the approach of Komninos et al. [15], which was developed for HCCI engines [16].

$$\dot{q}_{i-1\rightarrow i} = -k_{tot}\frac{\partial T}{\partial r_n} \cong -k_{tot}\frac{T_i - T_{i-1}}{(t_i + t_{i-1})/2} \tag{1}$$

The temperature gradient $\frac{\partial T}{\partial r_n}$ between two zones is discretized to a temperature difference, whereas $-k_{tot}$, T_i, and t_i are denoted as the total thermal conductivity, the zone temperature, and the zone thickness, respectively. The total thermal conductivity consists of a laminar and a turbulent part, which are calculated by estimating laminar and turbulent Prandtl numbers and gas mixture viscosities. Kominos et al. adapted this approach from Yang and Martin [17] and described it in more detail in [15, 16].

To assure a uniform pressure within all zones, each zone is connected to its adjacent zones within Cantera as valves, allowing a pressure difference dependent mass flow from one zone to another. The crevice zone and the outermost zone also transfer heat and mass to their adjacent zones, but also to the cylinder wall as a global heat transfer. The gas mixture in each zone of the single-zone and the multi-zone model is homogeneous and the piston movement induces volume changes.

For each timestep, the energy and species conservation equations are solved. For the single-zone model, the global heat transfer coefficient is calculated with the semi-empirical Woschni correlation [18] and the coefficients are taken from our recent work [10]. For the multi-zone model, a slightly modified version of the Woschni equation was used, according to Chang et al. [19], who measured the in-cylinder heat fluxes in an HCCI engine experimentally.

Our multi-zone model was then validated against our own measurements with methane/DME mixtures at $\phi = 1.9$ from [10] by using a zone distribution which led to a good agreement between the simulated fuel conversion, work, and heat output and the experimental data.

At TDC, the core zone and the crevice zone are set at 30% and 1.5% of the dead volume, respectively. The remaining volume is distributed between the five other zones so that the volume decreases from the inner to the outermost zone. The bore d and the height of the dead volume h_{TDC} of the cylinder are thus divided into the thickness of the outer zones t_i and the thickness of the core zone $t_{core,x}$ Eq. (2) and $t_{core,y}$, respectively Eq. (3).

$$d = 2\sum t_i + 2t_{core,x} \tag{2}$$

$$h_{TDC} = 2\sum t_i + 2t_{core,y} \tag{3}$$

The core zone thickness in direction of the piston movement $t_{core,y}$ is given by the cylinder height at top dead center h_{TDC} and the sum of the thickness of the outer zones Eq. (4). The thickness $t_{core,x}$ corresponds to the half diameter of the core zone and is kept constant.

$$t_{core,y} = \frac{h_{TDC} - 2\sum t_i}{2} \tag{4}$$

The thickness t_i of each outer zone is calculated according to Eq. (5).

$$t_i = L \bullet R^{N-(i+1)} \tag{5}$$

Here, $N = 7$ is the number of zones, R is set to 0.45 and L is solved within the model to assure that Eq. (3) is fulfilled. The zone volumes change with height changes due to the piston movement. The resulting volume distribution is visualized in Fig. 1.

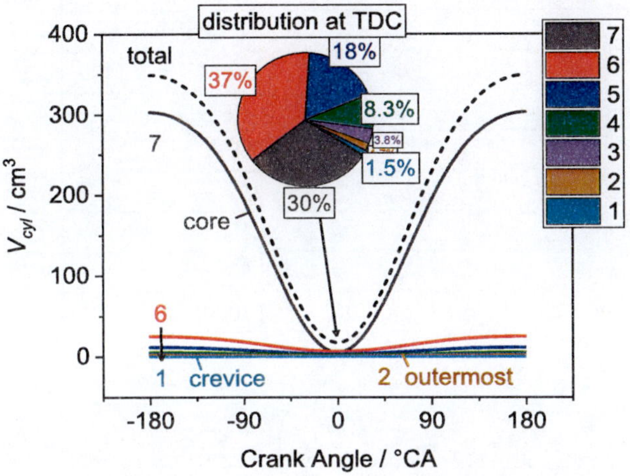

Fig. 1. Volumes of the seven zones of the multi-zone model as a function of crank angle and volume distribution at TDC (pie chart). At TDC, the core zone accounts for 30% of the dead volume, whereas the crevice zone accounts for 1.5%.

It was also proven that the MZM reproduced the methane conversion found experimentally in our previous work [10] within ±1.9%.

A representative natural gas is the fuel in this work, and it consist of 90% methane (CH_4), 9% ethane (C_2H_6), and 1% propane (C_3H_8). To compute the chemical kinetics of DME (CH_3OCH_3) and ozone (O_3), the Burke mechanism [20] for C1-C3 species is extended by the Zhao sub-mechanism [21] for ozone kinetics.

Table 1 presents the investigated engine properties and operation parameters. The equivalence ratio is varied between 1.5 and 2.5 and all other conditions remain constant. For the calculation of the equivalence ratio the additives are considered, and the amount of air is adjusted accordingly.

The fuel conversion X_i, product yields Y_i, and product selectivities S_i are evaluated to assess the influence of the additive choice on the chemical output of the polygeneration system. In the corresponding definitions Eqs. (6), (9), and (10), the stoichiometric coefficient, the entering molar flows, and the exiting molar flows are denoted as ν_i, $\dot{n}_i^{'}$, and $\dot{n}_i^{''}$, respectively. The total fuel conversion X_t in Eq. (8) considers the conversion of all fuel and additive species, weighted according to their entering mole fractions $x_i^{'}$ Eq. (7) in the fuel-additive mixture.

$$X_i = 1 - \left(\frac{\dot{n}_i^{''}}{\dot{n}_i^{'}}\right) \tag{6}$$

Table 1. Engine properties and operation parameters.

Description		Value	Unit
Rotational speed	n	1500	1/min
Compression ratio	ε	20	–
Bore	d	65	mm
Stroke	s	100	mm
No. of cylinders	Z	4	–
Displacement	D	1.327	cm^3
Equivalence ratio	Φ	1.5–2.5	–
Intake temperature	T_{in}	50	°C
Intake pressure	p_{in}	1	bar
Coolant temperature	T_C	100	°C
Th. conductivity wall	λ_W	53	W/(m K)
Convection coeff. coolant	α_C	3000	$W/(m^2 K)$

$$x_t^{'} = \sum\left(\frac{\dot{n}_i^{'}}{\sum \dot{n}_i^{'}}\right) = \sum x_i^{'} \tag{7}$$

$$X_t = \sum\left(x_i^{'} \bullet X_i\right) \tag{8}$$

$$Y_i = \frac{\dot{n}_i^{''} - \dot{n}_i^{'}}{\dot{n}_i^{'} \bullet x_t^{'} \bullet \nu_i} \tag{9}$$

$$S_i = \frac{Y_i}{X_t} \tag{10}$$

To evaluate the influence of the additive production on the exergetic efficiency, the HCCI engine is embedded in a system which also includes the additive production: the ozone generator and a black box model of the DME production process. Figure 2 displays a flow diagram of those three systems *add*, *poly*, and *sys*, their interconnections, and their corresponding exergy flows.

The exergetic efficiency ε_{add} of the DME production process ε_{DME} is estimated as 43.5% according to Zhang et al. [13] and the exergetic efficiency of the ozone generator ε_{ozone} is calculated within the Python model Eq. (11).

$$\varepsilon_{add} = \left.\frac{\dot{E}_{add}}{\dot{E}_{in,add}}\right|_{add=DME,ozone} \tag{11}$$

The ozone generator is supplied with air and power and provides a mixture of air and ozone. The power supply, the incoming air mass flow, and the ozone concentration in the output gas are taken from the data sheets of the Fujian Newland Entech Company [22] to calculate the exergy flows $\dot{E}$.

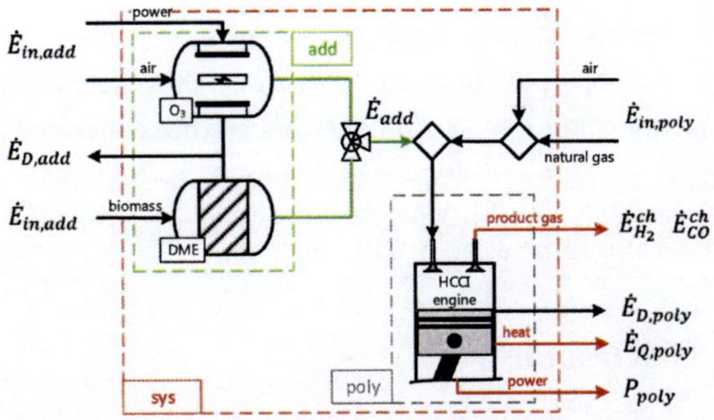

Fig. 2. Process flow diagram of the system sys, divided into two subsystems *add* and *poly*. The HCCI engine is supplied with either ozone or DME as additive.

To compare the exergy-share of DME and ozone, the exergy input ratio ε^*_{add} is defined as the ratio of the exergy flow of the additive to the total exergy flow of the engine input Eq. (12).

$$\varepsilon^*_{add} = \frac{\dot{E}_{add}}{\dot{E}_{in,poly} + \dot{E}_{add}} \tag{12}$$

Furthermore, the exergetic efficiency of the HCCI engine ε^A_{poly} and of the overall system ε^A_{sys} are defined according to Eqs. (13) and (14).

$$\varepsilon^A_{poly} = 1 - \frac{\dot{E}_{D,poly}}{\dot{E}_{in,poly} + \dot{E}_{add}} \tag{13}$$

$$\varepsilon^A_{sys} = 1 - \frac{\dot{E}_{D,poly} + \dot{E}_{D,add}}{\dot{E}_{in,poly} + \frac{\dot{E}_{add}}{\varepsilon_{add}}} \tag{14}$$

The exergy destruction $\dot{E}_D$ in the HCCI engine is estimated with the Gouy-Stodola theorem Eq. (15), the irreversible entropy production rate ($\dot{S}_{irr,poly}$) is calculated using the second law of thermodynamics.

$$\dot{E}_{D,poly} = T_{env} \bullet \dot{S}_{irr,poly} \tag{15}$$

Without full fuel conversion, the aforementioned efficiencies must be evaluated critically, since, as an extreme, the evaluated efficiency would be 100%, if no conversion takes place, which is not aimed. To assess the conversion to the wanted products only, a second definition for the exergetic efficiency is introduced for *poly* Eq. (16) and for *sys* Eq. (17), denoted with the superscript *B*.

$$\varepsilon^B_{poly} = \frac{\dot{E}^{ch}_{H_2} + \dot{E}^{ch}_{CO} + P_{poly} + \dot{E}_{Q,poly}}{\dot{E}_{in,poly} + \dot{E}_{add}} \tag{16}$$

$$\varepsilon^B_{sys} = \frac{\dot{E}^{ch}_{H_2} + \dot{E}^{ch}_{CO} + P_{poly} + \dot{E}_{Q,poly}}{\dot{E}_{in,poly} + \frac{\dot{E}_{add}}{\varepsilon_{add}}} \tag{17}$$

Here, only useful exergy stream outputs are considered: the chemical exergy flow of synthesis gas $\dot{E}^{ch}_{H_2}$ and $\dot{E}^{ch}_{CO}$, the power output P_{poly}, and the exergy flow of the heat transferred to the cooling water $\dot{E}_{Q,poly}$. Species that are not converted towards synthesis gas are thus neglected. Both definitions have advantages and disadvantages, since the second definition now neglects the possibility to re-use the unconverted educts in a following cycle. As a compromise, both will be shown.

3 Results and Discussion

First, the necessary amounts of ozone and DME for a stable combustion phasing of 7.4 ± 0.2 °CA after top dead center (°CA aTDC), using the SZM, are evaluated and compared. An ignition shortly after reaching top dead center is most favorable, the sensitivity of the combustion phasing on the exact amount of additive was investigated systematically by varying the additive amounts by ±20% at $\phi = 1.9$. The reference case for ozone represents the amount of ozone that was needed in the work of Schröder et al. [10] to substitute DME completely by ozone.

Second, these amounts are reduced or increased when reducing or increasing ϕ, respectively, and the equivalence ratio dependent product selectivities and exergetic efficiencies are discussed. All single-zone model results shown in this chapter are an average of four consecutive engine cycles. The multi-zone model results are taken from the fourth cycle.

Figure 3 illustrates the pressure traces for the DME and the ozone case at $\phi = 1.9$.

The relatively high compression ratio of 20 leads to maximum pressures of 53 bar and 59 bar before, and 174 bar and 167 bar after ignition, for DME and ozone, respectively. A similar CA50 is obtained with about 11 times less ozone compared to DME in the total mixture. In contrast to DME, ozone does not increase the molar heat capacity of the mixture and supplies radicals at lower temperatures and thus earlier crank angles. Because of the lower heat capacity, the temperature and pressure increase during the compression stroke is higher and less additive is needed. Figure 3 also indicates that small amounts of ozone shift the combustion phasing noticeably stronger, although the relative increase is stronger for DME: to shift the combustion phasing by one crank angle degree (considering the black cases as reference), additional 1.3% or rather 283 ppm DME must be added, whereas 2.5% or 50 ppm additional ozone is needed.

For DME, a small combustion phasing range is found, in which the ignition is unstable and only every second cycle ignites after intermediate species, such as formaldehyde and hydrogen peroxide, from the previous cycle are accumulated, which will be discussed later in this chapter. This effect is seen for the condition where the DME amount of the reference case is reduced by 4.5% or 950 ppm, marked in red in Fig. 3a. A pressure increase is noticeable, but the combustion phasing is very late, and the averaged pressure trace is not smooth, because every cycle differs.

The kinetic effects of ozone and DME on fuel-rich HCCI engine operated with neat methane was analyzed in our previous work [10]. We found that ozone decomposes very early and is completely converted at −40 °CA. Its decomposition yields atomic and molecular oxygen, the atoms react with methane, initiating the conversion. Furthermore, a buffer molecule was found to be important, previously: methyl hydroperoxide

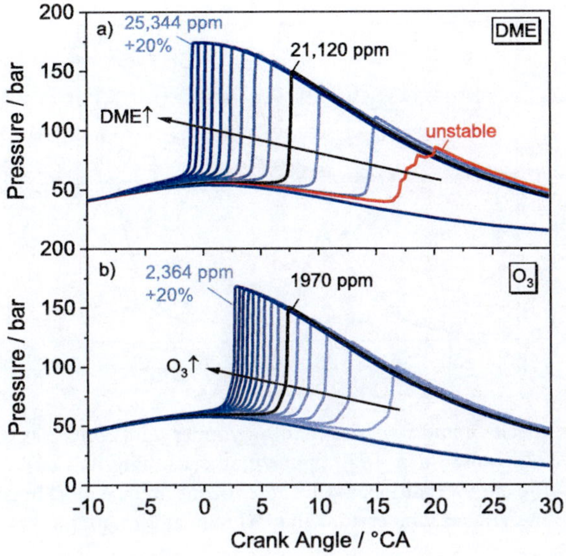

Fig. 3. Averaged pressure traces (SZM, four consecutive cycles) as a function of crank angle and additive amount in the total mixture for a) DME and b) ozone at $\phi = 1.9$. The reference additive amount (black line) is varied by ±20% (blue lines). For DME, an unstable ignition is found for the second lowest DME amount (red line).

(CH_3OOH). This relatively unstable molecule is formed for a brief period and increases the OH radical concentration later, due to decomposition at higher temperatures after further compression. Thus, it promotes the conversion of methane and the formation of formaldehyde (CH_2O) and hydrogen peroxide (H_2O_2), which are crucial precursors for methane ignition.

On the basis of these findings, the following section deals with the influence of the additional ethane and propane in natural gas mixtures on the intermediate species and the interaction with ozone. In this work, the engine is operated at higher rotational speed (+905 min^{-1}), higher compression ratio (+10) and lower intake temperature (−100 °C) compared to [10] in order to investigate conditions for the subsequent exergetic analysis, which are nearer to nowadays' engine parameters. This leads to a higher cylinder charge and shorter residence times.

In Fig. 4 two operating conditions are compared: the ozone reference case with 1970 ppm ozone in the total mixture (black line in Fig. 4b) and the ozone mixture with 1570 ppm ozone (−20%, dashed lines), the latter condition does not ignite (flat line in Fig. 4b).

Due to the 2.5 times shorter residence time and the lower intake temperature compared to the work of Schröder et al. [10], ozone is barely decomposed at −40 °CA. However, the higher compression ratio leads to a steeper temperature increase and thus ozone decomposition starts shortly after −40 °CA and is complete at −25 °CA. This applies for both cases and thus both ozone amounts. As a result, CH_3OOH, CH_2O and H_2O_2 are formed as described for the methane case before, but also ethyl hydroperoxide

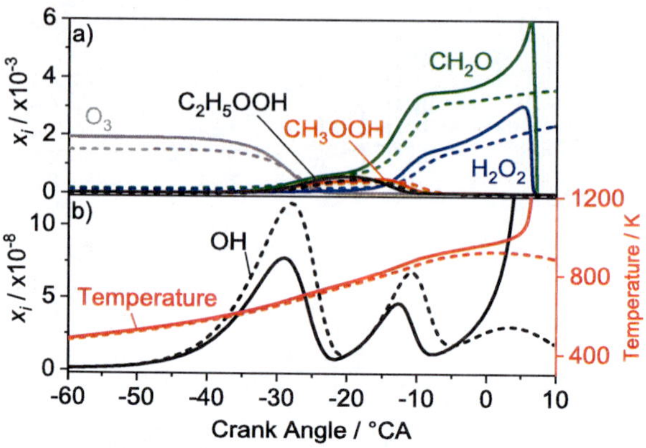

Fig. 4. Intermediate species mole fractions and in-cylinder temperature as a function of crank angle at $\phi = 1.9$ (SZM). Solid lines: 1970 ppm ozone cause ignition; dashed lines: 1570 ppm ozone (-20%) is insufficient for ignition. a) Illustrates ozone and the most important intermediate species, b) shows the in-cylinder temperature and OH radical formation.

(C2H5OOH). If the ozone amount is too low (dashed lines), CH_2O and H_2O_2 are accumulated in each cycle, the OH formation is increased, and the formation of CH_3OOH is comparable. OH radicals are formed in this first step by H-abstraction from the natural gas components, which react with oxygen atoms from the ozone decomposition. The OH radicals are then mainly reacting with the natural gas constituents, leading to a peak. Nevertheless, the second OH peak, which results from CH_3OOH decomposition to CH_3O and OH, is shifted by 2 °CA, because the temperature increases slightly slower. This leads to the effect that the expansion after reaching top dead center quenches the reactions and prevents the ignition of the mixture, illustrated by low OH radical amounts and no CH_2O or H_2O_2 conversion. However, the OH radical mole fraction is slightly higher in the non-igniting case, since CH_2O and H_2O_2 are accumulated, which lead to additional OH radical formation. Nevertheless, this effect is not sufficient for achieving ignition.

In Fig. 5, the case with 1970 ppm ozone which leads to ignition for natural gas, is compared to a methane case with the same ozone amount.

It is seen that the natural gas mixture starts to ignite, while methane is not converted yet. In the methane case the CH_3OOH formation shows a distinctive maximum at -22 °CA, whereas in the natural gas case, the formation is reduced at that point, and the maximum is found later at -14.4 °CA. In addition, ethene and propene are formed from twostep H abstractions from ethane and propane by OH radicals and ethyl hydroperoxide, as well as ethene and propene are formed as further buffer molecules. Kaczmarek et al. [12] showed that ethene and propene are typical intermediate species for the fuel-rich combustion of natural gas and that ethane and propane do not influence the CH_3OOH formation. This leads to a lower CH_3OOH mole fraction compared to the methane case, but due to the exothermal H-abstraction of ethane and propane, which leads to an intermediate C2H5OOH formation, the in-cylinder temperature and OH radical concentration are high enough for ignition. The initial ethane and propane mole

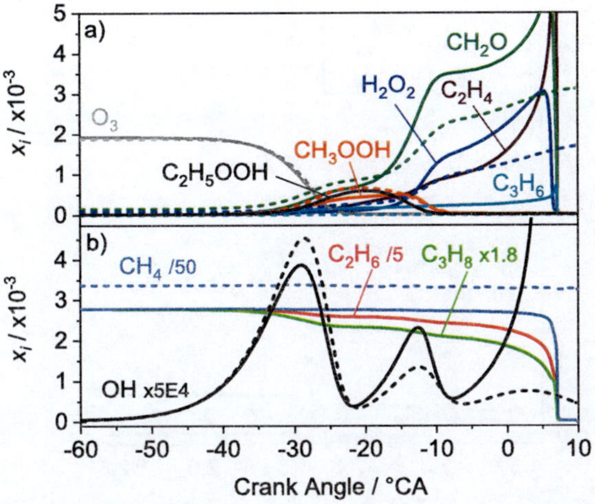

Fig. 5. Intermediate species mole fractions as a function of crank angle at $\phi = 1.9$ (SZM). Solid lines: natural gas; dashed lines: methane. a) Illustrates ozone and the most important intermediate species, b) shows the fuel conversion and OH radical formation.

fractions in the mixture correspond to 14,180 ppm and 1,580 ppm, respectively. If neat methane is to be ignited instead, these additives have to be replaced by only 640 ppm ozone, which emphasizes the effectiveness of ozone as an ignition promoter. The product yields are not affected by combustion phasing and are thus not further discussed here.

In the subsequent section, the necessary ozone and DME amounts for the equivalence ratio range of 1.5 to 2.5, to achieve ignition at CA50 = 6.5 ± 0.6 °CA, and the influence on the exergetic efficiencies are discussed. Figure 6a shows the necessary mole and mass fractions of DME and ozone as a function of equivalence ratio. Figure 6b illustrates the resulting exergy contribution of the additive to the exergy entering the engine ε^*_{add} and the efficiency of the additive production ε_{add}.

With increasing equivalence ratio, the additive mole and mass fractions needed for ignition increase for both additives. However, the DME-related gradients are larger: 1.0 mol-%/ϕ for DME and only 0.065 mol-%/ϕ for ozone. Up to 2.7 mol-% and 4.5 mass-% of DME in the total mixture are needed. This is about 9.9 to 11.6 times more compared to the ozone case, increasing with ϕ. The difference of the exergy ratios ε^*_{add} are even higher, as Fig. 6b illustrates. DME contributes 21% to the exergy input, whereas ozone is only accountable for a negligible contribution of 0.2%. On the contrary, the calculated exergetic efficiency of the ozone generator is poor with a rather constant value of 5.5%. At $\phi = 2.5$, for instance, 4.28 kW of electrical energy must be supplied to generate an ozone mass flow of 0.26 kg/h. The multi-zone model predicts similar figures for the DME case, for which it was validated. On the contrary, when the additive is switched from DME to ozone, much higher ozone amounts or higher intake temperatures are needed compared to the single-zone model. In Fig. 6, the multi-zone model results with ozone are obtained with an intake temperature of 100 °C, which is 50 °C higher than in the SZM case; this leads to an ignition with the same amount of ozone as in the single-zone model. Although ozone decomposes early in the compression stroke, it needs about 38°

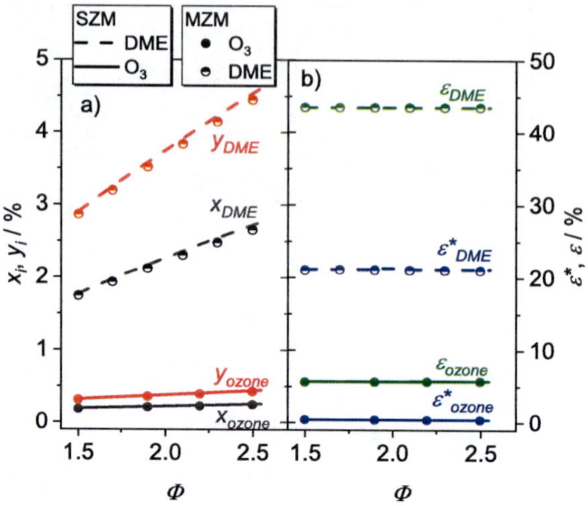

Fig. 6. a) Intake additive mole fractions x_i in black and mass fractions y_i in red, and b) intake exergy ratio ε^*_{add} and exergetic efficiency of the additive production ε_{add} as a function of equivalence ratio at CA50 = 6.5 ± 0.6 °CA (solid lines: ozone, dashed lines: DME; filled symbols: ozone multi-zone model with T_{in} = 100 °C (+50 °C), semi-filled symbols: DME multi-zone model).

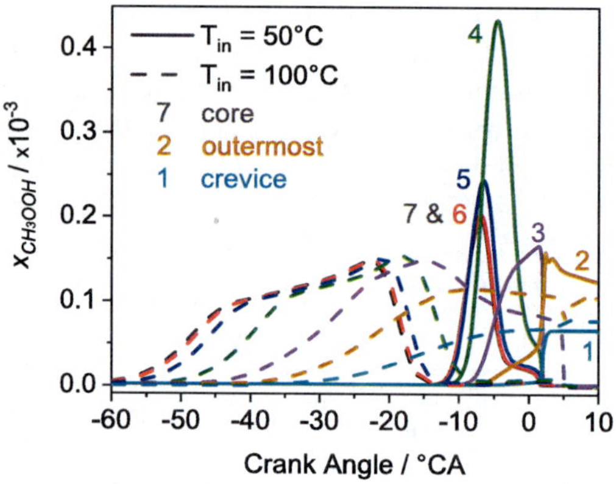

Fig. 7. Methyl hydroperoxide (CH_3OOH) mole fraction as a function of crank angle for all seven zones of the multi-zone model (ϕ = 1.9). Solid lines: intake temperature of 50 °C, no ignition. Dashed lines: intake temperature of 100 °C, ignition.

crank angle for full decomposition in the core zone ($\phi = 1.9$, $T_{in} = 50$ °C). This leads to a relatively late CH_3OOH formation and decomposition compared to the single-zone model. In the single-zone model, as shown in Fig. 4a and Fig. 5a, CH_3OOH is completely converted at −8 °CA. By contrast, in the multi-zone model, CH_3OOH is converted in the four innermost zones until +2 °CA, and no ignition is achieved. With an increased intake temperature of 100 °C, CH_3OOH is converted again at −8 °CA, comparable to the single-zone model, and the mixture ignites. This is illustrated in Fig. 7.

In contrast to ozone, DME converts late in the compression stroke and in the core zone: it is converted nearly twice as fast, which makes DME less sensitive on the heat and mass transfer from the core zone to its adjacent zones. This shows that due the complexity of the heat and mass transfer within the multi-zone model, this model must be validated against ozone measurements explicitly and investigated further in future work.

However, as seen from Fig. 8, the system's resulting exergetic efficiency decrease is minor compared to the DME case. In the figure, a comparison of the single-zone and multi-zone model is also included.

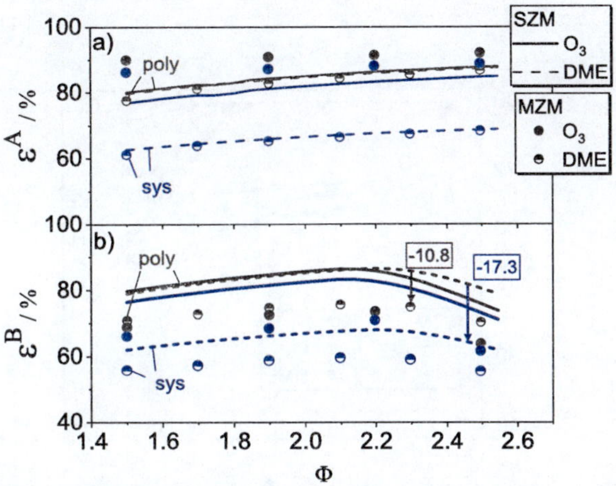

Fig. 8. Exergetic efficiency of the HCCI engine (*poly*, grey) and the entire system (*sys*, blue) as a function of equivalence ratio ϕ at CA50 = 6.5 ± 0.6 °CA (solid lines: ozone, dashed lines: DME; filled symbols: ozone multi-zone model with T_{in} = 100 °C (+50 °C), semi-filled symbols: DME multi-zone model). Top: definition A, bottom: definition B.

The exergetic efficiency of the HCCI engine (*poly*) reaches a value of 87.5%, regardless of which additive is used. Figure 8a also indicates that when the conversion is not considered in the efficiency definition, there is no distinctive difference between the SZM and MZM results. If the whole system is regarded, including the production efficiency of the additive, and not only the engine, further differences are recognized. The strong efficiency reduction by the DME production by up to 17.3 percentage points at ϕ = 2.5 (Fig. 8b, blue arrow) is remarkable, whereas the ozone production with the quite inefficient generator only leads to a reduction by 2.7 percentage points.

When the efficiency is defined only with the intended products, as done in definition B, Fig. 8b results from it. The efficiencies of ozone and DME still reach high values of 86.1% and 86.3%, respectively, but at lower equivalence ratios. At higher equivalence ratios, methane remains in the product gases, leading to a drop. The maximum efficiency for DME addition is slightly shifted to higher equivalence ratios compared to ozone addition: 2.2 compared to 2.15 in the ozone case.

The efficiency reduction, when the efficiency of additive production is considered (*sys*), is noticeable as well. In the DME case, the maximum efficiency decreases by 18.6

percentage points to 67.7%. With ozone there is a smaller reduction by 3.3 percentage points to 82.8%.

At $\phi = 2.2$ the conversion starts decreasing with decreasing amount of oxygen and thus the efficiency decreases as well. At $\phi = 2.3$, the efficiency of the HCCI engine (*poly*) calculated with the multi-zone model is up to 10.8 percentage points lower than those predicted by the single zone model (Fig. 8b, grey arrow). This is mainly an effect of the reduced conversion due to colder regions in the cylinder, e.g. crevice zones and zones adjoining the cylinder walls. With reduced conversion, the efficiency decreases since less chemical energy of the fuel is converted to work, heat, and synthesis gas.

In Fig. 9, the conversions predicted by the single-zone model and multi-zone model are compared.

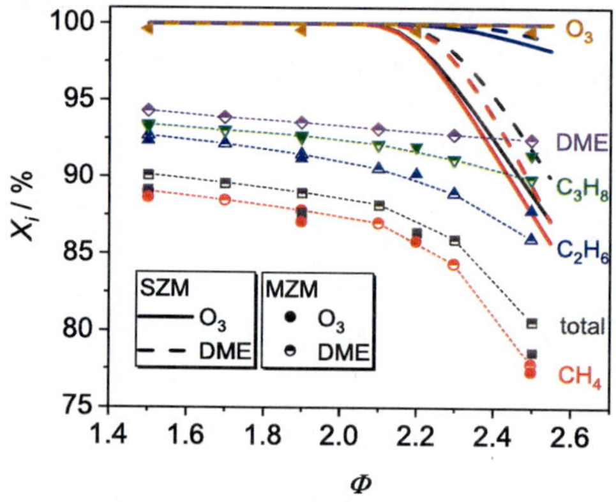

Fig. 9. Fuel and DME conversions as a function of equivalence ratio at CA50 = 6.5 ± 0.6 °CA (solid lines: ozone, dashed lines: DME, filled symbols: ozone multi-zone model with $T_{in} = 100$ °C (+50 °C), semi-filled symbols: DME multi-zone model).

In the single-zone model, propane and DME are always completely converted, whereas the ethane conversion decreases slightly at equivalence ratios above 2.2. The methane conversion is the lowest, so that the total conversion is 90.0% for DME and 87.2% for ozone at $\phi = 2.5$. The natural gas/DME mixture is more reactive at TDC, since DME is not fully converted until ignition occurs. This leads to a slightly higher overall fuel conversion compared to ozone and thus higher exergetic efficiencies at higher equivalence ratios, as seen in Fig. 8b.

The results of the multi-zone model shall be compared further to the single-zone-model. The multi-zone model predicts up to 12 percentage points smaller total conversions; values of 80.6% to 90.1% are estimated and thus the efficiencies are reduced. DME is converted by 94.3% to 92.4%, decreasing with ϕ – mainly because of the unreactive crevice zone, which, for example at $\phi = 1.9$, contains 3.5% to 9% of the total cylinder mass during combustion. This mass stems from mass transfers from outer, cooler zones when the inner zones ignite and induce a pressure increase. On the contrary, ozone is always converted completely, as emphasized before. The MZM calculations with ozone provide no additional insights since the outcome does not differ significantly from the single zone model results and are not further discussed in the following.

The efficiency reduction for DME addition at $\phi = 2.5$ is illustrated in more detail in Fig. 10, where the exergy stream outputs for the SZM and MZM are compared.

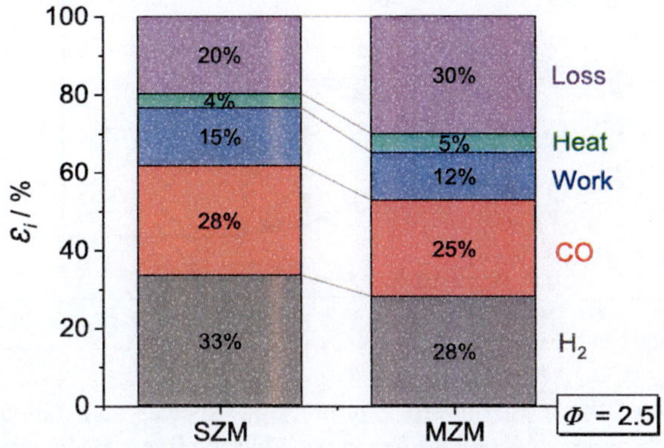

Fig. 10. Segmentation of the system's exergy stream outputs; comparison of the single zone model on the left and the multi-zone model on the right at $\phi = 2.5$ in the DME case.

The lower conversion in the MZM leads correspondingly to lower heat, work, and synthesis gas exergy streams in the system output. Nevertheless, the changes are quite uniform for the different useful outputs.

To determine promising operating conditions, it is also imperative to evaluate the equivalence ratio dependent product selectivities and compare them for both additives. In Fig. 11, the equivalence ratio dependence of the selectivities of CO, CO_2, H_2, H_2O, C_2H_2 and C_2H_4 are shown, again from both models. The main finding is that the predicted selectivities mainly depend on the equivalence ratio but only to a minor extent on the additive or the model used.

The acetylene (C_2H_2) and ethene (C_2H_4) selectivities reach 1.4% and 2.6% at the maximum investigated equivalence ratio of 2.5. Higher selectivites are expected at equivalence ratios of 5 and higher [23], but since synthesis gas is the target chemical here, these regions are not further examined. The synthesis gas selectivites increase with increasing equivalence ratio and reach a maximum at the equivalence ratio, where the highest efficiencies were found. Again, the maximum selectivities in the DME case are shifted slightly to higher equivalence ratios, compared to ozone. The maximum achievable selectivities of CO and H_2 are 88.4% and 66.5%, respectively. The MZM predicts slightly smaller synthesis gas selectivities and instead a small increase of the water selectivity. In comparison with methane, the natural gas selectivities do not differ noticeably for any investigated case. In practice, the natural gas composition does thus not influence the outcome of the process and must not be considered, when at least 90% of the natural gas consists of methane. However, for combustion control it is surely important to consider natural gas composition changes, since ignition is sensitive on this composition. Therefore, for combustion control, the additive mass flow must be adjusted on the same time-scale as the natural gas composition varies.

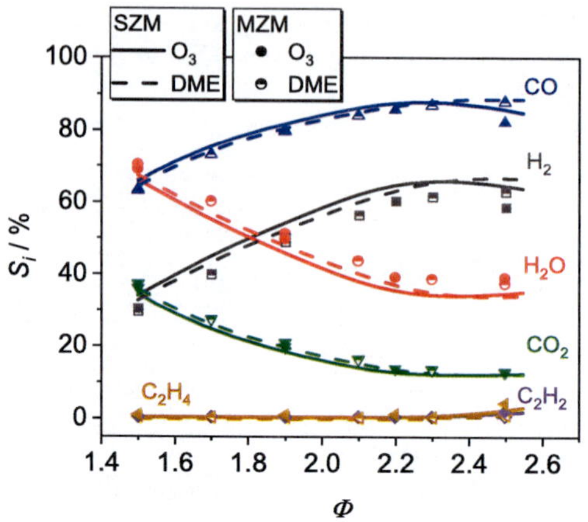

Fig. 11. Selectivities of the most important product gas species as a function of equivalence ratio ϕ at CA50 = 6.5 ± 0.6 °CA (solid lines: ozone, dashed lines: DME, filled symbols: ozone multi-zone model with T_{in} = 100 °C (+50 °C), semi-filled symbols: DME multi-zone model).

Finally, the decision about the most favorable additive for polygeneration in fuel-rich HCCI engines is not based on the product gas composition, but on the effort in producing the additive and its fuel conversion enhancement properties.

4 Conclusions

Fuel-rich HCCI engines for polygeneration of work, heat, and synthesis gas can be operated with natural gas with the help of additives such as ozone or DME. In this work, the influence of ozone on natural gas ignition and the exergy input of DME and ozone were investigated theoretically, and the influence of the additive production on the exergetic efficiency was evaluated.

The single-zone model results showed that natural gas is much more reactive than methane. Ethane and propane convert partially in the compression stroke and lead to ethene, propene, and OH radicals. However, the ethane and propane conversions do not favor, but slightly reduce the formation of methyl hydroperoxide, which is an important buffer molecule for fuel-rich methane ignition. Instead, ethyl hydroperoxide is intermediately formed as a second buffer molecule. If the same ignition timing shall be achieved with neat methane, ethane and propane can be substituted by only 640 ppm additional ozone, which emphasizes the efficiency of the ozone decomposition.

Compared to ozone, the necessary DME molar and mass fractions are up to 11 times higher, and its exergy contribution to the total mixture is even 95 times higher: DME contributes 21% to the exergy input of the engine, whereas 1700 ppm ozone only contribute 0.2%.

The exergetic efficiency of the ozone generator was determined and reached small values of 5.5%, but since ozone is used in small quantities the system's efficiency declines by only 3.3 percentage points and reasonable values of 82.8% are possible. The DME production reduces the system's efficiency by unfavorable 17 percentage points. The multi-zone model results show that the efficiency is strongly dependent on the fuel conversion and thus non-converted fuel should be recycled within the polygeneration system to maintain high efficiencies – especially at very fuel rich conditions.

The single-zone model predicted the product selectivities accurately and provided helpful species profiles for kinetic analyses. Additionally, if the exergetic efficiency is calculated with the Gouy-Stodola theorem, the resulting exergetic efficiencies only differed about 1 percentage point from the ones calculated with the multi-zone model. If the fuel conversion and the exergetic efficiency considering the conversion must be predicted accurately, the multi-zone model is much more suitable. The fuel conversion calculated by the multi-zone model matched our experimental values by $\pm 1.9\%$ and the exergetic efficiency was found to be 10.8 percentage points lower in the DME case. Since the multi-zone model has not been validated against ozone experiments so far, these results were obtained with a 50 °C higher intake temperature. Therefore, those conditions will be used for validation of the multi-zone model against natural gas/ozone experiments in a future work.

To conclude, ozone is a more favorable additive compared to DME – not only based on the necessary mass flows but on exergetic efficiency as well. These results encourage profound investigations of ozone kinetics for fuel-rich combustion technologies that require reactive fuel additives.

Acknowledgments. This research was funded by Deutsche Forschungsgemeinschaft (DFG), grant number 229243862 (AT24/13–3) within the framework of the DFG research unit FOR 1993 'Multi-functional conversion of chemical species and energy'. The authors gratefully acknowledge the DFG for the financial support.

References

1. Karim, G.A., Wierzba, I.: The production of hydrogen through the uncatalyzed partial oxidation of methane in an internal combustion engine. Int. J. Hydrogen Energy **33**, 2105–2110 (2008). https://doi.org/10.1016/j.ijhydene.2008.01.051
2. Lim, E.G., et al.: The engine reformer: Syngas production in an engine for compact gas-to-liquids synthesis. Can. J. Chem. Eng. **94**, 623–635 (2016). https://doi.org/10.1002/cjce.22443
3. Banke, K., Hegner, R., Schröder, D., Schulz, C., Atakan, B., Kaiser, S.A.: Power and syngas production from partial oxidation of fuel-rich methane/DME mixtures in an HCCI engine. Fuel **243**, 97–103 (2019). https://doi.org/10.1016/j.fuel.2019.01.076
4. Wiemann, S., Hegner, R., Atakan, B., Schulz, C., Kaiser, S.A.: Combined production of power and syngas in an internal combustion engine – experiments and simulations in SI and HCCI mode. Fuel **215**, 40–45 (2018). https://doi.org/10.1016/j.fuel.2017.11.002
5. Kaczmarek, D., Atakan, B., Kasper, T.: Investigation of the partial oxidation of methane/n-heptane-mixtures and the interaction of methane and n-heptane under ultra-rich conditions. Combust. Flame **205**, 345–357 (2019). https://doi.org/10.1016/j.combustflame.2019.04.005
6. Hegner, R., Werler, M., Schießl, R., Maas, U., Atakan, B.: Fuel-rich HCCI engines as chemical reactors for polygeneration: a modeling and experimental study on product species and thermodynamics. Energy Fuels **31**, 14079–14088 (2017). https://doi.org/10.1021/acs.energyfuels.7b02150

7. Kebriaei, M., HalvaeiNiasar, A., Ketabi, A.: A new pulsed power generator topology for corona discharge. In: 2016 7th Power Electronics and Drive Systems Technologies Conference (PEDSTC), 16–18 Feb 2016, pp. 577–581. IEEE, Tehran, Iran (2016)
8. Keum, S., Kuo, T.W.: Damköhler number analysis on the effect of ozone on auto-ignition and flame propagation in internal combustion engines. In: Volume 1: Large Bore Engines; Fuels; Advanced Combustion, 4 November 2018, V001T03A002. ASME, San Diego, California (2018)
9. Sayssouk, S., Nelson-Gruel, D., Caillol, C., Higelin, P., Chamaillard, Y.: Towards control of HCCI combustion by ozone addition: a mathematical approach to estimate combustion parameters. In: IFAC-PapersOnLine, vol. 49, pp. 361–368 (2016). https://doi.org/10.1016/j.ifacol.2016.08.054
10. Schröder, D., Banke, K., Kaiser, S.A., Atakan, B.: The kinetics of methane ignition in fuel-rich HCCI engines: DME replacement by ozone. In: Proceedings of the Combustion Institute, vol. 38, pp. 5567–5574 (2020).https://doi.org/10.1016/j.proci.2020.05.046
11. Duan, X., et al.: Effects of natural gas composition and compression ratio on the thermodynamic and combustion characteristics of a heavy-duty lean-burn SI engine fueled with liquefied natural gas. Fuel **254**, 115733 (2019). https://doi.org/10.1016/j.fuel.2019.115733
12. Kaczmarek, D., et al.: Plug-flow reactor and shock-tube study of the oxidation of very fuel-rich natural gas/DME/O2 mixtures. Combust. Flame **225**, 86–103 (2021). https://doi.org/10.1016/j.combustflame.2020.10.004
13. Zhang, X., Solli, C., Hertwich, E.G., Tian, X., Zhang, S.: Exergy analysis of the process for dimethyl ether production through biomass steam gasification. Ind. Eng. Chem. Res. **48**, 10976–10985 (2009). https://doi.org/10.1021/ie900199e
14. Goodwin, D.G., Speth, R.L., Moffat, H.K., Weber, B.W.: Cantera: an object-oriented software toolkit for chemical kinetics, thermodynamics, and transport processes (2021). https://www.cantera.org. Version 2.4.0
15. Komninos, N.P., Rakopoulos, C.D.: Heat transfer in HCCI phenomenological simulation models: a review. Appl. Energy **181**, 179–209 (2016). https://doi.org/10.1016/j.apenergy.2016.08.061
16. Komninos, N.P., Hountalas, D.T., Kouremenos, D.A.: Development of a new multi-zone model for the description of physical processes in HCCI engines. In: SAE Technical Paper 2004-01-0562. https://doi.org/10.4271/2004-01-0562
17. Yang, J., Martin, J.K.: Approximate solution—one-dimensional energy equation for transient, compressible, low mach number turbulent boundary layer flows. J. Heat Transf. **111**, 619–624 (1989). https://doi.org/10.1115/1.3250727
18. Woschni, G.: A universally applicable equation for the instantaneous heat transfer coefficient in the internal combustion engine. In: SAE Technical Paper 670931. https://doi.org/10.4271/670931
19. Chang, J., et al.: New heat transfer correlation for an HCCI engine derived from measurements of instantaneous surface heat flux. In: SAE Technical Paper 2004-01-2996. https://doi.org/10.4271/2004-01-2996
20. Burke, U., et al.: An ignition delay and kinetic modeling study of methane, dimethyl ether, and their mixtures at high pressures. Combust. Flame **162**, 315–330 (2015). https://doi.org/10.1016/j.combustflame.2014.08.014
21. Zhao, H., Yang, X., Ju, Y.: Kinetic studies of ozone assisted low temperature oxidation of dimethyl ether in a flow reactor using molecular-beam mass spectrometry. Combust. Flame **173**, 187–194 (2016). https://doi.org/10.1016/j.combustflame.2016.08.008

22. Fujian Newland Entech Company, Ozone Portfolio (2017). https://newlandentecheurope.com/wp-content/uploads/2017/06/Newland-EnTech-Europe_Ozone.pdf. Accessed 24 Feb 2021
23. Atakan, B., et al.: Flexible energy conversion and storage via high-temperature gas-phase reactions: The piston engine as a polygeneration reactor. Renew. Sustain. Energy Rev. **133**, 110264 (2020). https://doi.org/10.1016/j.rser.2020.110264

Enhancement of Blowout Limits in Lifted Swirled Flames in Methane-Air Combustor by the Use of Sinusoidally Driven Plasma Discharges

Maria Grazia De Giorgi[1(✉)], Sara Bonuso[1], Ghazanfar Mehdi[1], Mohamed Shamma[2], Stefan Raphael Harth[2], Nikolaos Zarzalis[2], and Dimosthenis Trimis[2]

[1] Department of Engineering for Innovation, University of Salento, Via per Monteroni, 73100 Lecce, Italy
mariagrazia.degiorgi@unisalento.it

[2] Engler-Bunte-Institute, Division of Combustion Technology, Karlsruhe Institute of Technology, Engler-Bunte-Ring 7, 76131 Karlsruhe, Germany

Abstract. This study focuses on the effects of continuous volumetric discharge of sinusoidal plasma actuator at 20 kHz coupled directly with methane-air premixed flame in the near field of the injector exit. A plasma actuator composed of a needle-type electrode placed at the center of the nozzle, connected with high-voltage, while the nozzle was acted as a grounded electrode with different input electrical power values was designed to enhance lean blowout performance in a swirl model combustor. The ionic wind induced by the electrical body force given by the flow ionization leads to velocity disturbance and subsequently affects the flame. To investigate the possible mechanism of the combustion control by the plasma through the aerodynamic effect high speed flow visualization was analyzed under quiescent conditions. Flow visualizations showed that the plasma discharge affects the flow dynamics near the burner exit. It was observed that by increasing the electrical power used for the actuation a recirculation zone is formed in the non-reacting flow field. Furthermore, comparative experiments between conventional and plasma-assisted combustion were carried out to analyze the combustion enhancement in terms of lean blowout performance. The effect of the input electric power of the plasma actuator was studied, and it was seen that at coupled plasma powers corresponding to less than 1% of the thermal output power, there is a significant improvement in the blow-out limit.

Keywords: Lifted flame · Low swirl combustor · Plasma assisted combustion · Lean blow-out

R. King and D. Peitsch (Eds.): AFCC 2021, NNFM 152, pp. 66–82, 2022.
https://doi.org/10.1007/978-3-030-90727-3_5

Nomenclature

A_{eff}	Effective area, [mm^2]
DBD	Dielectric barrier discharge
f	Frequency, [Hz]
FFT	Fast Fourier transforms
HV	High voltage
I(t)	Applied current signal, [A]
LBO	Lean blow-out
$\dot{m}_{air}$	Air mass flow rate, [kg/hr]
$\dot{m}_{fuel}$	Fuel mass flow rate, [kg/hr]
NTP	Non thermal plasma
$P_{elec.}$	Electrical Power consumed to produce the plasma, [W]
PIV	Particle Image Velocimetry
SN	Swirl number, [-]
SNR	Signal to noise ratio
V(t)	Applied voltage signal, [V]
V_{pp}	Peak to peak voltage, [V]
V_{max}	Maximum voltage used to produce the plasma, [V]
λ	Wavelength, [nm]
Φ_0	Equivalence ratio without plasma actuation, [-]

1 Introduction

Jet engines are facing serious challenges of re-ignition at high altitudes due to typical environmental conditions such as low atmospheric temperature and pressure. In recent years, plasma assisted actuators have attracted attention towards the ignition of aero-engines [1–3]. To address this issue, different plasma actuation techniques were considered, such as laser-induced plasma (LIP) [4,5], plasma torch [6], microwave discharge ion thrusters [7], gliding arc discharge [8], AC plasma discharge [9], and Dielectric barrier discharge (DBD) [10]. However, Non-thermal plasma (NTP) is a well-known technology due to its ability of actively control the flow dynamics of burners [11,12] and aircraft [13]. Among all NTPs, DBD showed great interest in aerospace applications due to real time operating conditions, low power utilization, no movement of parts, very easy to control, cheap to build, and light weight [14].

In NTPs, vibrational, rotational, and electronic temperatures are very different and the electron number density and neutral gas temperature are relatively low. However, it has a high electron temperature (1–100 eV), therefore, it is more kinetically active because of the rapid formation of neutral radicals and excited species via excitation reactions, electron impact dissociation reactions, charged transfer reactions, and subsequent energy relaxation [15,16]. These electron impact processes are mainly dependent on electron energy. Hence, the augment of ignition and combustion process by NTP heavily depends on plasma properties such as electron number density and electron temperature.

Since plasma produces free radicals, excited molecular species, charged species, electrons, long lifetime intermediate neutrals, ionic wind, fuel fragments, large density gradient, heat, and Lorentz and Coulomb forces. It affects ignition and combustion via three major characteristics: 1. Thermal properties (plasma has increased the temperature and accelerated the fuel oxidation and chemical reactions by following the Arrhenius law. 2. Kinetic properties (Plasma is produced charged species/ions and high energy electrons that will produce active radicals (i.e. O, H, and OH) via ion impact reactions, direct electron impact dissociation and recombination reactions (i.e. O2+ and N2+) and collision dissociation when reacted with vibrationally excited species (i.e. N2(v)) and electronically excited species (i.e. N2*, O2(Δg). Moreover, plasma also produced catalytic intermediate species (i.e. NO) and long lifetime reactive (i.e. O3) which accelerated low temperature fuel oxidation. 3. Transport properties (plasma broke down the large fuel fragments into small fuel molecules, and changed the fuel diffusivity, which ultimately amending the combustion [17]. The kinetic modeling of CH_4/air mixture was numerically and experimentally investigated by Aleksandrov et al. [18]. They observed that the active particles produced through a plasma discharge, accelerate the ignition process.

The ignition kinetics of CH4/O_2-Argon mixture at temperature ranges of 1230 K to 1719 K and pressure of 0.3 bar to 1.1 bar have been analyzed numerically and experimentally by subjecting high voltage nano-second plasma discharge in the shock tube were analyzed by Giorgi et al. [19], and Kosarev et al. [20]. It has been proved that ignition delay time was considerable reduced by using of nano-second plasma. To study the nascent kinetics of non-equilibrium pulsed discharge plasma, Uddi et al. [21] have performed the experimental measurements of time-resolved absolute atomic oxygen concentrations in air, methane-air, and ethylene-air mixture subjected to nano-second plasma by using two photon absorption laser induced fluorescence (TALIF) technique. Zhu et al. [22] studied the effects of gliding arc plasma featured with knife shaped electrode using atmospheric pressure. They observed that gliding arc plasma has the capability of producing a high number of OH radicals in turbulent flow. Lee et al. [23] investigated the rotating gliding arc plasma and concluded that hydrogen sensitivity reached almost 100% and plasma chemistry was improved. He et al. [24] also studied the effects of rotating gliding arc on flame extinction and combustion efficiency in swirling combustors. Improvement in lean blowout limits was observed when the inlet velocity lies in between 60 m/s to 120 m/s. Previously, it was observed that plasma assisted combustion (PAC) has great impacts on a swirl combustor. Neophytou et al. [25] investigated representative research work in this area. They performed the experimental work considering high altitude ignition conditions using lean direct injection combustion chamber. The flame kernel was quickly reduced and slowly increased after a critical point under successful ignition until the self-sustained combustion is achieved. It is also quite resembled with the theory of Chen and Ju's [26] for the minimum flame kernel. In general, the flame has faced three kinds of problems, re-ignition, partial flame out and full flame out. The flame out process is mainly influenced by the temperature of air.

In past, some research has been carried out considering advanced optical diagnostics equipment for swirl flame stabilization [27,28]. Besides it, simulation studies of swirl combustion are also quite essential to analyze the critical conditions of swirl flame ignition [29,30]. However, only a few experimental studies performed the ignition and flame blowout limits using plasma assisted ignition in swirled lifted flames. Therefore, the complete understanding of plasma ignition and blowout limits are still at an exploratory stage. In this study, a plasma actuator was implemented by using a central needle high voltage electrode and the nozzle as a grounded electrode was developed and equipped in a rectangular burner. The high voltage sinusoidal generator was used to drive the plasma actuator. Electrical characterization of induced plasma in presence of flame was conducted. The flow dynamics and lean blow out limits at different input electrical power conditions were analyzed.

2 Experimental Setup

2.1 Burner Geometry and Plasma Actuator Device

The experimental setup consists of a single cup burner with square section area of 105 mm width and 360 mm height. To allow good optical access for a wide variety of measurement techniques, the burner is bounded by four sides transparent quartz glass with 2 mm thickness. The air and methane are fed to the bottom of the burner shown in Fig. 1 and schematically in Fig. 3 through pressure regulators and two flow meters that controlled the gas flow rates.

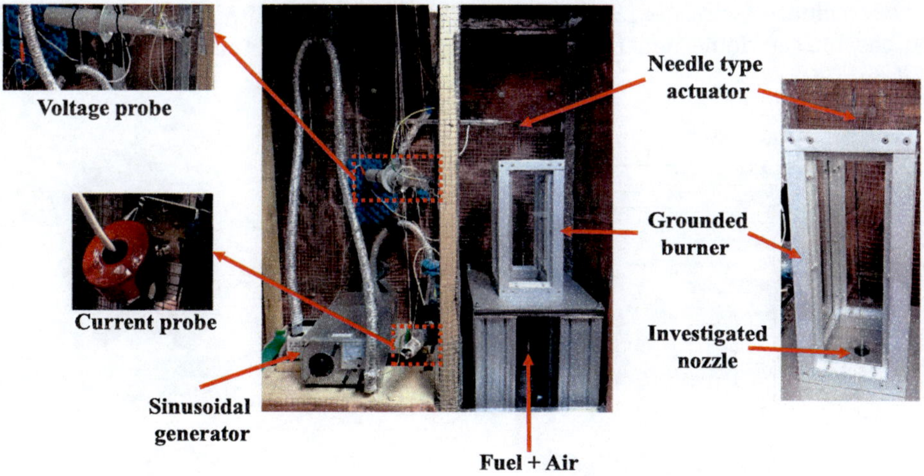

Fig. 1. Experimental setup of needle type plasma actuated burner.

To generate lifted swirled flames, an atmospheric nozzle based on the air-blast concept has been used with an effective area of $A_{eff} = 319\,\text{mm}^2$ to mimic the realistic geometrical scale for aero-engines similar to the investigated nozzle by Kasabov et al. [34].

This nozzle consists of five parts as depicted in Fig. 2; two swirlers to induce rotary movement to the air before entering the burner. The lower swirler is referred to as the primary swirler which consists of eight tangentially inclined vanes and the vane trailing edges are at an angle of 45° relative to the tube axis in a swirl. This swirl can be quantified as swirl number of SN = 0.76 and defined as:

$$SN = \frac{\dot{D}_i}{\dot{I}_i R_i} \tag{1}$$

where $\dot{D}_i$ is the angular momentum flux provided by the primary air channels, $\dot{I}_i$ is the axial momentum flux of the primary swirler and R_i denotes the characteristic length defined ad the inner radius of the lip at the smallest cross section.

The secondary air channels which namely as secondary swirler consists of twelve straight vanes with zero trailing angle. Thus, the air flowing through this secondary swirler is not orientated tangentially and therefore doesn't add any additional swirl to the flow (SN = 0).

These two swirlers are separated by prefilmer, where primary and secondary air streams meet at the lip of the prefilmer and start mixing with the fuel. The gaseous fuel is fed to the nozzle through gas inlet (lower part in Fig. 2) and the fuel is transferred to the prefilmer through eight holes in the primary swirler. The secondary swirler is bounded from the top by the diffuser which mounted on the burner dome and the combustion takes place further downstream the diffuser.

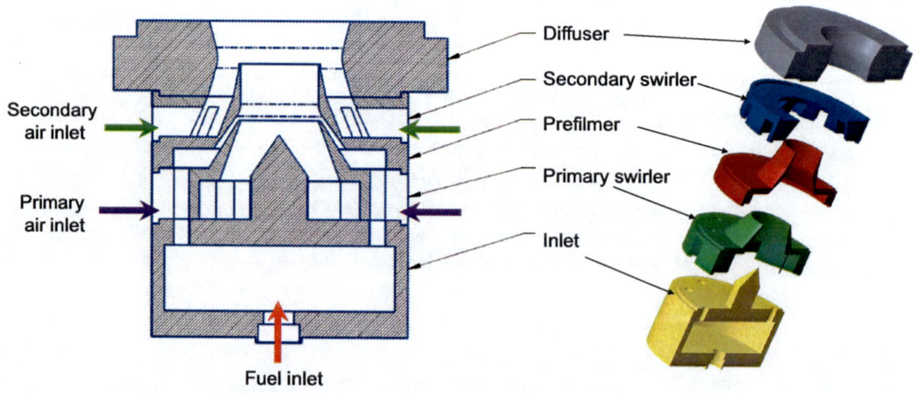

Fig. 2. Scheme of the investigated nozzle type.

To prevent interference of electromagnetic fields produced during the discharge process a Faraday cage was built around the test section. In particular, the actuator configuration consists of a tungsten needle having a length of 46 mm, with a diameter equal to 1 mm and a blunt tip with a 90° angle that acts as a high voltage electrode (HV) and the grounded electrode (GND) represented by the nozzle, as shown in Fig. 3.

Tests were performed with the actuator off and on at different working conditions and a wide range of fuel and airflow rates. All experiments were performed at a laboratory ambient temperature of about 298 K, humidity of about 60% and pressure of 1 bar.

2.2 Electrical Characterization

The experimental set-up used for the electrical characterization consisted of a high voltage (HV) amplifier (PVM500 Plasma Resonant and Dielectric Barrier Corona Driver), a high voltage probe (Tektronix P6015A), a current transformer (Bergoz Current Transformer CT-D1.0-B), and an oscilloscope (R&S3000). The voltage probe can measure DC voltages up to 20 kVRMS and pulses up to 40 kV (peak, 100 ms duration). The 75 MHz bandwidth enables to capture fast, high-voltage signals and it is characterized by an attenuation ratio equal to 1:1000, as shown in Fig. 3. The HV amplifier supplies the central needle HV electrode with a sinusoidal voltage waveform V(t), characterized by a frequency of 20 kHz, different amplitudes, and different input electrical power. The HV probe, current probe and oscilloscope were used to retrieve the voltage-current characteristic curves (as a function of time, t) and electrical power. The HV probe was located on the HV connector side, and the current transformer was located on the grounded side. Both the HV probe and current transformer were connected to the oscilloscope, and the corresponding signals were recorded on a dedicated PC with an accuracy given by the following formula:

$$Measurement\ accuracy = \pm(DC gain\ accuracy \times |Reading\ net - offset| + Offset\ accuracy) \quad (2)$$

where;

- DC gain accuracy is: 1.5% for the input sensitivity >5 mV/div (voltage case), and 2% for the input sensitivity ≤5 mV/div to ≥1 mV/div (Current case).
- Reading is the value of voltage or current measured by oscilloscope.
- Net offset is the offset value set in each test.
- Offset accuracy = ± (0.5% × $|offset|$ + 0.1 div × input sensitivity + 0.5 mV)
- Input sensitivity is in the range of 0.5 mV/div to 10 V/div at 1 MΩ.

The connection between probes and oscilloscope allowed simultaneous measurement of the applied voltage and the current flowing in the discharge, respectively. A single acquisition of the oscilloscope captured 2500 data points at a

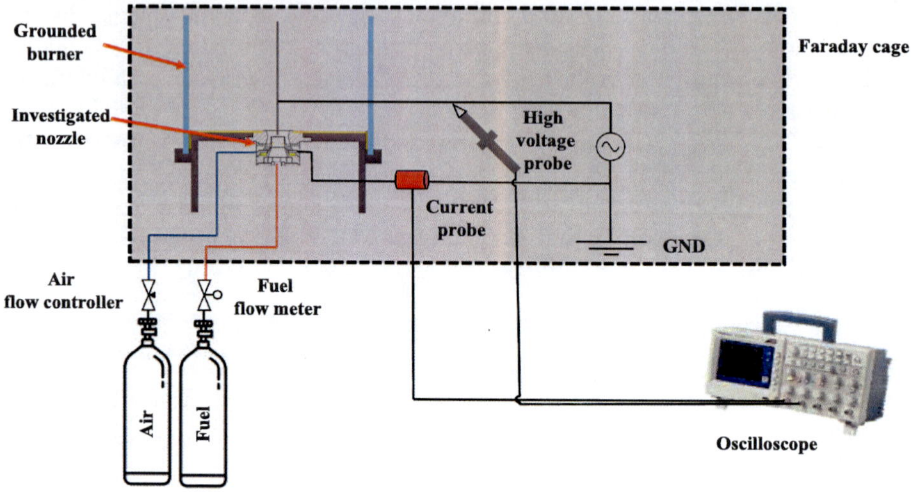

Fig. 3. Experimental set-up for the electrical characterization.

sampling rate of 25 MHz, equivalent to two time periods (T) for each signal. For each input voltage, 128 single acquisitions were recorded and averaged and the mean values of the applied voltage signal V (t) and of the current signal I (t) were obtained.

In order to increase measurement accuracy, for each test case, the V(t) and I(t) curves were recorded for n times throughout the experiment. The mean value of the respective power dissipation was considered representative of each test. The electric power dissipation $P_{elec.}$ is calculated as follows:

$$P_{elec.} = \frac{1}{2T}\int_0^{2T} I(t)V(t)\mathrm{d}t \tag{3}$$

For the power calculation, a numerical integration was performed using the trapezoidal rule and the corresponding uncertainty of each power measurement was estimated by standard uncertainty analysis methodology [31].

3 Results and Discussions

3.1 Electrical Characterization

Through the experimental campaign, quiescent experiments without flame and reacting flow experiments were performed. The complete details of air and fuel composition, along with the plasma actuation conditions and mean electrical power delivered to the flame as shown in Table 1. Figure 4 showed the applied voltage and current characteristics (on the left) and the instantaneous electrical power (on the right).

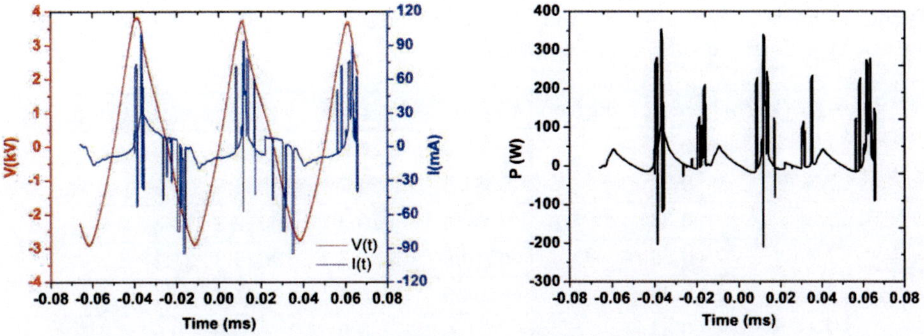

(a) $\dot{m_{air}}$ = 7.3 kg/hr; $\dot{m_f}$ = 0.33 kg/hr; V_{pp} = 6604.8 V; P = 29.37 W

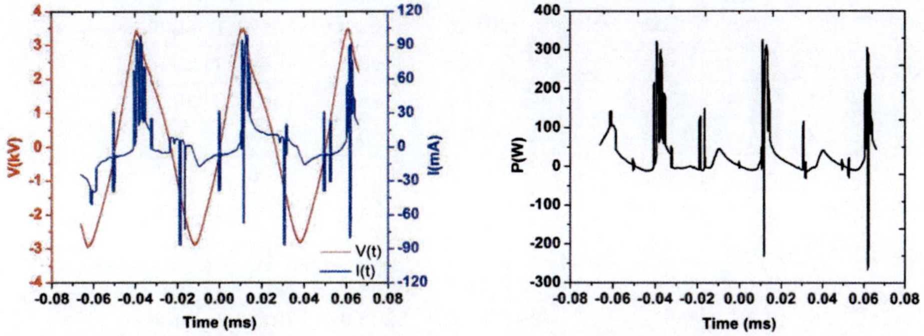

(b) $\dot{m_{air}}$ = 9.8 kg/hr; $\dot{m_f}$ = 0.33 kg/hr; V_{pp} = 6348.2 V; P = 30.7 W

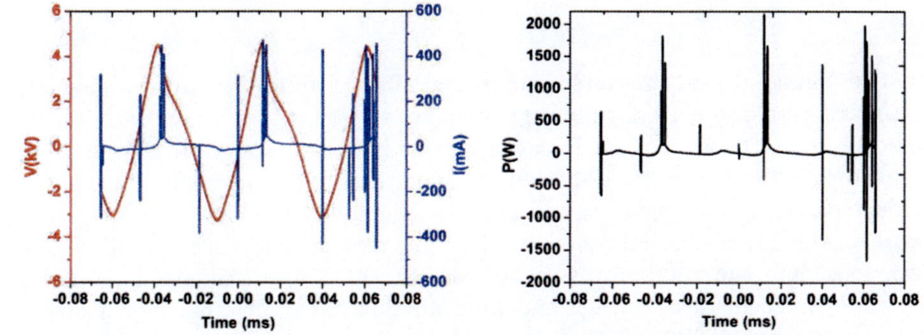

(c) $\dot{m_{air}}$ = 12.2 kg/hr; $\dot{m_f}$ = 0.33 kg/hr; V_{pp} = 7889.1 V; P = 33.03 W

Fig. 4. Voltage, current waveforms and instantaneous electrical power at fixed fuel flow rate of 0.33 kg/hr by varying air flow rates of 7.3 kg/hr (a), 9.8 kg/hr (b) and 12.2 kg/hr (c), and electrical power P.

Table 1. Fuel/air composition and electrical characteristics of plasma actuation for the reacting flow and quiescent conditions.

$\dot{m_{air}}$ (kg/hr)	$\dot{m_{fuel}}$ (kg/hr)	V_{pp} (V)	V_{max} (V)	f (Hz)	$P_{elec.}$ (W)
7.3	0.33	0.0	0.0	0.0	0.0
7.3	0.33	5581.6	2924.7	20038.8	6.9
7.3	0.33	7562.5	3919.2	20012.8	18.3
7.3	0.33	6910.9	3630.3	19629.3	21.1
7.3	0.33	6604.9	3717.9	19905.3	29.4
9.8	0.33	0.0	0.0	0.0	0.0
9.8	0.33	5554.4	2903.2	19964.4	6.4
9.8	0.33	8011.1	4162.3	19771.1	19.5
9.8	0.33	5714.3	3100.1	18965.3	25.0
9.8	0.33	6348.2	3475.4	19679.6	30.7
12.2	0.33	0.0	0.0	0.0	0.0
12.2	0.33	5565.5	2907.6	20082.3	6.1
12.2	0.33	8373.2	4394.6	19881.0	20.18
12.2	0.33	5900.5	3197.5	18849.4	30.8
12.2	0.33	7889.1	4607.9	20088.1	33.03
0.0	0.0	10082.2	5206.9	20133.7	14.2
0.0	0.0	19566.2	10216.0	20004.2	53.8

The voltage signal presents a sinusoidal shape, while the current waveform presents several large-amplitude current spikes superimposed on a low-amplitude current signal. These current spikes appear while the plasma discharge occurs and indicate the filamentary discharges. The discharge current is made of two evident parts by both positive-and negative-going cycle, which is related to the two types of micro-discharges. Some high current spikes in the positive-going cycle exist and the maximum value of the current is approximately 400 mA in the case of the highest voltage and high air flow rate. This can be associated with the streamer discharge which is filamentary and is capable of releasing energy for a short time, while the glow-like discharge which is developing in the negative-going cycle causes several current pulses of reduced amplitude. The plasma microdischarge cause to a change in the electrical impedance within the actuator and this leads to a current spike. These spikes present an ultrashort duration and a random time interval, less than the sinusoidal cycle period. The low-amplitude current shows a nonsinusoidal waveform, in which the frequency of the applied voltage can be identified. This low-amplitude current is given by the cold capacitive actuator displacement current within the actuator, corona discharge currents occurring between spikes and additional currents associated with the plasma processes, such as variable surface charge movements and volume charge equalizations. The signal's current peak to the average value ratio

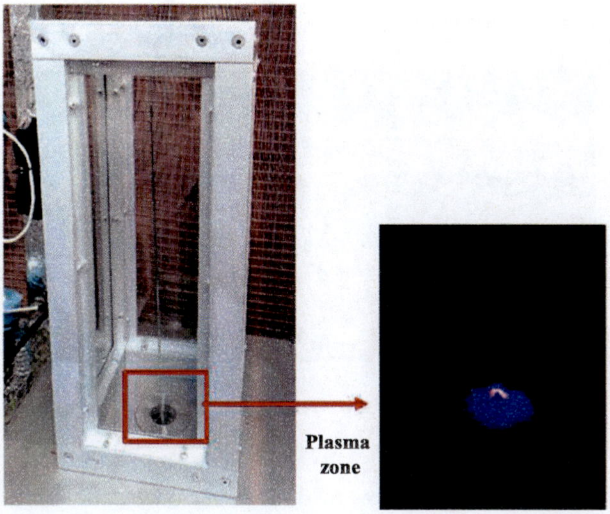

Fig. 5. Plasma generated by the plasma actuator in proximity of the burner exit.

("crest factor") has a very dynamic range, hence the data acquisition rate and resolution required to resolve the spikes should be high and have a great impact on accuracy of the power estimation. Since the power is directly related with the current, these high amplitude spikes are also visible in the instantaneous electrical power waveform. It is also possible to visualize the volumetric discharge of the plasma which appears as a purple cone extending beyond the radial edge of the high voltage electrode as shown in Fig. 5.

3.2 Flow Visualization and PIV Lab in Quiescent Mode

Smoke flow visualizations were performed by using the high speed CCD camera MEMRECAM GX-3 equipped with a Nikkor 60 mm f/2.8d A/F objective and an interference filter (wavelength, $\lambda = 532$ nm). The images were taken at three different frequency and resolutions: a) 50 Hz for 10 s of acquisition time, 1280 $\times$ 960 ppi; b) 1000 Hz for 4 s of acquisition time, 640 $\times$ 480 ppi; c) 5000 Hz for 3 s of acquisition time, 320 $\times$ 240 ppi. The behaviour of vortices and the average flow field induced by the actuator has been studied using the computer-based image processing approach called Digital Particle Image Velocimetry (DPIV) implemented in the open source "PIVlab - Time-Resolved Digital Particle Image Velocimetry" code from Thielicke and Stamhuis [35]. Incense smoke (particle average diameter of 0.3 μm) has been used for filling the test chamber and Laser 532 nm 300 mW mod. MGL-III-532 was used to illuminate the visualisation plane. The input electrical power for the quiescent flow is reported in Table 1.

Figure 6 shows an example of the acquired instantaneous smoke flow visualization image in the case of input electrical power equal to 53 W after 10 s from the start of the actuation. For the analysis of the induced flow field, the

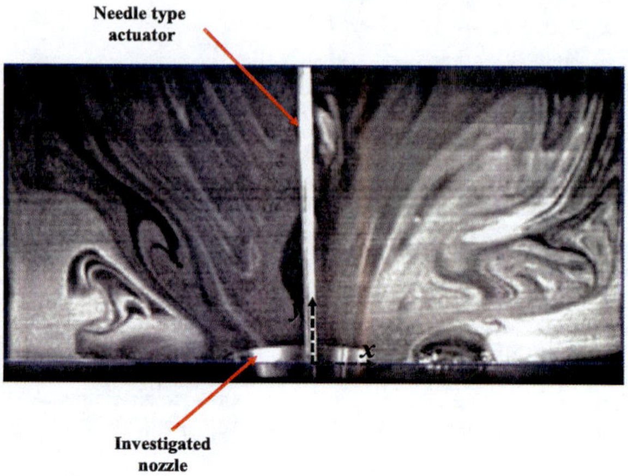

Fig. 6. Image of the instantaneous flow field induced under quiescent conditions without flow by the plasma actuator under quiescent conditions and input electrical power of 53 W acquired after 10 s from the start of the actuation.

acquired images were cropped to analyze only the region of interest (ROI), then to improve the contrast between the object and the background. So a ROI of 100 mm × 100 mm is set. The Contrast Limited Adaptive Histogram Equalization (CLAHE) method was used enabled with 20 pixels window size [32] and as the PIV algorithm, the fast Fourier transform (FFT-based) cross-correlation method was applied [33]. The interrogation window size set to 64 × 64 pixels in the first pass and then was reduced to 32 × 32 pixels in the second pass and 16 × 16 in the third passage, with a 50% overlapping for each pass and adaptive interrogation window. Non-overlapping image sequencing (A–B) was chosen for each run of image processing. Finally, improperly matched vectors were removed and vectors having a signal-to-noise ratio (SNR, ratio of the highest peak to the second highest peak in each individual correlation map) over 3 were analysed.

Figure 7 shows the instantaneous smoke flow visualizations, after subtraction of time-average intensity, acquired at a frequency equal to 50 Hz. It is evident the different flow structures in proximity of the electrodes and the burner exit for an input electrical power approximately equal to 14 W and the one of 53 W. The induced flow field was mainly determined by the strength of the plasma actuation.

As confirmed also by the PIV visualization in Fig. 8, at a low amplitude voltage the induced flow is more attached to the HV needle, which suggests that the momentum transfer between ions and neutral particles mainly happens near the ionization zone around the needle. While increasing the electrical power the vortexes were generated and a recirculation zone is created on the right side of the box. The maximum time-average velocity magnitude is approximately 0.01 m/s for both cases, however when the excitation increased there is a variation in the flow direction, the velocity distribution changed and negative radial velocity

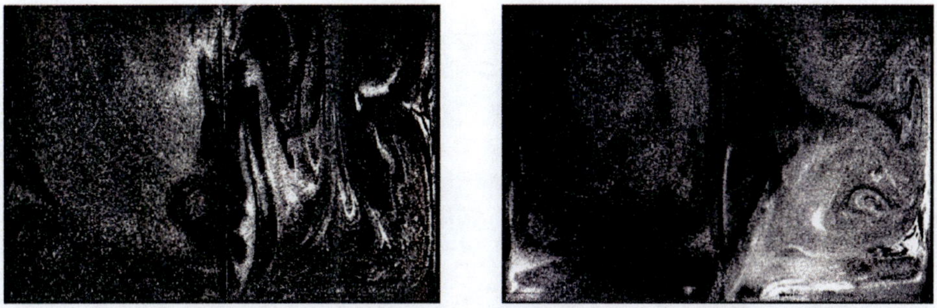

Fig. 7. Instantaneous smoke flow field after subtraction of the background given time-average intensity; where (a) at 35% of amplitude (P = 14 W) and (b) at 70% of amplitude (P = 53 W).

values as well as reverse axial velocity were present. The induced majority airflow covers an area about 5 cm in height and 5 cm in width. The flow recirculation can enhance fuel/air mixing which can in turn enhance the combustion process.

3.3 Extension of Lean Blow Out (LBO) Limits

Stably range is very important parameter for safety and has direct influence on the pollutant emissions. The LBO point is defined as the flame extinguishes. During these tests, a stable condition was set and then the fuel flow slowly reduced until the flame went out. The LBO stability of plasma flame has analyzed using different actuation power and compered non-plasma flame as shown in direct photographs in Fig. 9. In conditions near to LBO, the plasma-assisted flame appears broader with respect to the clean case without actuation, in particular a boost in flame length and an increase in the luminous intensity has been found as long as the applied electrical power increases.

Figure 10 shows how the LBO changes with varying the HV amplitude, hence the plasma power. A noticeable extension of LBO limit was observed when plasma is on.

As the amplitude increases, the equivalence ratio when the flame blows out decreases for all the air mass flow rate values. The maximum reduction of the equivalence ratio at lean blow out in percentage terms is of the order of 52% for the maximum plasma HV amplitude when the air mass is 9.8 kg/hr and the power is 30.7 W. In terms of fuel flow rate, for $\dot{m_{air}} = 14.6$ kg/hr and P = 40.5 W the blow off fuel flow rate is 0.17 kg/hr compared to 0.35 kg/hr without actuation.

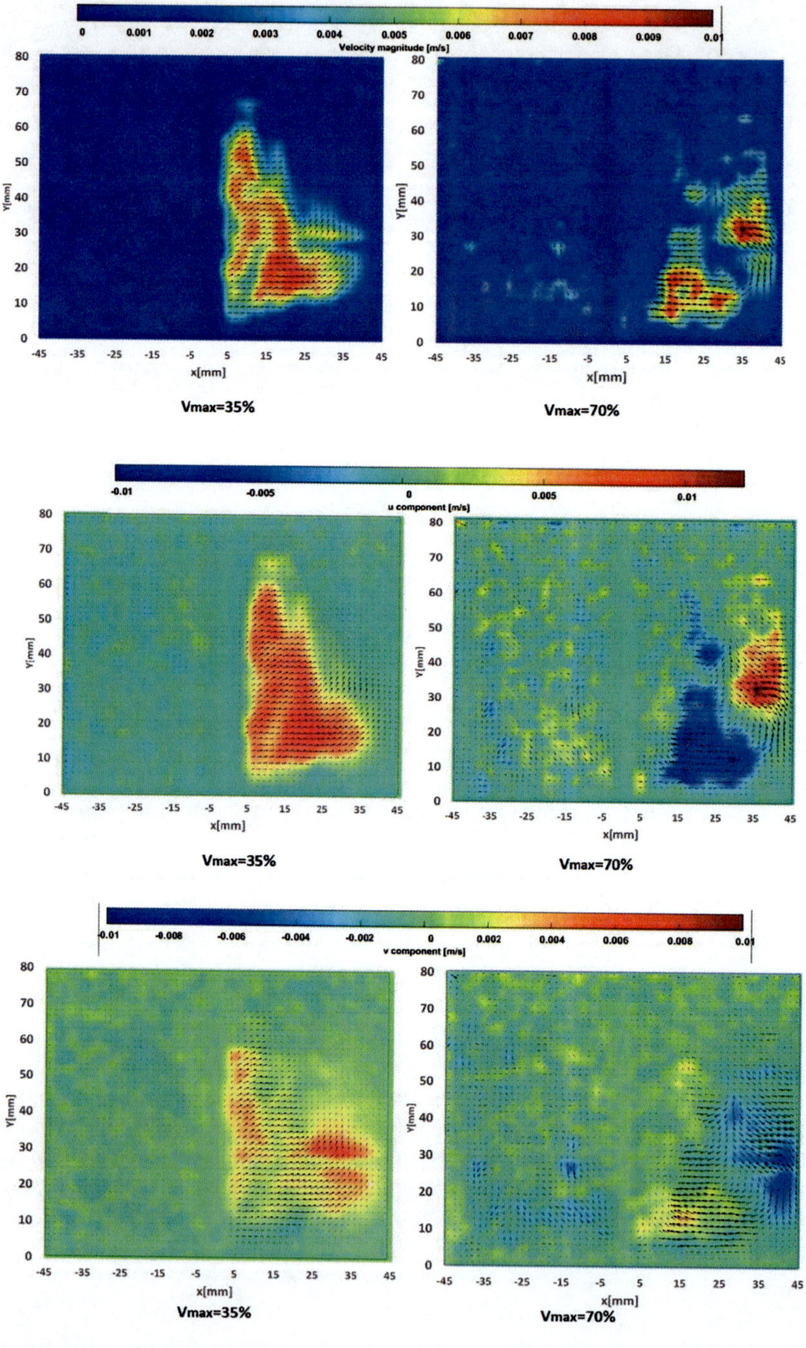

Fig. 8. Time averaged velocity magnitude (Vmag), radial (Vx), and axial (Vy) components at different plasma actuation at maximum amplitude equal to 35%, P = 14 W (on the left), 70%, P = 53 W (on the right).

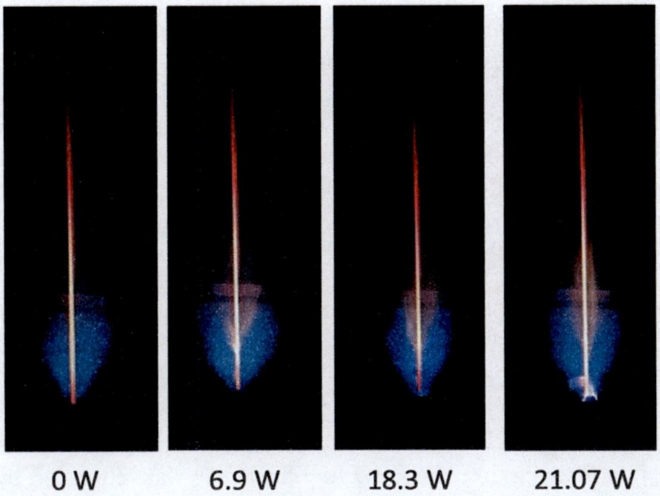

Fig. 9. Picture of flame without (left image) and in presence of plasma actuation at different HV amplitude at conditions near the LBO; $\dot{m_{air}} = 7.3\,\text{kg/hr}$ and $\dot{m_f} = 0.33\,\text{kg/hr}$.

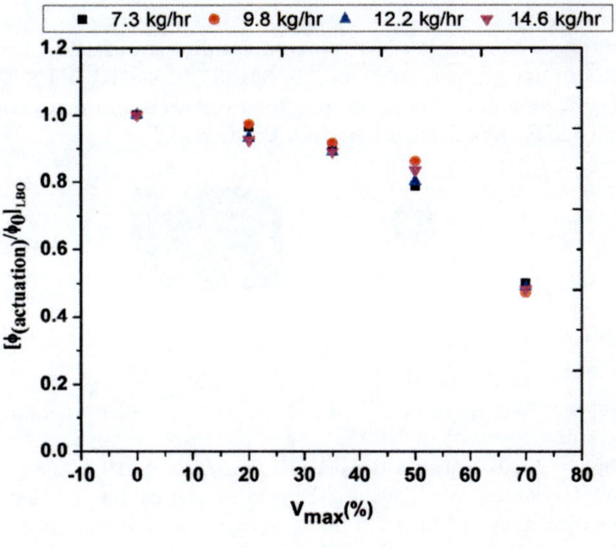

Fig. 10. Lean blow out limits at different HV amplitude and different mass flow rate.

4 Conclusions

A ringneedle plasma actuator power by a sinusoidal wave high voltage has been incorporated into the low swirl burner to improve the flame ignition and lean blowout performance. The study found that despite the low energy consumption in terms of electrical power, the plasma is effective in controlling flow and combustion. This study focuses on the flow induced by plasma excitation in the area near the burner outlet. The analysis of the aerodynamics under quiescent conditions showed an impact of the electrical power on the generated flow structures and in particular the presence of recirculation in the case of higher electrical power.

However, plasma excitation still plays a very significant role in the flame dynamics by influencing the fluid dynamics characteristics of the reacting case. In the reacting flow fields, the DBD plasma generates not only an electrical effect but also active species, which affect the dynamic behavior of the flame, particularly under lean flame conditions. It was found that the use of plasma actuation with an input electrical power less than 1% of the thermal power, the extension of LBO limits was more than 50%. In particular, a relevant extension of LBO limit was observed in presence of plasma discharges. At the highest air flow rate ($\dot{m_{air}} = 14.6\,\text{kg/hr}$) the LBO fuel flow rate was extended by 7.4% with an electrical power P = 8 W, by 11.1% with P = 20.6 W, by 16.66% with P = 35.2 W and by 51.85% with P = 40.5 W. Lowering the air flow rate the effect of the plasma actuation was more relevant and the LBO limits were significantly extended, at the air flow rate equal to 9.8 kg/s the maximum reduction of the LBO fuel flow rate was of 53% with P = 30.7 W.

Acknowledgment. This project has received funding from the Clean Sky 2 Joint Undertaking (JU) under grant agreement No. 831881 (CHAiRLIFT). The JU receives support from the European Union's Horizon 2020 research and innovation programme and the Clean Sky 2 JU members other than the Union.

References

1. Wei, B., Wu, Y., Liang, H., Su, Z., Li, Y.: Flow control on a high-lift wing with microsecond pulsed surface dielectric barrier discharge actuator. Aerosp. Sci. Technol. **96**, 105584 (2020). https://doi.org/10.1016/j.ast.2019.105584
2. Liu, F., Yan, H., Zhan, W., Xue, Y.: Effects of steady and pulsed discharge arcs on shock wave control in Mach 2.5 flow. Aerosp. Sci. Technol. **93**, 105330 (2019). https://doi.org/10.1016/j.ast.2019.105330
3. Sun, G., Wu, Z., Li, H., Zeng, L.: Discharge voltage characteristic in ablative pulsed plasma thrusters. Aerosp. Sci. Technol. **86**, 153–159 (2019). https://doi.org/10.1016/j.ast.2019.01.017
4. Cai, Z., Zhu, J., Sun, M., Wang, Z.: Effect of cavity fueling schemes on the laser-induced plasma ignition process in a scramjet combustor. Aerosp. Sci. Technol. **78**, 197–204 (2018). https://doi.org/10.1016/j.ast.2018.04.016
5. Cai, Z., Zhu, J., Sun, M., Wang, Z., Bai, X.-S.: Ignition processes and modes excited by laser-induced plasma in a cavity-based supersonic combustor. Appl. Energy **228**, 1777–1782 (2018). https://doi.org/10.1016/j.apenergy.2018.07.079

6. Zhu, H., Hao, W., Li, C., Ding, Q., Wu, B.: Application of flow control strategy of blowing, synthetic and plasma jet actuators in vertical axis wind turbines. Aerosp. Sci. Technol. **88**, 468–480 (2019). https://doi.org/10.1016/j.ast.2019.03.022
7. Tani, Y., Tsukizaki, R., Koda, D., Nishiyama, K., Kuninaka, H.: Performance improvement of the μ10 microwave discharge ion thruster by expansion of the plasma production volume. Acta Astronaut. **157**, 425–434 (2019). https://doi.org/10.1016/j.actaastro.2018.12.023
8. Feng, R., Li, J., Wu, Y., Zhu, J., Song, X., Li, X.: Experimental investigation on gliding arc discharge plasma ignition and flame stabilization in scramjet combustor. Aerosp. Sci. Technol. **79**, 145–153 (2018). https://doi.org/10.1016/j.ast.2018.05.036
9. Wang, W., Chen, Q., Mao, X.: AC plasma retarded flame spread over thin solid fuels in a simulated microgravity environment. Aerosp. Sci. Technol. **92**, 139–147 (2019). https://doi.org/10.1016/j.ast.2019.05.051
10. De Giorgi, M.G., Motta, V., Suma, A.: Influence of actuation parameters of multi-DBD plasma actuators on the static and dynamic behavior of an airfoil in unsteady flow. Aerosp. Sci. Technol. **96**, 105587 (2020). https://doi.org/10.1016/j.ast.2019.105587
11. De Giorgi, M.G., et al.: Plasma assisted flame stabilization in a non-premixed lean burner. Energy Proc. **82**, 410–416 (2015). https://doi.org/10.1016/j.egypro.2015.11.825
12. De Giorgi, M.G., et al.: Lean blowout sensing and plasma actuation of non-premixed flames. IEEE Sens. J. **16**(10), 3896–3903 (2016). https://doi.org/10.1109/JSEN.2016.2538970
13. Rosocha, L.A., Kim, Y., Anderson, G.K., Abbate, S.: Combustion enhancement using silent electrical discharges. Int. J. Plasma Environ. Sci. Technol. **1**(1), 8–13 (2007)
14. Ruma, M., Ahasan, H., Ranipet, H.B.: A survey of non-thermal plasma and their generation methods. Int. J. Renew. Energy Environ. Eng. **04**(01), 006–012 (2016)
15. Hicks, A., Norberg, S., Shawcross, P., Lempert, W.R., Rich, J.W., Adamovich, I.V.: Singlet oxygen generation in a high pressure non-self-sustained electric discharge. J. Phys. D Appl. Phys. **38**(20), 3812 (2005). https://doi.org/10.1088/0022-3727/38/20/007
16. Raizer, Y.P.: Gas discharge physics. Springer, Barcelona (1991)
17. Ju, Y., Sun, W.: Plasma assisted combustion: dynamics and chemistry. Prog. Energy Combust. Sci. **48**, 21–83 (2015). https://doi.org/10.1016/j.pecs.2014.12.002
18. Aleksandrov, N.L., Kindysheva, S.V., Kukaev, E.N., Starikovsjaya, S.M., Starikovskii, A.Y.: Simulation of the ignition of a methane-air mixture by a high-voltage nanosecond discharge. Plasma Phys. Rep. **35**(10), 867–882 (2009). https://doi.org/10.1134/S1063780X09100109
19. De Giorgi, M.G., Ficarella, A., Fontanarosa, D., Elisa Pescini, E., Suma, A.: Investigation of the effects of plasma discharges on methane decomposition for combustion enhancement of a lean flame. Energies **13**, 1452 (2020). https://doi.org/10.3390/en13061452
20. Kosarev, I.N., Aleksandrov, N.L., Kindysheva, S.V., Starikovskaia, S.M., Starikovskii, A.Y.: Kinetics of ignition of saturated hydrocarbons by nonequilibrium plasma: CH4-containing mixtures. Combust. Flame **154**(3), 569–586 (2008). https://doi.org/10.1016/j.combustflame.2008.03.007

21. Uddi, M., Jiang, N., Mintusov, E., Adamovich, I.V., Lempert, W.R.: Atomic oxygen measurements in air and air/fuel nanosecond pulse discharges by two photon laser induced fluorescence. Proc. Combust. Inst. **32**, 929–936 (2009). https://doi.org/10.1016/j.proci.2008.06.049
22. Zhu, J., et al.: Sustained diffusive alternating current gliding arc discharge in atmospheric pressure air. Appl. Phys. Lett. **105** (2014). https://doi.org/10.1063/1.4903781
23. Lee, D.H., Kim, K.T., Cha, M.S., Song, Y.H.: Optimization scheme of a rotating gliding arc reactor for partial oxidation of methane. Proc. Combust. Inst. **31**, 3343–3351 (2007). https://doi.org/10.1016/j.proci.2006.07.230
24. He, L., Chen, Y., Deng, J., Lei, J., Fei, L., Liu, P.: Experimental study of rotating gliding arc discharge plasma-assisted combustion in an aero-engine combustion chamber. Chin. J. Aeronaut. **32**, 337–346 (2019). https://doi.org/10.1016/j.cja.2018.12.014
25. Neophytou, A., Mastorakos, E.: Simulations of laminar flame propagation in droplet mists. Combust. Flame **156**, 1627–1640 (2009). https://doi.org/10.1016/j.combustflame.2009.02.014
26. Chen, Z., Ju, Y.: Theoretical analysis of the evolution from ignition kernel to flame ball and planar flame. Combust. Theory Model. **11**, 427–453 (2007). https://doi.org/10.1080/13647830600999850
27. Taamallah, S., Shanbhogue, S.J., Ghoniem, A.F.: Turbulent flame stabilization modes in premixed swirl combustion: physical mechanism and Karlovitz number-based criterion. Combust. Flame **166**, 19–33 (2016). https://doi.org/10.1016/j.combustflame.2015.12.007
28. Liu, H., Zhao, J., Shui, C., Cai, W.: Reconstruction and analysis of non-premixed turbulent swirl flames based on kHz-rate multi-angular endoscopic volumetric tomography. Aerosp. Sci. Technol. **91**, 422–433 (2019). https://doi.org/10.1016/j.ast.2019.05.025
29. Zhou, L.X.: Comparison of studies on flow and flame structures in different swirl combustors. Aerosp. Sci. Technol. **80**, 29–37 (2018). https://doi.org/10.1016/j.ast.2018.06.032
30. Wang, K., Li, F., Zou, P., Lin, X., Mao, R., Yu, X.: Effect of the fuel-air flow velocity on heat release rate of swirling non-premixed methane flames. Aerosp. Sci. Technol. **95** (2019). https://doi.org/10.1016/j.ast.2019.105465
31. Pescini, E., Francioso, L., De Giorgi, M.G., Ficarella, A.: Investigation of a micro dielectric barrier discharge plasma actuator for regional aircraft active flow control. Plasma Sci. IEEE Trans. **43**(10), 3668–3680 (2015). https://doi.org/10.1109/TPS.2015.2461016
32. Zuiderveld, K.: Contrast limited adaptive histogram equalization. Graph. Gems, 474–485 (1994)
33. Thielicke, W.: The Flapping Flight of Birds - Analysis and Application University of Groningen (2014). https://doi.org/10.1016/j.cja.2018.12.014. http://irs.ub.rug.nl/ppn/382783069
34. Kasabov, P., Zarzalis, N., Habisreuther, P.: Experimental study on lifted flames operated with liquid kerosene at elevated pressure and stabilized by outer recirculation. Flow Turbul. Combust. **90**, 605–619 (2013). https://doi.org/10.1007/s10494-013-9444-1
35. Thielicke, W., Stamhuis, E.J.: PIVLab-time-resolved digital particle image velocimetry tool for MATLAB. Published under the BSD license, programmed with MATLAB 7, no. 0.246 (2014): R14

Part II: Combustor-Turbine-Integration: Pressure Fluctuations and Turbulence

Computational Simulation of an Exhaust Plenum Charged by a Multi-tube Pulsed Detonation Combustor

Maikel Nadolski[1], Mohammad Rezay Haghdoost[2], Kilian Oberleithner[2], and Rupert Klein[1](✉)

[1] FB Mathematik & Informatik, Freie Universität Berlin, Berlin, Germany
{nadolski,rupert.klein}@math.fu-berlin.de
[2] Laboratory for Flow Instabilities and Dynamics, Institute of Fluid Dynamics and Technical Acoustics, Technische Universität Berlin, Berlin, Germany
mohammad.rezayhaghdoost@campus.tu-berlin.de, oberleithner@tu-berlin.de

Abstract. An efficient computational framework for the numerical simulation of multi-tube pressure-gain combustors for gas turbine applications is introduced. It is based on the open-source AMReX software platform (https://amrex-codes.github.io/amrex/), enhanced to allow for the flexible coupling of multiple computational domains with possibly different dimensionality. Here, six pulsed detonation tubes represented by an efficient one-dimensional model are coupled at their outlets to a threedimensional plenum reservoir. The plenum outlet can be partially blocked to simulate the flow resistance of the first stage of a turbine. This paper presents a validation of this computational setup against measurements obtained from a laboratory experiment with unblocked plenum, and a numerical simulation study of the flow generated by periodic sequential firing of the six detonation tubes into a plenum with a partially blocked exit cross section simulating the flow resistance of a turbine.

Keywords: Multi-domain–multi-dimension CFD · Pressure gain combustion · Pulsed detonation combustors

1 Introduction

Established technology for the most efficient gas turbines relies on combustion via premixed turbulent deflagration. Gas turbine efficiency has been optimized over decades, and even small improvements come at extensive development costs. Pressure gain combustion processes have been identified more recently to bear considerable potential for substantial increases in gas turbine efficiency. In this context, pulsed detonation combustion represents one of the promising alternatives Stathopoulos et al. (2015); Pandey and Debnath (2016).

A pulsed detonation combustor (PDC) operates in a cycle of (i) scavenging and charging, (ii) combustion by an initial deflagration phase followed by deflagration-to-detonation-transition (DDT) and complete burnout by a detonation wave, and

R. King and D. Peitsch (Eds.): AFCC 2021, NNFM 152, pp. 85–102, 2022.
https://doi.org/10.1007/978-3-030-90727-3_6

(iii) expansion of the high-pressure combustion products at the combustor exit. One certain contribution to the expected efficiency gain relative to deflagrative combustion arises from chemical energy conversion occurring at much higher temperatures in the detonation wave as the leading shock of the detonation very strongly compresses the combustible gas prior to chemical heat release. The associated cyclic gasdynamic compression, however, implies a strongly pulsating flow at the combustor exit which can be detrimental for both turbine efficiency and longevity if it were to be received by the turbine in this form.

One approach to controlling the intensity of pulsations received by the turbine consists of inserting an intermediate buffer volume, called the "turbine plenum" or, in short, "plenum" below, in between the combustor exits and the turbine entry plane. Such a plenum will not only disperse the pressure peaks induced by the PDCs but, with a suitable design of combustors, plenum, and their coupling interface, it may also serve as a reservoir that holds a substantially higher mean pressure than the compressor exit pressure. This latter relative pressure gain from compressor exit to turbine plenum may further contribute to the overall engine efficiency if it can be realized without impeding the scavenging of the combustor.

The design of the turbine plenum therefore deserves particular attention, and this is the backdrop of the present study. Motivated by the need to explore different combustor-plenum design combinations, we have developed a flexible and efficient computational framework for related numerical simulations. Its key ingredients are

- an explicit finite volume solver for the compressible Euler equations featuring
- space-time adaptivity,
- generic coupling to reaction kinetics software libraries,
- cut-cell representation of complex domain geometries,
- multi-block domain decomposition, and
- coupling of subdomains with different spatial dimensions.

In previous studies we have validated parts of this computational simulation framework against experimental measurements focusing on the qualitative correctness of coarse-grid simulations (Nadolski et al. 2018), on the detailed representation of shock induced flow structures near the exit of a shock or combustion tube (Rezay Haghdoost et al. 2020a), and on the geometrically more complex case of the flow through a shock divider (Rezay Haghdoost et al. 2020b). Here we aim to validate a simplified model of a PDC combustor implemented within this framework against experimental measurements of a six-tube PDC coupled to an annular plenum as displayed in Fig. 1 (see Rezay Haghdoost et al. 2021, for a detailed description).

Two operation scenarios are considered in this context. In the first scenario, for model validation, only a single combustor is fired. Between two such shots, a long period of scavenging is allowed for, so as to re-establish a quasi-steady inert gas flow through the combustor and the plenum. Only just before the next shot, the inflow stream is charged with fuel over a given time interval, such that about 75% of the tube's length is filled with the desired combustible mixture and the next shot can be fired. The second scenario involves the operation of all

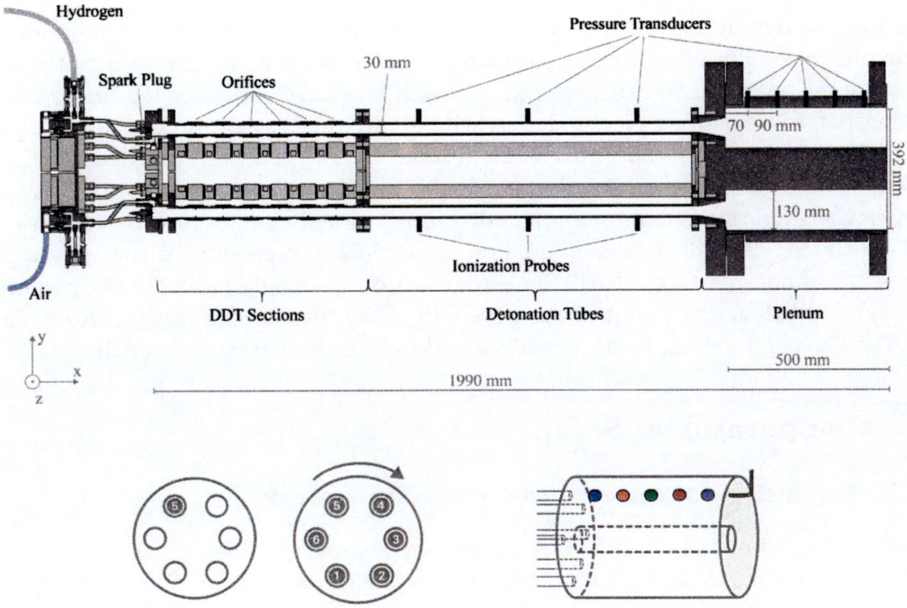

Fig. 1. Top: Test rig for experimental investigations of six pulsed detonation tubes coupled to an annular plenum. The plenum exit can be partially blocked by variable end plates. Bottom right: Sketch of the annular plenum showing the five pressure transducers placed at regular intervals along the plenum wall downstream of the fifth combustor (marked grey in the most left panel). The same color coding of the transducers will be used in later displays of the measured pressure time series. The two panels on the left indicate the two different operating modes (single- and multi-tube operation).

six PDC tubes, which are continuously scavenged by a prescribed fresh air mass flux. The tubes are charged with fuel as in the single-shot experiment, and fired in a continuous sequence one after its previous neighbor in intervals of 10 ms. This scenario is used here for a first computational study of how a relatively large plenum may help to attenuate the extremely strong pulsations induced by the PDC tubes before they reach the plenum exit. An extensive experimental study which includes ion probe measurements within one reference tube (marked grey in the bottom left panel of Fig. 1) to follow the combustion front evolution as well as pressure measurements along this tube and its prolongation into the plenum, though yet without sizeable blocking of the plenum exit, is available in (Rezay Haghdoost et al. 2021).

The realistic simulation of deflagration-to-detonation transition requires the capability to faithfully simulate multi-dimensional turbulent deflagrations in complex geometries, including chemical kinetic effects leading to auto-ignition in flow-induced hot spots and their very local interaction with compression waves (see, e.g., Gamezo et al. 2008; Gray et al. 2017; Bengoechea et al. 2018). As, in this project, we are interested not in these detailed processes within the combustion tube but rather in the impact of the highly dynamic PDC combustor outflow on the gas dynamics of the plenum, we represent the combustors by simplified one-dimensional models to be described in Sect. 3 below. The free parameters of this model are adjusted to reasonably match the measurements for single-tube firing before we move to a simulation with multi-tube operation of the system.

The remainder of the paper is structured as follows: Sect. 2 provides additional information regarding the numerical simulation system. Section 3 compares the single-tube experiments and computational simulations for validation of the latter. Section 4 presents a first numerical exploration of the effects of increasing blockage of the turbine plenum exit on the gasdynamics of the plenum flow, including a discussion of the possibility of a sustained mean pressure gain across the combustors. Section 5 offers additional discussion and conclusions.

2 Computational Setup

2.1 Extensions of the AMReX CFD Framework

For the implementation of the flow solver, we have adopted the open source software framework AMReX (https://amrex-codes.github.io/amrex/) for patch-wise adaptive structured grid calculations. The development of this platform is spearheaded and maintained by Lawrence Berkeley National Laboratory (LBNL), National Renewable Energy Laboratory (NREL), and Argonne National Laboratory (ANL). The framework provides high-level functionality for the handling of a hierarchy of refined grid patches and for the implementation of highly efficient code that can be dispatched to a variety of high-performance computing architectures.

On this platform we have implemented an explicit second order finite volume compressible Euler solver that uses Strang splitting to account for both multiple space dimensions and chemical reactions. The gasdynamics solver is a MUSCL scheme (van Leer 1979; Munz 1986). It employs the HLLE numerical flux (Einfeldt 1988) with an adaptive utilization of the correction in the advected variables (HLLE(M)) as described by Berndt (2014), and with slope limiting in the characteristic variables. Complex flow domain geometries are accounted for through a variant of the cut-cell technique for use with directional operator splitting as developed in (Klein et al. 2009; Gokhale 2018; Nadolski 2021). The solver is second order in global norms and subject to a reduction to first order at curved solid walls treated by the cut cell scheme.

For the present study, we are interested mainly in efficient simulations of the gas dynamics in the plenum, whereas the details of the combustion process in the PDC chambers are less relevant. It seems justified, therefore, to represent the combustion chambers by quasi-onedimensional simulations, while accounting for three-dimensional effects only for the flow in the turbine plenum. To achieve this within the AMReX framework, we have extended its capabilities accordingly. Prior to our developments, AMReX required the user to fix the number of spatial dimensions for a simulation, and to introduce one large rectangular master mesh that covered the entire flow domain. The computational grid patches on which the actual calculations were to be carried out had to be placed within – and spatially referenced relative to – this grid super structure.

This principal grid structure has been generalized in this work in two ways. First, we now allow for a general multiblock grid structure so that several rectangular patches may form the basic reference grid. Any two members of this basic structure that touch each other may do so across rather general rectangular subsets of the planes (or lines) forming their surfaces. The second change concerns situations in which the flow on some of these patches can be expected to be approximately uniform in one or more of the cartesian directions. The system now allows for a reduced-dimensional representation of that flow, even though it may couple to a full-dimensional patch somewhere along its surface. The logics of coupling two subdomains with different grid dimensions essentially follows the general adaptive mesh refinement (AMR) strategy in that the lower-dimensional grid patch is treated as a particular coarsened version of an underlying full-dimensional grid of higher resolution. Technical details of these modifications are described in (Nadolski 2021), and the technology has been made publicly available as part the AMReX distribution.

We are thus in a position to simulate the six PDC tubes seen in the experimental setup (Fig. 1) in one space dimension, while pursuing fully three-dimensional simulations of the plenum flow. To avoid too abrupt a change from the one- to the three-dimensional representations, the last 100 mm of the combustor lengths are covered by the three-dimensional grid, so that the coupling interfaces lie in an axial section within which the assumption of a nearly one-dimensional flow in the combustion tubes is still justified. The according computational grid for the domain represented in three space dimensions is shown in Fig. 2, which presents the setup of the computational grid as used in the subsequent simulations, with spacial cell sizes $\Delta x = \Delta y = \Delta z = 1.81$ mm.

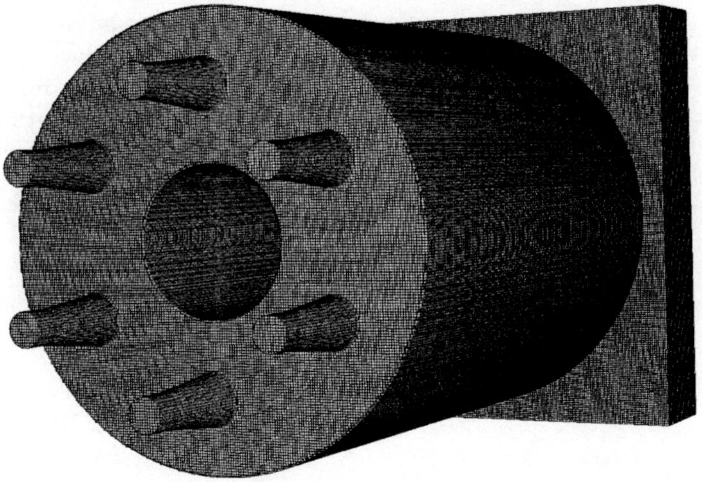

Fig. 2. Grid arrangement for the computational simulation of the gas dynamic processes in the plenum part of the text rig shown in Fig. 1

As we are interested in leading order effects of the gas dynamics in the plenum, we solve the compressible Euler equations for a chemically reacting gas only on the one-dimensional combustor subdomains, while representing the dynamics in the plenum by non-reacting flow with the equation of state obtained from the chemical/thermodynamics library FLAMEMASTER (Pitsch 2021), which is also used for the reactive flow simulation within the onedimensional combustor submodels.

2.2 One-Dimensional Model of the PDC Tube

To avoid the necessity of a detailed, inherently multi-dimensional simulation of deflagration-to-detonation transition (DDT) in the combustors, we utilize a simplified one-dimensional computational model to generate the strong shock outflow into the turbine plenum. This model consists of a one-dimensional computational domain of length L representing the PDC tube (see Fig. 3). At the left end of the tube, the mass flux $\dot{m}$ is prescribed, and this approximates the forced choked inflow applied in the associated experiments in (Rezay Haghdoost et al. 2021) rather well. This mass flux consists of just air or of a stoichiometric hydrogen-air mixture, with equivalence ratio $\phi_{\mathrm{hyd}} = 1.0$. Prior to ignition, the inflow is switched from clean air to the mixture for 26 ms, which leads to a charged length L_{chrg} of the tube of, on average, about 1 100 mm.

The deflagration-to-detonation transition in a real PDC requires a certain run-up distance before the detonation is established. We mimick this process here by a sudden internal energy addition within a width of δx_{ign} centered on location x_{ign}, and by replacing the charged unburnt gas to the left of $x_{\mathrm{ign}} - \frac{1}{2}\delta x_{\mathrm{ign}}$ with burnt gas conditions obtained by constant volume combustion. The local energy addition is modelled by imposing a triangular-shaped temperature profile within this interval with a maximum temperature of T_1 and minimum temperature T_0 as sketched in Fig. 3. This temperature rise occurs at constant density and thus induces a pressure rise and associated strong pressure waves departing to the left and right from the ignition region. It also triggers immediate chemical heat release in the rightward travelling wave wherever the temperature exceeds the ignition threshold of the hydrogen-air mixture. This procedure reproducibly induces the emergence of a detonation wave, and allows us to capture the essential dynamics of the pressure waves that enter the plenum after exiting the PDC chambers.

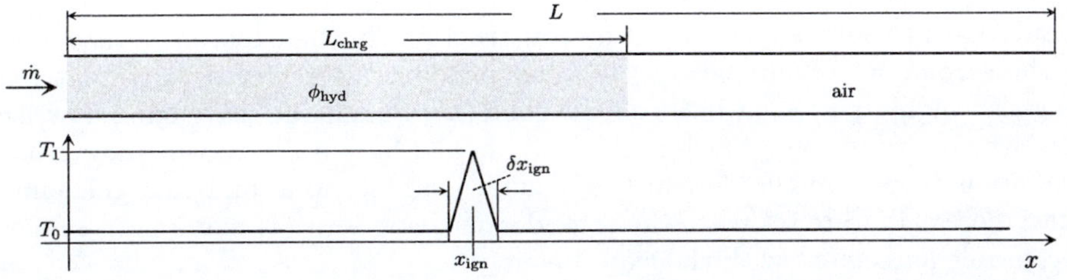

Fig. 3. Sketch of the one-dimensional model of a PDC chamber. $\dot{m}$: mass flux imposed at the entry of the tube; (L_{chrg}, ϕ_{hyd}): length of subdomain charged with hydrogen-air mixture and related equivalence ratio; $(x_{ign}, \delta x_{ign}, T_0, T_1)$: center and width of ignition subdomain, minimum and maximum temperatures of the triangular heating profile.

The model's free parameters are, $(\dot{m}, L, L_{chrg}, x_{ign}, \delta x_{ign}, T_0, T_1, \phi_{hyd})$. For lack of space, we do not present a comprehensive parameter investigation in this paper. Rather, we demonstrate in an exemplary fashion the influence of the inflow mass flux $\dot{m}$ and of the ignition location x_{ign} below, with the other parameters fixed at values that have turned out to yield satisfactory results in a more extensive study (Nadolski 2021), see Table 1.

3 Validation Based on Single-Tube–Single-Shot Tests

For the single-tube–single-shot experiments and numerical simulations, the plenum exit is unblocked and only a single PDC tube out of a total of six is fired. Pressure time series are monitored by the five pressure transducers located on the outer wall of the annular plenum directly downstream of the firing tube at equal spacing (see Fig. 1). The focus is on reproducible single shots separated in time sufficiently to avoid any influence of previous shots on the next one. In a first very straightforwardly implementable series of numerical runs, each simulation is set up separately and with the gas at rest initially. Thus, independence

Table 1. Parameters of the one-dimensional PDC model that are fixed in the present validation study w.r.t. the remaining parameters, i.e., imposed mass flux $\dot{m}$ and hot spot location x_{ign}.

Total PDC tube length	L	1 490	mm
Length of charge	L_{chrg}	1 100	mm
Width of ignition spot	δx_{ign}	50	mm
Lower hot spot temperature	T_0	300	K
Upper hot spot temperature	T_1	2 000	K
Equivalence ratio of charge	ϕ_{hyd}	1.0	

of subsequent shots is guaranteed by construction. In the experiment as well as in the second set of simulations, this independence is achieved by a relatively long scavenging period in between the shots. Although the scavenging flow has a relatively low average Mach number of about Ma = 0.13, we will show below that its influence on the flow in the plenum is substantial, and that the somewhat more elaborate set up of the second simulation series is required to achieve acceptable agreement with the experiment.

Figure 4 shows the results of the pressure measurements. The color coding corresponds to that used to indicate the different pressure transducers in Fig. 1. These measured pressure traces (solid lines) provide the reference for the subsequent numerical sensitivity study (dashed lines).

The simulations that generated the two columns of results in Fig. 4 differ in terms of the imposed mass flux $\dot{m}$ (see Fig. 3). For the left column, the gas was at rest initially, i.e., $\dot{m} = 0$, whereas for the right column, the same mass flux as measured in the experiment was imposed and before charging and firing a shot we let the system settle into a nearly steady state. The rows in the figure represent different locations x_{ign} of the hot spot in the tube. If we let $k = 1, 2$ label the rows, then $x_{\text{ign,k}} = (2k - 1) \cdot 100\,\text{mm}$.

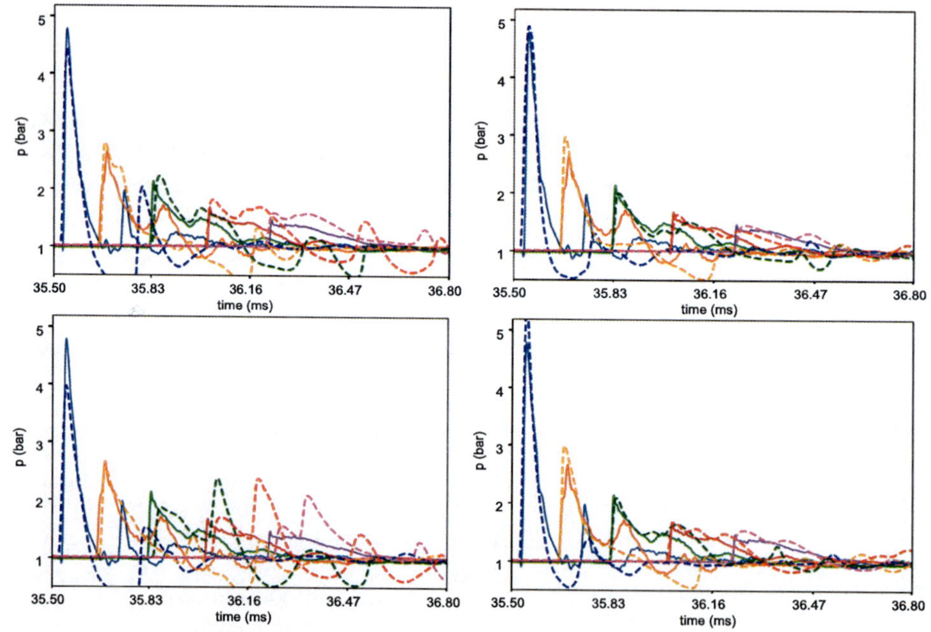

Fig. 4. Measured (solid) and simulated (dashed) pressure traces at the five locations along the plenum wall with color coding as indicated in Fig. 1. Left column: Individual shots into gas at rest; right column: Individual shots into scavenging flow charged for 26 ms before firing, corresponding to a length of the charged gas of approximately 1.1 m. Ignition imposed at $(2k - 1) \cdot 100\,\text{mm}$ for the result shown in the kth row.

The output for experiment and simulations has been shifted along the time axis to guarantee that the first steep pressure rise for the first transducer matches well between the experiment and all the simulations. Both the left and right columns reveal a significant effect of the hot spot location. In the left column (zero initial mass flux), e.g., the pressure trace taken at the fourth transducer location (red dashed curve) develops a strong secondary pulse as x_{ign} increases. This secondary pulse is entirely absent from the solid red line, i.e., from the results of the corresponding measurement. In this sense, the computational signal for hot spot position $x_{\text{ign}} = 0\,\text{mm}$ would seem most promising, were it not for a significant discrepancy in the trace from the first transducer location (blue curves). In fact, both panels on the left show a roughly equal time lag between the simulated first and second maximum of the blue dashed curves, and this time lag is substantially longer than that seen in the experiment. Thus, when the flow is at rest initially a reasonable fit between the measured and simulated pressure traces cannot be achieved by varying the hot spot location alone.

The panels in the right column of Fig. 4, for which the imposed mass flux leads to a Mach number of the scavenging flow of Ma ≈ 0.13, do not feature the strong secondary hump in the red (or green) dashed curves as seen in the left column. Thus, this secondary mode is traced back to the absence of a mean flow here. Moreover, the time lag between the first and second pressure maxima of the first transducer (blue curves) shows a clear dependence on the hot spot location. In the second panel the time lag between the maxima agrees very well with that seen in the measurement. Since other features of the traces also agree reasonably well, we take the massflux corresponding to the column on the right and the hot spot location of $x_{\text{ign}} = 300\,\text{mm}$ as the model parameters to be used in the subsequent multi-tube simulations.

The only remaining sizeable discrepancy between the measured and simulated traces concerns the minimum between the first two maxima, which is much lower in the simulations than it is in the measurements. This has been traced back to the way we model the initiation of the detonation wave. In the experiment, an accelerating turbulent deflagration pushes the flow to the transition point, so that the volume in the first part of the tube between tube entry and the establishing detonation is essentially filled with burnt gas at an elevated pressure and the still non-reacting gas in front of the flame has been set in motion by the prior flame-induced gas expansion. In contrast, in the simulation the detonation is triggered by hot spot ignition in a stream of gas that is moving at the nearly constant velocity imposed by the forced inflow mass flux, and the gas between tube entry and ignition location undergoes an instantaneous isochoric reaction. As a consequence, the pre-frontal velocities seen by the emerging detonations as well as the pressures in the respective first sections of the tube differ to some extent between simulation and the experiment. Further improvements of the model readily come to mind but are considered beyond the scope of this paper, because the principal dynamics in the plenum is captured sufficiently well by the present computational setup to enable a first assessment of the dominant gasdynamic effects in the plenum.

4 Multi-tube Firing with Partially Blocked Plenum Exit

Here we pursue a computational simulation study to investigate the response of a multi-tube array of PDC combustors firing in a regular sequence into an annular plenum in analogy with that of the experimental setup of Fig. 1. A partial blockage of the plenum exit simulates the flow resistance of an adjacent turbine (see Fig. 5).

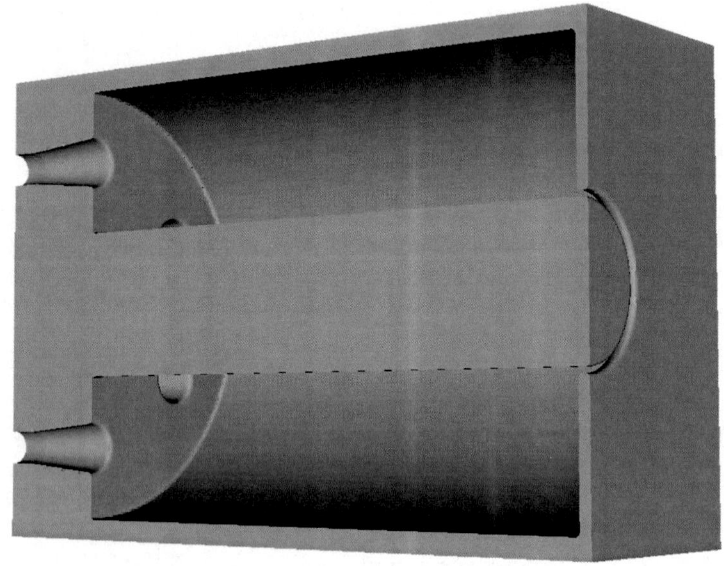

Fig. 5. View of the plenum with a 98% blocked exit mimicking the flow resistance of a turbine.

In this simulation, the six combustors receive a constant imposed mass flux of fresh air or reactive mixture from their entry that induces an average inflow Mach number of $\mathrm{Ma} = 0.13$. One tube fires 10 ms after its next counter-clockwise neighbor, when viewed in the direction of the inflow. Prior to simulating the firing of a tube, it is charged with a stoichiometric hydrogen-air mixture over approximately 1.1 m of its length, corresponding to a time interval of fuel injection of 26 ms, see lower right panel of Fig. 7.

4.1 Plenum Pressure for Forced Inert Gas Flow

A first evaluation concerns the effect of blocking the plenum exit for an inert gas flow generated by imposing the reference mass flux uniformly at the entries of the six combustors. Figure 6 shows pressure traces monitored at the locations of the five pressure transducers in the plenum when the system is started from a medium at rest under ambient conditions and the entry mass flux is imposed instantaneously at time $t = 0$. After a short time lag, a shock wave arrives at the first transducer, and then the system develops an intense wave activity that leads to the effective mean pressure rise needed to force the incoming mass flux through the remaining crossection of the plenum exit. The left panel in the figure corresponds to a blockage ratio of 96%, while the panel on the right was obtained with the 98% blockage also used for later runs with sequentially firing tubes.

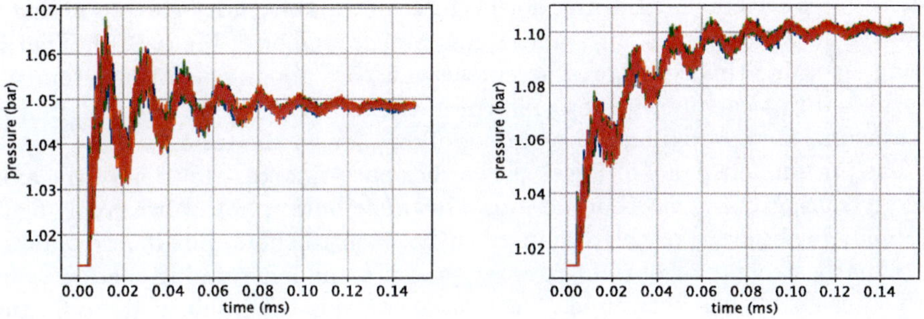

Fig. 6. Increased average plenum pressure needed to force the given mass flux through the partially blocked plenum exit (see Fig. 5). Left: 96% blocking ratio; Right: 98% blocking ratio as used for the multi-tube sequential firing test case.

A similar effect is to be expected in case of reacting flow. Yet, in this case, the chemical heat release will raise temperatures and force the gas to strongly expand. The prescribed mass flux can then pass the plenum exit only if it attains much higher exit flow velocity, and this requires considerably higher excess pressure in the plenum relative to the environment. See also the discussion in the next paragraph.

Table 2. Sequential firing conditions

Firing sequence	next neighbors, clockwise, every 10 ms	
Average length of charge	L_{chrg}	1 100 mm
Charge equivalence ratio	ϕ	1.0

4.2 Mean Plenum Pressure and End-to-End Total Pressure Differences

Here we discuss aspects of the fluid dynamics within the combustors and the downstream plenum under sequential firing conditions. The six tubes are fired one after the other in time intervals of 10 ms. Figure 7 documents the typical cycles of the dynamics within one of the combustors (tube 0) after initial transients have essentially settled. The space-time temperature distribution (top left panel) reveals that between two shots pressure waves are reflected repeatedly between the forced left and open right ends of the tube. This leads to an approximate equilibration of the pressure and the re-establishment of a nearly stationary scavenging flow.

The bottom right panel of Fig. 7 shows the space-time distribution of the fuel mass fraction. The inflow is charged with fuel over the last 26 ms before the tube is fired. During this time span, the tip of the charge distribution moves to a distance of, on average, 1 100 mm downstream of the tube entry corresponding to a flow velocity of about 42 m/s and a flow Mach number of Ma $\approx$ 0.13. That is, during the same time span pressure waves can pass about seven times the tube length, and this corroborates our interpretation of the temperature distribution as revealing the approximate pressure equilibration in the tube.

The bottom left panel of Fig. 7 shows the space-time distribution of the axial velocity during the same time period. This distribution not only reveals again the oscillatory nature of the dynamics in between to shots, but it also shows a particularly strong pair of right-running pressure and left-running suction waves right after a shot has been fired. The suction wave is the result of the reflection of the strong detonation-induced shock at the sudden area increase from the end of the tube to the full plenum crossection.

The space-time distribution of pressure (Fig. 7, top right panel), due to the linear scaling of the color code and the extreme pressure spikes arising in the detonation wave, mainly highlights the short time intervals during which the tube is fired and the combustion driven wave passes the length of the tube. To document that our simplified model of the DDT process actually does generate detonation waves, Fig. 8 shows the pressure distributions as a function of space for six output time slices at and after one of the shots in tube 0 at time intervals of 0.2 ms. It is seen how an overdriven detonation, recognizeable by the typical Zel'dovic-von Neumann-Döring (ZND) pressure spike, is generated right after the energy addition (dotted line [with spike at $x \approx -0.975$ m]), that – as long as there is any combustible left – it settles into a mode more akin to a Chapman-Jouguet (CJ) detonation (dashed and long dash-dotted lines), and that it turns into a decaying shock wave towards the right end of the tube (short dash-dotted, solid, and dotted [with spike at $x \approx -0.2$ m] lines).

The same mass flux is forced here through the system as it is in the inert flow case discussed above, but the PDC tubes are now fired sequentially as described in Table 2. The entropy of the burnt gas is considerably larger than that of the inert gas before, so that the burnt gas tends to expand relative to the conditions in the inert flow, and the volume flux through the plenum exit, and hence the exit velocity, increases. This is accompanied by a mean pressure rise as displayed in Fig. 9. Instead of the moderate pressure increase by about 10% for inert gas flow seen in Fig. 6, the pressure now rises above ambient pressure more substantially. Indeed, after the first transient between 50 and 350 ms the remaining effective pressure rise amounts to an increase by about 25%.

Note that this pressure increase arises across then entire system comprising the combustors and the exit plenum. Therefore, a comparably elevated pressure

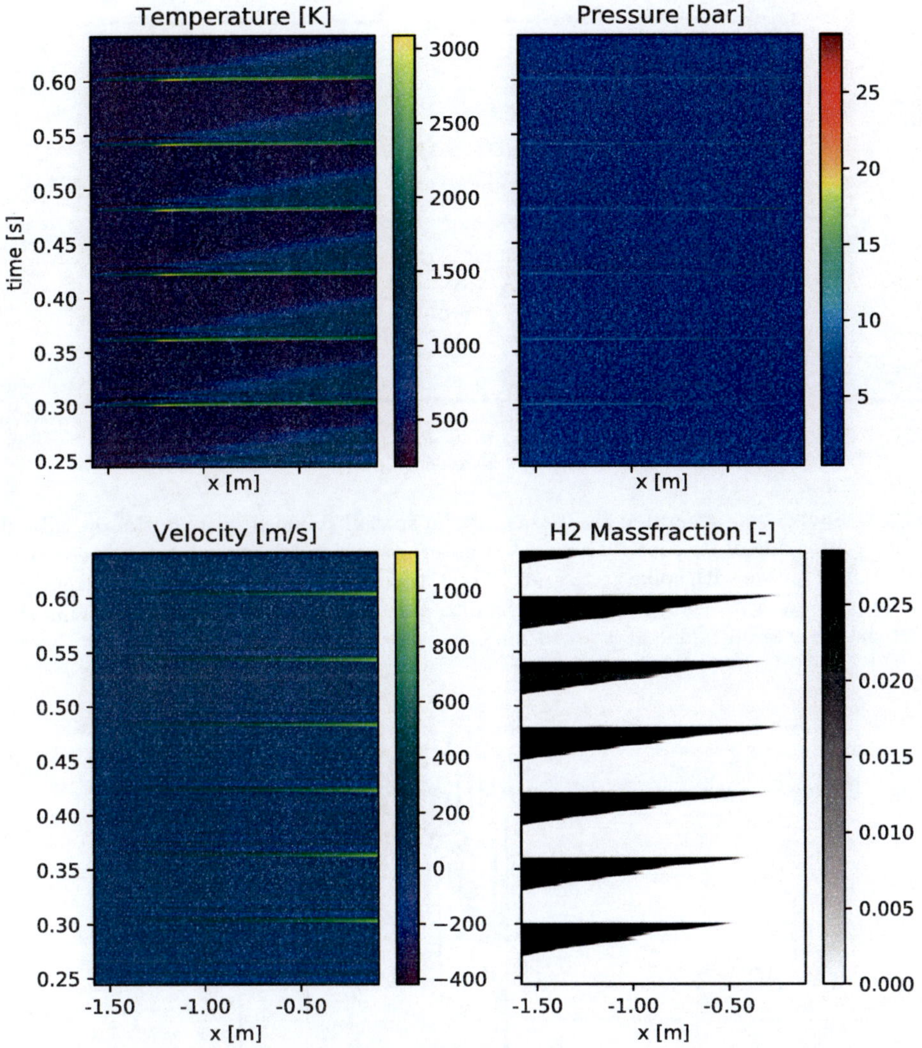

Fig. 7. Space-time traces of the gasdynamics within one of the combustors of the multi-tube configuration.

has to be imposed at the combustor entries to establish the required scavenging flow, and the observed pressure increase does not correspond to a pressure gain from the compressor to the turbine plenum as referred to in the discussion of the second potential efficiency gain in the introduction. To corroborate this, consider the comparison of the time series of area-averaged total pressures at the combustion chamber entry (blue curve) and in the smallest crossection of the plenum exit (orange curve) in Fig. 10. After initial transients have settled, the time and area averaged total pressures at the combustor inlet and plenum

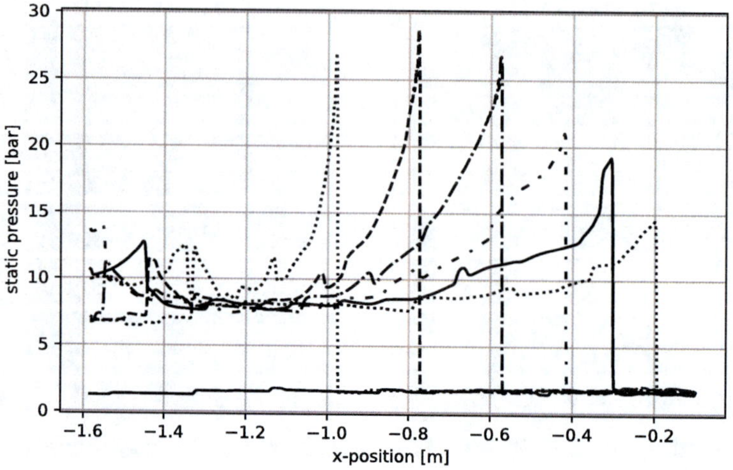

Fig. 8. Snapshots, separated by 0.2 ms, of the spatial pressure distribution in tube 0 shortly after one of its shots. The right-running wave develops first into an overdriven detonation (curve with spike at $x \approx -0.775$ m), then attenuates to a near-CJ detonation mode (spike at $x \approx -0.575$ m), and turns into a decaying shock wave when leaving the fuel-charged region (spike at $x \approx -0.2$ m).

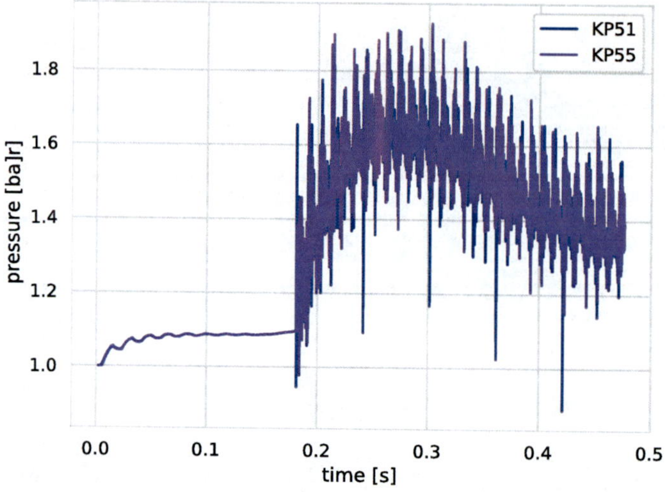

Fig. 9. Pressure traces monitored at the locations of the experimental pressure sensors 1 (KP51) and 5 (KP55) in the plenum (see Fig. 1) under the sequential firing conditions from Table 2.

exit are approximately the same. In fact, realizing an effective pressure gain across a PDC chamber seems impossible if the combustor is to be scavenged by a total volume flux that is comparable to or larger than its own volume, as is the case here. Confer again the bottom right panel of Fig. 7 and its discussion

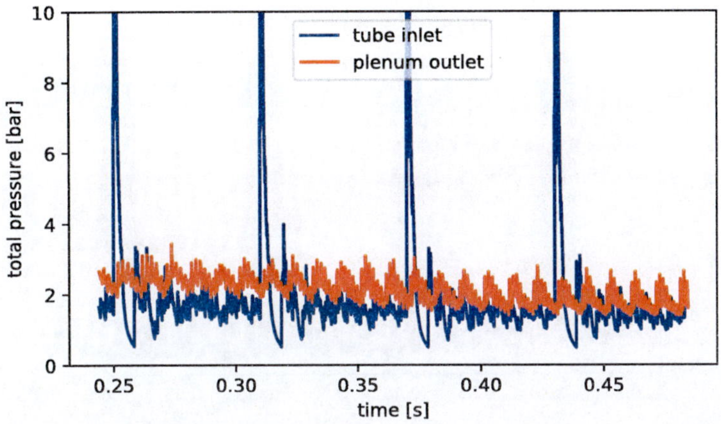

Fig. 10. Time series of area-averaged total pressures at the entry of combustor 0 (left), and within the plenum outlet crossection (right). Note that the total pressure spikes in the combustor reach up and beyond 15 bar. Their traces are truncated in this graph so as to better resolve the fluctuations of much lower amplitude that remain when the pressure pulses directly driven by the chemical heat release have passed.

above. Unless the scavenging flow is close to sonic, pressure waves will oscillate many times across the length of the tube during the scavenging process, leading to an effective pressure equilibration.

A possible remedy could lie in the controlled use of gasdynamic effects involving large amplitude pressure waves in the plenum as realized in studies of the "shockless explosion combustion" (SEC) process (Berndt 2016; Berndt and Klein 2017). An exploration of this idea for PDC applications is beyond the scope of the present work. See, however, (Wolff 2019) for a sketch of a related concept. In any case, even without such a mean pressure rise across the combustors, efficiency gains of a PDC-driven gasturbine relative to a classical turbine with deflagrative combustion should still arise from the first effect mentioned in the introduction, i.e., from chemical reactions occurring at substantially higher temperature in the PDC case.

The turbine plenum's primary purpose is a partial equilibration of pressure to avoid excessive fluctuations of the thermodynamic state and flow velocities within the turbine entry section. Figure 11 shows again the time series of the area-averaged total pressure within the plenum exit crossection (orange) as in Fig. 10, and this is compared here with the time series of the spatial variance of the total pressure in the same crossection. Clearly, the temporal variability of the area-averaged total pressure exceeds the total pressure's spatial variances by a factor of at least two, so that the former provides a good estimate for the total pressure fluctuations in the plenum exit crossection. We conclude, therefore, that the very high pressure spikes induced by detonation combustion, with observed maximum pressures as high as 25 bar, are strongly dispersed within the plenum as intended.

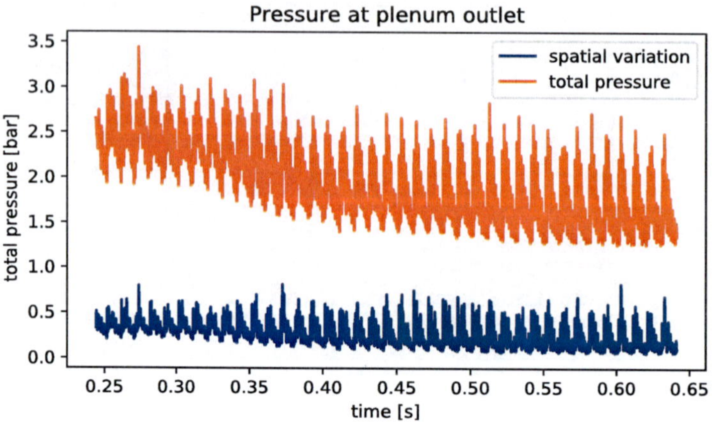

Fig. 11. Time series of the spatial variance of total pressure (blue) and of the total mean pressure (orange) in the smallest crossection of the plenum exit.

5 Conclusions

This manuscript documents the development of an efficient simulation model for multi-tube pulsed-detonation-combustors coupled to a plenum reservoir based on the reactive Euler equations. This system represents the combustion tubes by effective onedimensional models while it provides a fully threedimensional representation of the flow in the plenum. The simulation model has passed validation tests against single-shot experiments within the same geometry.

Simulations with sequential firing of the six combustors with realistic firing frequencies have revealed that the need for substantial scavenging of the tubes between shots leads to pressure equilibration between combustor inlet and plenum. As a consequence, an effective mean pressure rise across the combustors cannot be achieved with this set-up. Alternative designs aiming to overcome this barrier are conceiveable but their study is beyond the scope of the present paper.

The plenum, even though of a relatively simple design that is easy to realize in the experimental context, does succeed in drastically reducing the total pressure variability at the plenum exit (equivalent conceptually to the turbine entry) relative to that at the combustor entries where the full pressure amplitude of the detonation waves is felt. The remaining pressure variability may still be challenging to handle with a standard turbine, so more elaborate designs of the plenum and, particularly, the plenum exit should be investigated in future work.

Aside from the time it takes for a tube to discharge after a shot into the plenum and the pressure to fall back to nearly the compressor pressure, the firing frequency of the current setup is limited by the need to substantially scavenge the combustors between two shots. This is necessary as part of the charge lengths is lost to the less efficient deflagration to detonation transition process. Therefore, and because options of minimizing the DDT-distance are limited, maximization of the efficiency gain promised by detonative combustion requires maximization of the charge length. When the charge length becomes comparable in magnitude to the entire length of the tube, the scavenging process is dominated by the advection time scale, which then controls the order of magnitude of the maximum firing frequency. This limitation can be overcome only by near sonic scavenging flow which will come, however, with much stronger dissipative losses.

An idea to overcome the second limitation described above, i.e., that with the current setup an effective pressure increase across the detonation tube is not realizeable, has been formulated by Wolff (2019). One might trigger a strong transversally travelling acoustic mode in the plenum be suitable firing sequences of the detonation tubes, and by letting the combustors enter the plenum at a suitable angle. When the phase of low pressure passes by a tube exit, synchronized with the scavenging flow, then the mean (total) pressure in the turbine plenum might still be higher than that at the combustor inlets or in a preceding compressor plenum.

References

Bengoechea, S., Gray, J., Reiss, J., Moeck, J., Paschereit, O., Sesterhenn, J.: Detonation initiation in pipes with a single obstacle for mixtures of hydrogen and oxygen-enriched air. Combust. Flame **198**, 290–304 (2018). https://doi.org/10.1016/j.combustflame.2018.09.017. https://www.sciencedirect.com/science/article/pii/S0010218018304085

Berndt, P.: On the use of the HLL-scheme for the simulation of the multi-species Euler equations. In: Fuhrmann, J., Ohlberger, M., Rohde, C. (eds.) Finite Volumes for Complex Applications VII - Elliptic, Parabolic and Hyperbolic Problems. Springer Proceedings in Mathematics & Statistics, vol. 78, pp. 809–816. Springer, Cham (2014). https://doi.org/10.1007/978-3-319-05591-6_81

Berndt, P.: Mathematical modeling of the shockless explosion combustion. Ph.D. thesis, Freie Universität Berlin, Berlin (2016)

Berndt, P., Klein, R.: Modeling the kinetics of the shockless explosion combustion. Combust. Flame **175**, 16–26 (2017)

Einfeldt, B.: On Godunov-type methods for gas dynamics. SIAM J. Numer. Anal. **25**, 294–318 (1988)

Gamezo, V.N., Ogawa, T., Oran, E.S.: Flame acceleration and DDT in channels with obstacles: effect of obstacle spacing. Combust. Flame **155**(1), 302–315 (2008). https://doi.org/10.1016/j.combustflame.2008.06.004. https://www.sciencedirect.com/science/article/pii/S0010218008001934

Gokhale, N.K.R., Nikiforakis, N.: A dimensionally split cartesian cut cell method for hyperbolic conservation laws. J. Comput. Phys. **364**, 186–208 (2018)

Gray, J., Lemke, M., Reiss, J., Paschereit, C.O., Sesterhenn, J., Moeck, J.P.: A compact shock-focusing geometry for detonation initiation: experiments and adjoint-based variational data assimilation. Combust. Flame **183**, 144–156 (2017). https://doi.org/10.1016/j.combustflame.2017.03.014

Klein, R., Bates, K.R., Nikiforakis, N.: Well-balanced compressible cut-cell simulation of atmospheric flow. Philos. Trans. Royal Soc. Lond. A Math. Phys. Eng. Sci. **367**(1907), 4559–4575 (2009). https://doi.org/10.1098/rsta.2009.0174, http://rsta.royalsocietypublishing.org/content/367/1907/4559. http://rsta.royalsocietypublishing.org/content/367/1907/4559.full.pdf

van Leer, B.: Towards the ultimate conservative difference scheme. J. Comput. Phys. **32**, 101–136 (1979)

Munz, C.D.: On the construction and comparison of two-step schemes for the Euler equations. In: Hirschel, E. (ed.) Notes on Numerical Fluid Mechanics, vol. 14, pp. 195–217. Vieweg Verlag, Braunschweig/Wiesbaden (1986)

Nadolski, M.: Ph.D. thesis, Freie Universität Berlin (2021)

Nadolski, M., Rezay Haghdoost, M., Gray, J., Edgington-Mitchell, D., Oberleithner, K., Klein, R.: Validation of numerical simulations by the exhaust flow of a PDC based high speed Schlieren. In: Notes on Numerical Fluid Mechanics and Multidisciplinary Design: Proceedings of AFCC 2018, vol. 141, pp. 237–253 (2018)

Pandey, K., Debnath, P.: Review on recent advances in pulse detonation engines. J. Combust., 4193034 (2016). https://doi.org/10.1155/2016/4193034

Pitsch, H.: Flamemaster: A C++ computer program for 0d combustion and 1d laminar flame calculations (2021). https://www.itv.rwth-aachen.de/downloads/flamemaster/

Rezay Haghdoost, M., Edington-Mitchell, D., Nadolski, M., Klein, R., Oberleithner, K.: Dynamic evolution of a transient supersonic trailing jet induced by a strong incident shock wave. Phys. Rev. Fluids **5**(073), 401 (2020a)

Rezay Haghdoost, M., et al.: Evaluation of shock dividers using numerical and experimental methods. AIAA J., 2020–0926 (2020b)

Rezay Haghdoost, M., et al.: Mitigation of pressure fluctuations from an array of pulse detonation combustors. J. Eng. Gas Turbines Power **143**(7) (2021). https://doi.org/10.1115/1.4049857

Stathopoulos, P., Vinkeloe, J., Paschereit, C.: Thermodynamic evaluation of constant volume combustion for gasturbine power cycles. In: 11th International Gas Turbine Congress, Tokyo, Japan, pp. 15–20 (2015)

Wolff, S.D.: Entwicklung von fehlererkennungs-, zustandsschätzungs-, regelungs- und ganzzahligen optimalsteuerungsmethoden für pulsierende, detonative brennkammern anhand eines akustischen ersatzsystems. Doctoral thesis, Technische Universität Berlin, Berlin (2019). https://doi.org/10.14279/depositonce-8148

Pressure Fluctuations in an Annular Plenum Downstream of a Multi-tube Pulse Detonation Combustor

Fabian Habicht[1(✉)], Fatma Cansu Yücel[1], Myles Bohon[2], Mohammad Rezay Haghdoost[3], Kilian Oberleithner[3], and Christian Oliver Paschereit[1]

[1] Chair of Fluid Dynamics, Institute of Fluid Dynamics and Technical Acoustics, TU Berlin, Straße des 17. Juni 135, 10623 Berlin, Germany
fabian.habicht@tu-berlin.de

[2] Chair of Pressure Gain Combustion, Institute of Fluid Dynamics and Technical Acoustics, TU Berlin, Straße des 17. Juni 135, 10623 Berlin, Germany

[3] Laboratory for Flow Instabilities and Dynamics, Institute of Fluid Dynamics and Technical Acoustics, TU Berlin, Straße des 17. Juni 135, 10623 Berlin, Germany

Abstract. The generation of large pressure fluctuations at the combustor outlet due to the periodic combustion process involving propagating detonation waves is a major drawback on the way of integrating a pulse detonation combustor (PDC) into a gas turbine. Recently, the attachment of an annular plenum downstream of a multi-tube PDC was proposed to allow for the attenuation of the pressure amplitudes. In this work, pressure data is recorded at various axial and azimuthal positions in the annular plenum allowing for a quantification of pressure fluctuations. Furthermore, a systematic study was conducted to evaluate the effect of the firing pattern and an outlet blockage on both the longitudinal change of the peak amplitudes and the pressure fluctuations throughout the entire cycle duration. The results suggest that a sequential firing pattern should be preferred over the simultaneous firing of multiple PDC tubes, as it results in the lowest pressure fluctuations at the plenum outlet.

Keywords: Pulse detonation combustor (PDC) · Annular plenum · Pressure fluctuations

1 Introduction

Pressure gain combustion has been in the scope of research in the past decades as a promising concept for increasing the thermal efficiency of gas turbines [10]. Among others, pulse detonation combustors (PDCs) revealed a promising concept to realize pressure gain combustion. The tubular combustor is periodically filled with a fuel–oxidizer mixture, which is ignited close to the combustor

R. King and D. Peitsch (Eds.): AFCC 2021, NNFM 152, pp. 103–115, 2022.
https://doi.org/10.1007/978-3-030-90727-3_7

head. Commonly, the integration of geometric obstacles is used to promote the deflagration-to-detonation transition (DDT). The initiated detonation wave then combusts the remaining mixture nearly instantaneously, resulting in a considerable rise in pressure. By conducting numerical simulations, Xisto et al. [11] demonstrated that the periodically fluctuating flow conditions downstream of a PDC induce significant variations in the incidence angle of the rotor blades, which results in a reduced turbine efficiency. Fernelius and Gorrell [1] found that the decrease in turbine efficiency mainly depends on the amplitude of pressure fluctuations, while the pulsing frequency only plays a secondary role. Therefore, several investigations were conducted in the past aiming for a reduction of these pressure oscillations. Schauer et al. [9] experimentally studied the interaction of a PDC with a centrifugal turbine. They stated high losses through the turbine stage expansion. Measurements by Rouser et al. [8] revealed highly unsteady flow conditions downstream of a multi-tube PDC, which led to a decrease in the cycle-averaged efficiency of an attached turbine. Moreover, they observed a decrease in the pressure amplitude for an increasing firing frequency of the PDC, which was found to be beneficial for the turbine operation. In addition to the operation frequency of a PDC, adjusting the firing pattern has been under investigation by Rasheed et al. [5]. They found that the choice of the firing pattern affected the pressure amplitudes at the turbine inlet. Specifically, a sequential firing was observed to result in reduced fluctuations compared with a simultaneous operation of the tubes. In consistence with this, asynchronous firing of a double-tube PDC was found to be beneficial in terms of turbine efficiency by Qui et al. [4]. Rezay Haghdoost et al. [7] proposed the integration of an annular plenum as an alternative concept to achieve reduced pressure amplitudes at the turbine inlet plane downstream of a PDC. They demonstrated a reduction by nearly 70% along the plenum axis as the shock waves exiting the combustion tubes diffracted in azimuthal direction.

In this work, pressure fluctuations in an annular plenum downstream of a multi-tube PDC are examined experimentally, which allows for identifying favorable operating conditions to reduce pressure fluctuations at the plenum outlet. In particular, the propagation and interaction of shock waves inside the annular plenum are analyzed and the longitudinal change of the maximum pressure amplitude along the plenum axis is quantified for various firing patternsand operation frequencies. Furthermore, the effect of an outlet blockage on the pressure evolution and the longitudinal change of pressure amplitudes is evaluated.

2 Experimental Setup and Measurement Procedure

The experiments presented in this work, were conducted on a multi-tube PDC test rig, which is sketched in Fig. 1. The test facility consists of six PDC tubes and a plenum, arranged in can–annular configuration. Each PDC tube is composed of an injection section, a DDT section, and a detonation section. In the injection section, the fuel is added to a continuous air flow through three parallel arranged solenoid valves (Bosch 0280158827) in a jet in cross flow configuration. The

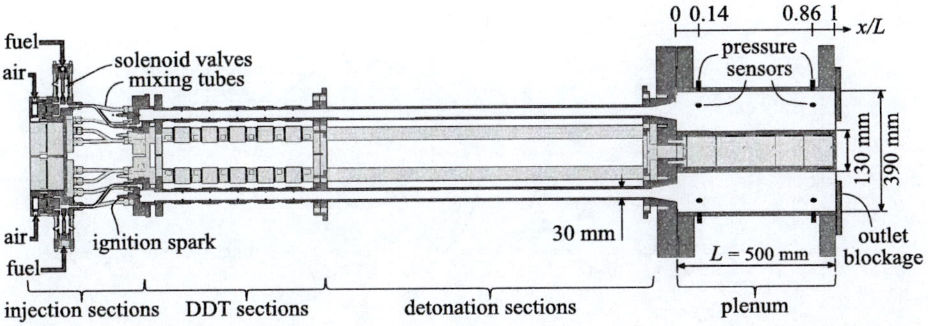

Fig. 1. Cross section of the multi-tube PDC consisting of six circumferentially arranged tubes

mixture is subsequently guided through three mixing tubes before it is radially injected into the combustion tube through three circumferentially distributed ports. A spark plug at the center of the flat combustor head allows for the controlled ignition of the injected fuel–air mixture. A time delay of 4 ms between the closing of the fuel valves and the spark discharge is applied to create an air buffer in the mixing tubes before ignition. By this, the mechanical and thermal stress on the injection section is minimized. Each DDT section is equipped with a series of orifice plates with a blockage ratio of 0.43 equally spaced at 85 mm to ensure reliable detonation initiation. Downstream of the DDT section, a straight tube with an inner diameter of 30 mm and a length of 0.8 m is attached to allow for the combustion of the injected fuel–air mixture by means of a propagating detonation wave. The transition from the PDC outlet to the annular plenum is realized by diverging nozzles increasing the cross section area of each combustion tube by a factor of 2.25. All six combustion tubes are arranged on a pitch-circle diameter of 260 mm, as sketched in Fig. 2a that visualizes a rear view of the test rig. The radial height of the annular channel is sized to match the distance of two neighboring tubes of 130 mm, while the tube outlets are located at half the channel height. Thus, the outer wall of the plenum has a diameter of 390 mm and the center body has a diameter of 130 mm. The axial length of the plenum is set to $L = 500$ mm. This length was chosen to investigate the propagation of the pressure waves emitted from the PDC tubs and to assess the attenuation of pressure amplitudes along the plenum axis, which can then be used to design an improved plenum geometry with a reduced axial length. In order to examine the effect of an outlet blockage on the longitudinal change of pressure amplitudes in the annular plenum, an orifice plate is installed at the plenum outlet. Three different orifice plates are used to achieve blockage ratios of $\beta \in [0.5, 0.7, 0.9]$, as shown in Fig. 2b. The applied blockage plates were designed to emulate the effect of a reduction of the annular cross section area, which would be required when attaching a suitable turbine with respect to the applied mass flux. Measuring the static pressure throughout the entire operation period revealed no increase in static pressure for any applied blockage. The flow through

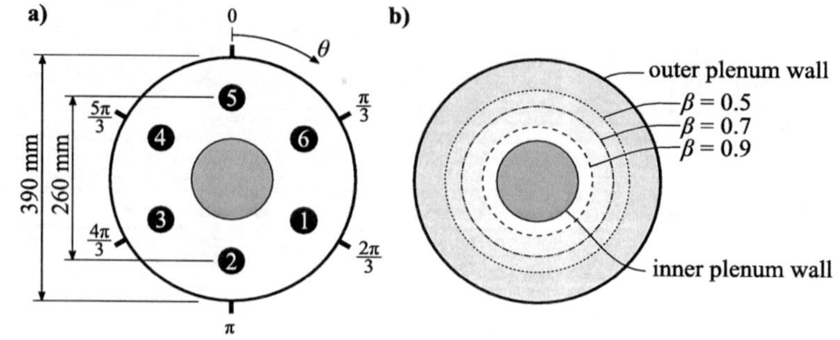

Fig. 2. Rear view of the annular plenum

the plenum exit is therefore assumed not to be choked at any time, which ensures comparable flow conditions for all conducted measurements. Two piezoresistive pressure transducers (Kulite DTL) were flush-mounted into the outer plenum wall at $x/L = 0.14$ and $x/L = 0.86$ downstream of PDC tube 5 ($\theta = 0$). An additional piezoresistive sensor was installed close to the plenum outlet on the opposite side of the plenum ($\theta = \pi$, $x/L = 0.86$). In combination with nine piezoelectric pressure transducers (PCB 112A05), this resulted in two arrays of six sensors, evenly distributed in circumferential direction, at $x/L = 0.14$ and $x/L = 0.86$ (see Fig. 1). As observed in previous experiments [7], all applied pressure transducers implied a measurement uncertainty of less than 2%. All pressure values, which are included in the following are to be considered as absolute pressure.

A continuous air flow rate of $\dot{m}_{\text{air}} = 900\,\text{kg/h}$ is provided for all measurements, equally distributed among six PDC tubes using a manifold. A constant mass flow rate with a maximum deviation of 0.7 % is assured by applying a closed-loop control, including the evaluation of the actual mass flow rate measured by a Coriolis mass flow meter (Endress+Hauser Promass 80A) and the adjustment of the position of an electric proportional valve (Bürkert 2712). The fuel mass flow rate during the injection period is controlled by setting the supply pressure by means of a dome-loaded pressure regulator (Swagelok RD8) to 5 bar. In previous investigations [2,3], this methodology was proven to allow for the injection of a stoichiometric hydrogen–air mixture with a nearly constant equivalence ratio along the combustor. An injection duration of 21 ms was applied, which results in the combustion tubes being entirely filled with reactive mixture. For each measurement, the PDC tubes are operated for ten consecutive cycles, with a firing frequency of 16.7 Hz ($t_{\text{cycle}} = 60\,\text{ms}$). Although this resulted in an operation duration of only 0.6 s, no systematic change in the recorded pressure signals over the conducted combustion cycles was observed. Thus, the recorded pressure signals and the following examination are expected to be representative for an arbitrary operation time.

Six different firing patterns, as illustrated in Fig. 3, are applied. The naming convention of the firing patterns is based on the number of simultaneously

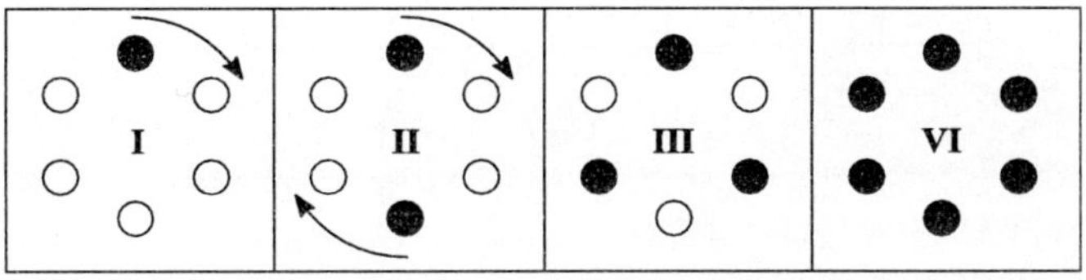

Fig. 3. Firing patterns

operated tubes. Firing pattern **I** implies sequential firing of all six tubes with a delay of 10 ms between two consecutive ignitions. Simultaneous ignition of two opposite tubes is obtained from applying firing pattern **II**. Here, the delay between two subsequent ignitions is increased to 20 ms. Alternating ignition of a set of three tubes is realized by pattern **III**. Firing pattern **VI** results in the simultaneous operation of all six PDC tubes. As the firing frequency of a single PDC tube remains constant for all patterns, the overall thermal power of the multi-tube PDC is independent of the applied firing pattern. By the variation of the number of simultaneously operated tubes n, the frequency at which shocks enter the plenum varies. This parameter is called the effective firing frequency f_{eff} and is calculated from

$$f_{\text{eff}} = \frac{N}{n} f_{\text{tube}}, \tag{1}$$

where $N = 6$ represents the total number of PDC tubes.

3 Results and Discussion

In this section the evolution of the pressure waves exiting the PDC tubes and subsequently entering the annular plenum is analyzed for different firing patterns and plenum outlet blockage ratios. For this, pressure signals measured at different azimuthal and axial positions at the outer plenum wall are evaluated in the following.

3.1 Examination of Pressure Signals

In this section, the main features of the shock wave propagation inside the plenum are examined by means of pressure signals at the outer plenum wall. In addition, the impact of the firing pattern is analyzed. Once a detonation wave reaches the mixture–air intersection at the PDC outlet, it can not sustain due to the absence of reactive mixture, and thus, transitions into a propagating shock wave. Subsequently the shock wave diffracts in the divergent nozzle due to the increase in the cross section area from the PDC tube to the annular plenum. The recorded pressure amplitude p at six azimuthal and two axial positions for the operation with firing pattern **I** are shown in Fig. 4 for an example cycle. As the examination of all available data (not shown for brevity) revealed very

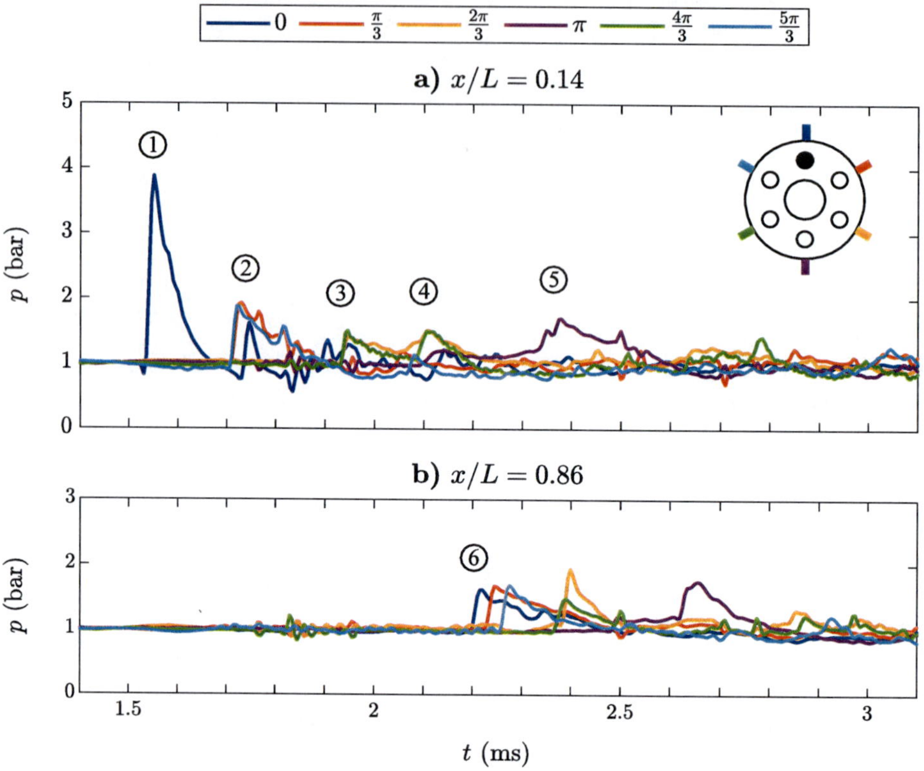

Fig. 4. Pressure histories subsequent to the firing of PDC tube 5 at two different axial positions. The PDC was operated with firing pattern **I** and $\beta = 0$.

similar pressure signals, the given example is considered to be representative. The time is given with respect to the spark discharge in the firing PDC tube.

At $t \approx 1.5\,\text{ms}$, a sharp increase in pressure is detected at $\theta = 0$ and $x/L = 0.14$ ①. The maximum measured amplitude of $p_{\text{max}} \approx 4\,\text{bar}$ is significantly smaller than the CJ pressure of $p_{\text{CJ}} \approx 15\,\text{bar}$, which denotes the pressure behind the detonation wave propagating through the stoichiometric hydrogen–air mixture at atmospheric conditions in the combustion tube. This decay in the pressure amplitude is attributed to two effects: i) the transition from a detonation front into a shock wave and ii) the increase in the cross section area from the PDC tube to the annular plenum. A second pressure peak is observed at this sensor position $t \approx 1.75\,\text{ms}$ ②, which can be explained by a pressure wave reflected at the center body [6]. Within the same period, the shock wave exiting the PDC tube propagates in azimuthal direction and evokes a sudden increase in static pressure at both $\theta = \frac{\pi}{3}$ and $\frac{5\pi}{3}$. As the pressure wave propagates further, it arrives at the azimuthal positions $\frac{2\pi}{3}$ and $\frac{4\pi}{3}$, where it induces a pressure increase at $t \approx 1.95\,\text{ms}$ ③. A second pressure peak is visible at these two positions at $t \approx 2.1\,\text{ms}$ ④. Simultaneously, a first increase in the pressure at $\theta = \pi$

is observed, which is followed by a second pressure rise at $t \approx 2.35\,\text{ms}$ ⑤. Subsequently, only minor pressure fluctuations are visible at $x/L = 0.14$. Close to the plenum outlet (Fig. 4b), the pressure signal of each sensor contains one distinct maximum, respectively. As expected, the propagating pressure wave is first detected downstream of the firing PDC tube at $\theta = 0$ ⑥. Subsequently, peaks in the pressure histories are observed according to the azimuthal distance of the sensors to the firing PDC tube. Interestingly, the maximum pressure amplitudes close to the plenum outlet are not observed downstream of the firing tube, but rather on the opposite side of the annulus at $\theta = \frac{2\pi}{3}$ and $\theta = \pi$. A discussion regarding the underlying mechanism leading to the peak pressure on the opposite side of the firing tube at the plenum outlet is given in [6]. For $x/L = 0.14$, the pressure signals at $\theta = \frac{\pi}{3}$ and $\frac{5\pi}{3}$ are very similar. The same statement holds for $\theta = \frac{2\pi}{3}$ and $\frac{4\pi}{3}$ at this axial position, which indicates a mostly axisymmetric propagation of the shock wave in both azimuthal directions. Nevertheless, at $x/L = 0.86$, notable deviations are observed for opposed azimuthal sensor positions with respect to the firing PDC tube, which have already been reported in [6]. We explain this asymmetric pressure evolution by the interaction of the shock wave exiting the PDC with acoustic modes inside the annular plenum. The existence of these mode shapes were verified in [2]. In particular, a number of rotating azimuthal mode shapes were identified, when firing pattern **I** was applied. When propagating through the plenum, the main pressure wave interacts with these acoustic modes, resulting a non-symmetric pressure evolution at $x/L = 0.86$.

To evaluate the interaction of multiple shock waves simultaneously exiting the PDC tubes, the recorded pressure signals for the operation with firing pattern **II** are shown in Fig. 5. As tubes 2 and 5 fire synchronously, two distinct pressure peaks with $p_{\max} \approx 4\,\text{bar}$ are detected at $x/L = 0.14$ for $\theta = 0$ and $\theta = \pi$, respectively ①. It should be noted that the recorded pressure signals at these azimuthal positions are not expected to be congruent, as they represent the pressure evolution due to two different detonation events. Furthermore, different sensors are installed with various measurement principles (piezoresistive at $\theta = 0$ and piezoelectric at $\theta = \pi$). In particular, a non-physical decrease in the measured pressure value for the piezoelectric transducer is assumed. Nevertheless, the recorded pressure traces can be used to analyze the propagation of the pressure waves in the annular plenum. Analogous to the findings from the operation with pattern **I**, the reflection of the shock wave at the center body results in a second pressure peak at these two sensors 0.25 ms after the detection of the first shock wave ②. As the two leading shock waves propagate in azimuthal direction, they cause a simultaneous increase in static pressure at all remaining sensor positions with $x/L = 0.14$. At $x/L = 0.86$, the pressure waves first arrive at $\theta = 0$ and $\theta = \pi$ ③ as they are located slightly closer to the outlets of the firing PDC tubes than the other sensors at this axial position. In contrast to the pressure signals in Fig. 4b, a second pressure peak is observed for firing pattern **II** (Fig. 5b) at each sensor position (④ and ⑤). Both peaks are associated with the propagating leading shock waves induced from the simultaneously firing of

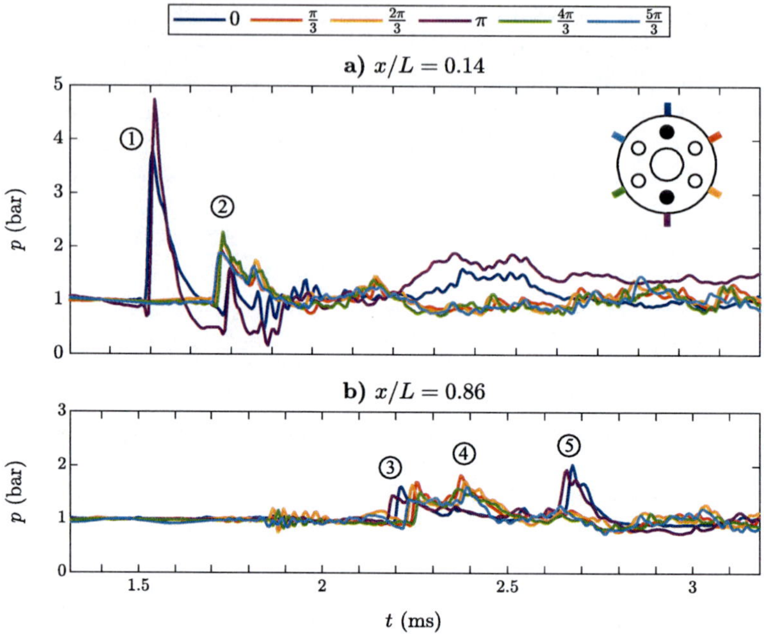

Fig. 5. Pressure histories subsequent to the firing of PDC tube 5 at two different axial positions. The PDC was operated with firing pattern **II** and $\beta = 0$.

two combustion tubes. These shock waves arrive at a given sensor position with a certain time delay due to the varying distance between the sensor position and the outlets of the two firing PDC tubes, respectively.

As discussed by Fernelius and Gorrell [1], the pressure amplitudes at the turbine inlet greatly affect the efficiency of a hybrid PDC–turbine system. However, not only the maximum peak pressure but also the distribution along the circumference is expected to impact the system performance. To allow for a quantitative examination of the azimuthal distribution of p_{max}, the cycle averaged pressure amplitudes $\bar{p}_{\text{max}}(\tilde{\theta})$ are calculated from

$$\bar{p}_{\text{max}}(\tilde{\theta}) = \frac{1}{N_{\text{c}} N_{\text{t}}} \sum_{k=1}^{N_{\text{c}}} \sum_{j=1}^{N_{\text{t}}} p_{\text{max},j,k}(\theta_j + \tilde{\theta}) \tag{2}$$

with the number of combustion cycles per tube $N_{\text{c}} = 10$ and the number of PDC tubes $N_{\text{t}} = 6$. $p_{\text{max},j,k}$ represents the maximum pressure after firing tube j for the k-th, while θ_j denotes the azimuthal position of the firing tube. The azimuthal distribution of the cycle-averaged maximum pressure amplitude $\bar{p}_{\text{max}}$ is visualized in Fig. 6 for all investigated firing patterns. The two lines (blue and orange) in each plot represent the values of $\bar{p}_{\text{max}}(\tilde{\theta})$ at the two different axial positions $x/L = 0.14$ and $x/L = 0.86$, where $\tilde{\theta}$ represents the azimuthal position with respect to the firing combustion tube.

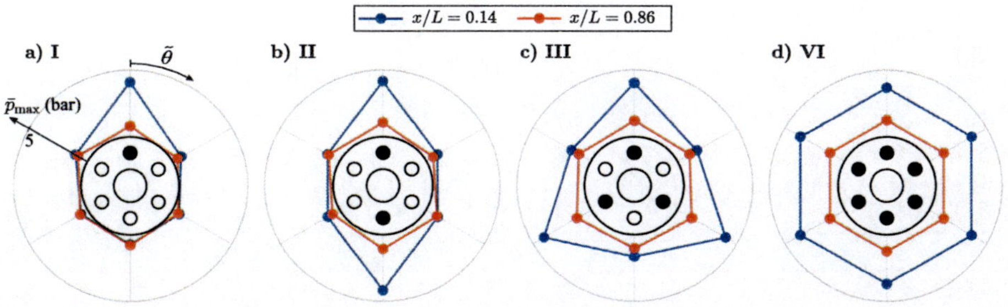

Fig. 6. Azimuthal distribution of maximum pressure amplitudes for all investigated firing patterns. The radial coordinate $\bar{p}_{\mathrm{max}}$ scale from 0 to 5 bar.

When applying firing pattern **I**, a large pressure amplitude is observed at $x/L = 0.14$ and $\theta = 0$ (Fig. 6a, blue line). This amplitude, however, is drastically attenuated along the plenum axis, resulting in a nearly homogeneous distribution of moderate pressure amplitudes close to the plenum outlet (Fig. 6a, orange line). For the simultaneous firing of two opposite tubes (pattern **II**) the amplitudes at $x/L = 0.86$ downstream of the firing tubes are increased (Fig. 6b, orange line). By further increasing the number of simultaneously firing tubes to three (pattern **III**) or even six (pattern **VI**), the pressure amplitudes at the plenum outlet increase simultaneously (Fig. 6c and d). This trend can be explained by the restricted expansion of the leading shock waves in azimuthal direction due to the coexistence of multiple shock waves, exiting the PDC from a number of simultaneously operated tubes.

3.2 Longitudinal Change of Pressure Amplitudes

The attenuation of pressure amplitudes along the plenum axis is the main purpose of integrating an annular plenum downstream of the PDC. In this work, the longitudinal change of pressure amplitudes is quantified by the pressure ratio Π, which is determined from the maximum pressure amplitudes at a given azimuthal position according to

$$\Pi = \frac{\bar{p}_{\mathrm{max}}(x/L = 0.86)}{\bar{p}_{\mathrm{max}}(x/L = 0.14)}. \tag{3}$$

Values of $\Pi > 1$ indicate a larger maximum pressure amplitude near the plenum outlet ($x/L = 0.86$) than recorded close to the plenum inlet ($x/L = 0.14$). In contrast to this, $\Pi < 1$ indicate a longitudinal attenuation in $\bar{p}_{\mathrm{max}}$. The resulting azimuthal distributions are visualized in Fig. 7 for all investigated firing patterns and two blockage ratios of the plenum outlet, $\beta = 0$ and 0.9, respectively.

Regardless of the blockage ratio and the firing pattern, the largest longitudinal attenuation in the pressure amplitude is observed at the azimuthal positions of the firing PDC tube(s), resulting in the smallest values of Π for the respective position(s). In addition, the plenum blockage ratio β was found to have no significant impact on Π at these positions. For sequential firing (pattern **I**),

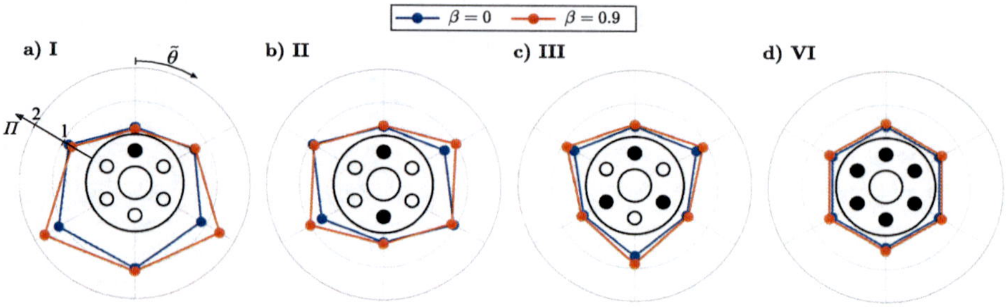

Fig. 7. Azimuthal distribution of Π for all investigated firing patterns and two blockage ratios. The radial coordinate scales from 0 to 2.

$\Pi > 1$ is observed at multiple azimuthal positions. This indicates larger pressure amplitudes at the plenum outlet compared with the plenum inlet at $\theta = \frac{2\pi}{3}$, π, and $\frac{4\pi}{3}$ in case the tube at $\theta = 0$ fires. As discussed in [7], the reflection of the leading shock at the plenum outer wall results in the formation of a Mach stem at $\theta \approx \pi$. When a plenum outlet blockage of $\beta = 0.9$ is introduced this pressure wave is reflected at the plenum outlet, resulting in a further increased pressure amplitude at $x/L = 0.86$. The azimuthal distribution of Π for firing pattern **II** visualizes a decrease in the pressure amplitude from the inlet to the outlet of the plenum for all considered azimuthal positions when firing two opposite PDC tubes simultaneously. The same findings can be deduced from the results for the operation with patterns **III** and **VI**. When comparing the azimuthal distributions of Π for the different patterns, the minimum value of Π in azimuthal direction is steadily observed downstream of the firing PDC tube(s): $\theta = 0$ for firing pattern **I**, $\theta = 0$ and $\theta = \pi$ for firing pattern **II**, and $\theta \in [0, \frac{2\pi}{3}, \frac{4\pi}{3}]$ for firing pattern **III**. For firing pattern **VI**, a homogeneous distribution of Π is obtained. When increasing the number of simultaneously operated tubes, this minimum value of Π increases gradually, which indicates a decreasing longitudinal attenuation of the maximum pressure amplitude along the plenum axis. In addition, the application of an outlet blockage results in a smaller reduction of pressure amplitudes at some distinct azimuthal positions.

In order to quantitatively compare the obtained pressure amplitudes close to the plenum outlet for all investigated firing patterns and blockage ratios, $P_{\max}$ is calculated as the average value of the maximum pressure amplitude $\bar{p}_{\max}(\tilde{\theta})$ in azimuthal direction for $x/L = 0.86$ according to

$$P_{\max} = \frac{1}{N_\theta} \sum_{i=0}^{N_\theta - 1} \bar{p}_{\max}(\theta_i), \tag{4}$$

with $\theta_i = \frac{2\pi i}{N_\theta}$. In this work, $N_\theta = 6$, as sensors were installed at six equidistant azimuthal positions. The obtained values are plotted in Fig. 8a for all applied firing patterns with respect to the blockage ratio of the plenum outlet. In addition, the circumferential average of the root–mean–square (RMS) value during the entire operation duration is determined as

$$P_{\mathrm{RMS}} = \frac{1}{N_\theta} \sum_{i=0}^{N_\theta - 1} p_{\mathrm{RMS}}(\theta_i). \tag{5}$$

The results are plotted in Fig. 8b.

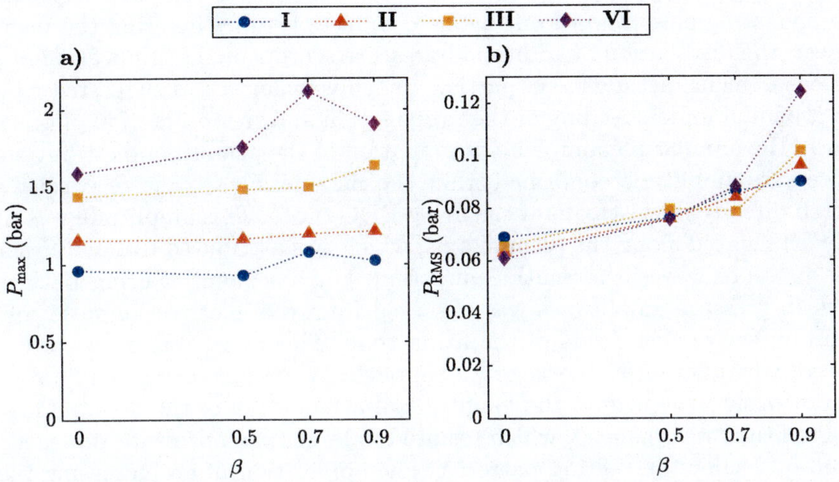

Fig. 8. (a) Cycle-averaged maximum pressure amplitude in azimuthal direction and (b) p_{RMS} at $x/L = 0.86$.

It can well be seen in Fig. 8a that an increasing number of simultaneously firing tubes significantly increases the maximum pressure amplitude along the circumference for all applied blockage ratios. When no constriction is applied to the plenum outlet ($\beta = 0$), a small reduction in the RMS value of pressure fluctuations is observed when increasing the number of simultaneously firing tubes (Fig. 8b). This trend, however, is reversed when the maximum blockage ratio of $\beta = 0.9$ is applied. In conclusion, the presented results suggest that sequential firing is favorable not only because it results in the largest longitudinal attenuation of peak amplitudes of the shock waves exiting the PDC, but also because of the smallest RMS value of pressure fluctuations, which is expected to be beneficial for the operation of an attached turbine.

4 Conclusion

Pressure fluctuations in an annular plenum downstream of a multi-tube pulse detonation combustor (PDC) were investigated experimentally. Each combustion tube was operated at a constant firing frequency of 16.7 Hz, while four different firing patterns were applied, defining the succession of firing tubes. In addition to measurements with an open plenum outlet, three different blockage plates were installed, resulting in blockage ratios of 0.5, 0.7 and 0.9 at the plenum outlet. The recorded pressure signals subsequent to a detonation event were evaluated for two arrays of six circumferentially distributed pressure sensors at different axial positions at the plenum outer wall. The observed pressure histories could well be explained by the propagation of pressure waves that originated from the propagating detonation front in the combustion tubes entering the plenum. However, the installation of additional pressure sensors or the application of planar measurement methods, e.g. particle image velocimetry, is suggested to gain more profound understanding of the propagation and the interaction of pressure waves in the annular plenum. The examination of the azimuthal distributions of the pressure amplitude confirmed, that the maximum peaks were consistently detected directly downstream of the firing PDC tubes. This amplitude was drastically attenuated along the plenum axis, which was attributed to the diffraction of the pressure wave in azimuthal direction. For a sequential firing pattern, a growth in pressure amplitudes was observed from the inlet to the outlet of the plenum on the opposite side of the firing tube. This effect was even more pronounced, when an outlet blockage was applied. When increasing the number of simultaneously firing tubes, the longitudinal attenuation of the maximum pressure amplitude was reduced, which resulted in larger peak pressure values at the plenum outlet for these firing patterns. The application of an increasing blockage ratio at the plenum outlet enhanced this deviation between the investigated firing patterns. The evaluation of the root–mean–square value of pressure fluctuations throughout the entire operation duration at the plenum outlet revealed a considerable increase in pressure fluctuations for an increasing number of simultaneously firing tubes when a blockage ratio of 0.9 was applied. In contrast to the results from the operation with an open plenum outlet, the findings from the examination with the largest applied blockage ratio revealed that the sequential firing of the PDC tubes is favorable to reduce pressure fluctuations at the inlet of a potentially attached turbine.

Acknowledgments. The authors gratefully acknowledge support by the Deutsche Forschungsgemeinschaft (DFG) as part of Collaborative Research Center SFB 1029 "Substantial efficiency increase in gas turbines through direct use of coupled unsteady combustion and flow dynamics" on projects A01 and C01.

References

1. Fernelius, M.H., Gorrell, S.E.: Predicting efficiency of a turbine driven by pulsing flow. In: Turbo Expo: Power for Land, Sea, and Air, vol. 50787, p. V02AT40A008. American Society of Mechanical Engineers (2017)
2. Habicht, F., Yücel, F.C., Rezay Haghdoost, M., Oberleithner, K., Paschereit, C.O.: Acoustic modes in a plenum downstream of a multitube pulse detonation combustor. AIAA J., 1–12 (2021)
3. Habicht, F.E., Yücel, F.C., Hanraths, N., Djordjevic, N., Paschereit, C.O.: Lean operation of a pulse detonation combustor by fuel stratification. J. Eng. Gas Turbines and Power **143**(5), 051009 (2021)

4. Qiu, H., Xiong, C., Zheng, L.: Experimental investigation of an air-breathing pulse detonation turbine prototype engine. Appl. Therm. Eng. **104**, 596–602 (2016)
5. Rasheed, A., Furman, A.H., Dean, A.J.: Pressure measurements and attenuation in a hybrid multitube pulse detonation turbine system. J. Propul. Power **25**(1), 148–161 (2009)
6. Haghdoost, M.R., Edgington-Mitchell, D., Paschereit, C.O., Oberleithner, K.: High-speed schlieren and particle image velocimetry of the exhaust flow of a pulse detonation combustor. AIAA J. **58**(8), 3527–3543 (2020)
7. Rezay Haghdoost, M., et al.: Mitigation of pressure fluctuations from an array of pulse detonation combustors. J. Eng. Gas Turbines Power (2020)
8. Rouser, K., King, P., Schauer, F., Sondergaard, R., Hoke, J.: Experimental performance evaluation of a turbine driven by pulsed detonations. In: 51st AIAA aerospace sciences meeting including the new horizons forum and aerospace exposition, p. 1212 (2013)
9. Schauer, F., Bradley, R., Hoke, J.: Interaction of a Pulsed Detonation Engine with a Turbine. Technical report, Air Force Research Lab Wright-Patterson AFB OH Propulsion Directorate (2003)
10. Stathopoulos, P.: Comprehensive thermodynamic analysis of the Humphrey cycle for gas turbines with pressure gain combustion. Energies **11**(12), 3521 (2018)
11. Xisto, C., Petit, O., Grönstedt, T., Rolt, A., Lundbladh, A., Paniagua, G.: The efficiency of a pulsed detonation combustor-axial turbine integration. Aerosp. Sci. Technol. **82**, 80–91 (2018)

Reduction of Pressure Fluctuations in an Annular Pulsed Detonation Combustor Mockup by Iterative Learning Control Using Eigenvector-Based Binary Solution Sets and Iterative Model Identification

Daniel Topalovic(✉), Florian Arnold, and Rudibert King

Chair of Measurement and Control, Technische Universität Berlin, 10623 Berlin, Germany
{daniel.topalovic,f.arnold,rudibert.king}@tu-berlin.de

Abstract. The approximation of constant volume combustion using a pulsed detonation engine is a promising approach to overcome the efficiency limitations of state-of-the-art gas turbines. It is assumed that the firing synchronization between different detonation tubes in an actual setup of such a machine significantly affects the overall performance. Therefore, to maintain reliable performance even under disturbances, the utilized firing pattern of the operation should be regulated using closed-loop control. A suitable control approach for such a machine that is cyclic in nature is iterative learning control (ILC). This contribution presents an advanced reformulation of a binary ILC method to minimize pressure fluctuations by altering the firing patterns of the detonation tubes. This method uses an eigenvector-based determination of the feasible binary solution set of control inputs to reduce the computational cost. Moreover, an ILC method using iterative model identification is introduced that minimizes pressure fluctuations under linear and non-linear system behavior, allowing operation even if a linear model insufficiently describes the system. Both methods are tested on an acoustic mockup of an annular pulsed detonation combustor.

Keywords: Iterative learning control · Firing pattern · Pulsed detonation combustion · Closed-loop control

1 Introduction

As recent efficiency improvements of state-of-the-art gas turbines have mainly been driven by evolutionary enhancements in the development of single components, the increase in efficiency has declined [1]. Significant improvement by continuing the single-component design strategy is therefore unlikely. Consequently, a change in the underlying thermodynamic cycle is required to push beyond the current efficiency limitations.

R. King and D. Peitsch (Eds.): AFCC 2021, NNFM 152, pp. 116–131, 2022.
https://doi.org/10.1007/978-3-030-90727-3_8

A transition from the commonly used Brayton cycle to the so-called Humphrey cycle could increase the efficiency of the underlying combustion process and hence increase overall machine efficiency [2]. This transition would require a change from isobaric combustion to isochoric combustion. Isochoric combustion can be approximated through different technical approaches—rotating detonation combustion (RDC) [3], shockless explosion combustion (SEC) [4], or pulsed detonation combustion (PDC) [5].

This contribution will focus on a gas turbine design that exploits the PDC concept. However, the methods shown can be similarly applied to an SEC-based setup. As described in [6], we assume that a manifold of PDC combustion tubes replaces the typical combustion chamber of a conventional gas turbine. The operation of an individual PDC combustion tube is cyclic in nature. To simplify, the cycle can be divided into three phases. In the first phase, the tube is filled with an ignitable air–fuel mixture. In the second phase, the gas is ignited to induce deflagrative combustion. This deflagrative combustion travels through the tube until a shock-focusing geometry is reached, at which point the deflagration to detonation transition occurs and a detonation front forms. The detonation front passes through the remaining mixture with supersonic speed and burns it, resulting in approximated isochoric combustion. In the third phase, the tube is purged.

The operation cycle of the PDC combustion tubes results in a non-steady but periodic flow upstream of the actual turbine. The impulse-like flows reaching the turbine can endanger the machine's overall efficiency and result in dynamic mechanical strain on the turbine. A possible solution for the safe and successful operation of the machine that ensures increased efficiency of the combustion is the homogenization of the pressure fluctuations in front of the turbine. This can be achieved either passively through buffering elements, such as a plenum between the firing tubes and the turbine, or actively using control, as applied to the mockup system of a PDC combustion setup in [7] or [8].

Even though the actual system behavior is nonlinear, it can be assumed that the effect of the pressure fluctuations induced through the periodically applied combustion can be approximated by superposing the pressure fluctuations of single firing events. Therefore, we can deduce that the pressure field in front of the turbine is a function of the synchronization between firing events of different tubes and that suitable synchronization of firing events, that is a firing pattern, can be used to minimize the pressure inhomogeneity in front of the turbine. Consequently, a proper representation of the firing pattern can be used as a control input for the system and for the mathematical description in the first step.

An input representation in a binary domain has proven to be favorable for easy and precise model identification with respect to the superposition assumption [13]. If, for a single tube, a discrete-time control input vector of length p is introduced, describing all time instants of one period, in a binary formulation, a 1 in this vector marks the time instant at which a single combustion tube fires. All remaining entries are zeros. However, this requires a control design in the binary domain. Further, due to the periodic nature of the underlying combustion, the system itself has a periodic input–output nature in terms of operation. This periodic system behavior can be exploited to control such a system by applying iterative learning control (ILC). An overview of different ILC approaches can be found in [9] and [10].

If unconstrained control inputs and a linear system dynamic described by a model are given, an optimal control input can be easily calculated. Unfortunately, this does not hold if constraints, such as the mentioned binary control input, must be respected. Because the constraint optimization problem cannot be solved analytically, a binary quadratic program (QP) has to be solved numerically in each iteration. Therefore, for applications with a fast dynamic, a high number of sampling instances, and a small cycle duration, the resulting computational cost of this approach is often too high to respect real-time requirements [8]. However, an ILC-based solution to this problem has been proposed and successfully applied to an acoustic mockup of a PDC combustor in [8]. The approach is based on an order reduction by partially omitting model information in each iteration of the ILC so that the solution space of the optimization problem shrinks. While the results are promising, the reduction scheme used in [8] is elementary. In the present contribution, we investigate and apply an enhanced reduction scheme with a more sophisticated composition of the solution space proposed in [8]. The composition of the solution space is based on the evaluation of eigenvectors of the quadratic objective of the underlying QP.

Additionally, as an ideal description of a PDC-based gas turbine seems unlikely using a linear model, a control approach for operating conditions that a linear model cannot approximate sufficiently well is proposed. A model-free ILC, which can also minimize the pressure fluctuations in front of the turbine, is introduced and applied in this contribution as an alternative or supplement to the binary ILC. The approach is based on transforming the control problem from the binary domain to the real-valued domain and subsequent iterative model identification that requires no model to be identified before the application. Both control algorithms are applied to the acoustic mockup test rig of an annular pulsed detonation combustor that was also used in [11], [12] and [13].

The remainder of this contribution is organized as follows. In Sect. 2, the acoustic mockup test rig is described and an artificial output nonlinearity is defined, which is used to determine the applicability of the model-free control approach. Section 3 describes the advanced binary eigenvector-based ILC, and Sect. 4 introduces the model-free ILC method. In Sect. 5, the results of the proposed methods are presented. Finally, Sect. 6 consists of conclusions and an outlook.

2 Experimental Setup

The test rig used in this contribution is an annular pulsed detonation combustor mockup. The setup represents a PDC-based gas turbine plenum that connects the combustion tubes with the actual turbine. The mockup test rig only exploits

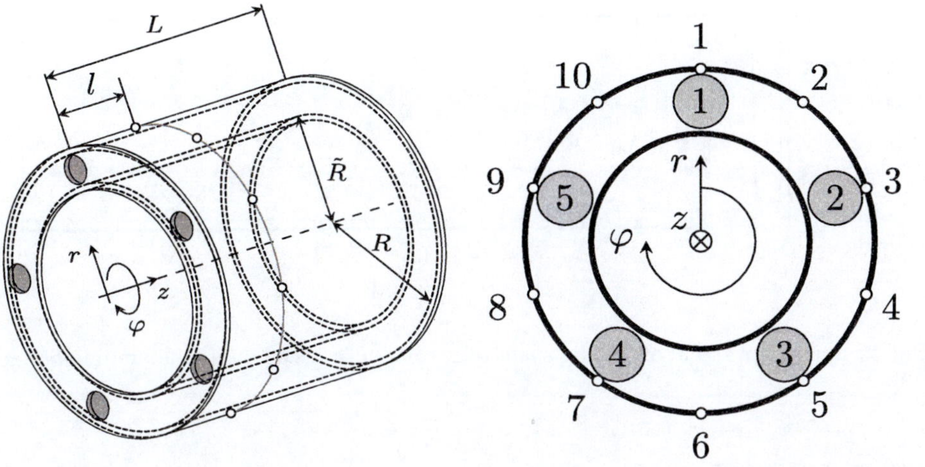

a) Geometry of the test rig **b)** Microphone and loudspeaker positions

Fig. 1. Geometry, sensor, and actuator position of the acoustic mockup test rig [13]

acoustic features to mimic the pressure field generated in the plenum. Therefore, no flow or temperature fluctuations are addressed.

While in an actual PDC-based gas turbine the pressure fluctuations in the plenum induced by combustion tubes can be expected to be much higher than in the mockup test rig, we still assume that superposition of pressure fluctuations holds and, to a certain degree, that linear acoustics sufficiently approximate the dynamic of the pressure field inside the plenum. Further, no additional differences of the acoustic properties of the setup compared to a real combustor-turbine inferface, e.g., due to the bend of guide vanes, are addressed in this setup, since the focus of its use is the proof of concept of the applied control methods and not a exact replication of a real setup. Therefore, the test rig allows only for a methodical study of control concepts that can be similarly applied to the PDC-based gas turbine. The test rig has already been used in previous studies regarding the misfire detection of a PDC-based gas turbine using artificial neural networks [11] and hypothesis testing [12]. Additionally, the calculation of optimal control for such a setup has been investigated [13].

Figure 1a) exemplifies the geometry of the test rig. The main feature is defined by two conically placed aluminum cylinders of length $L = 0.6$ m with an inner radius of the outer cylinder of $R = 0.4$ m and an outer radius of the inner cylinder of $\tilde{R}$ =0.32 m. Both cylinders are mounted between two annular aluminum plates. The plate at $z = 0$ m represents the side of the plenum that is connected to the combustion tubes and has five holes where the actuation elements are mounted. The plate at $z = 0.6$ m represents the turbine side of the plenum. As gas turbines are usually designed in a way that the first turbine stage is flown through sonically [14], the sound-reflecting plate suitably imitates the boundary conditions at the turbine side of the plenum. The choice of the annular geometry

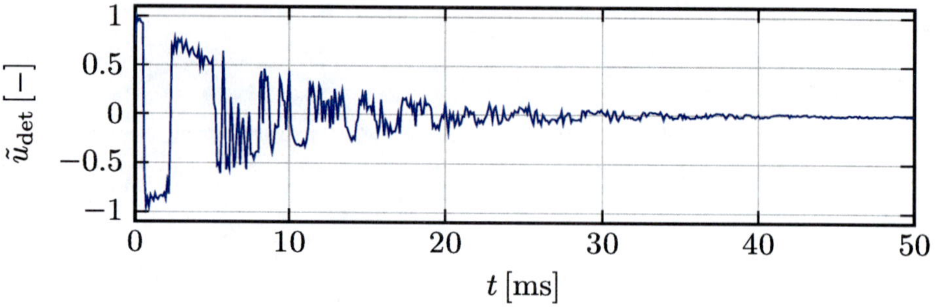

Fig. 2. Excitation signal of a firing event over time normalized on its maximum voltage [13]

of the plenum reflects the need to connect the turbine with the compressor through a shaft in an actual PDC-based gas turbine.

As no actual PDC-based combustion tubes could be used in this test rig, the acoustic excitation of the pressure field inside the plenum is realized by loudspeakers. In the setup, $n_u = 5$ TI-100 8 Ω Visaton loudspeakers are used that are mounted at the combustion side plate at the positions $\varphi \in \{0°, 72°, 144°, 216°, 288°\}$, as can be seen in Fig. 1b). In principle the number of loudspeakers is arbitrary. The shown methods can be transferred to any number of actuators.

In an experiment, whenever a firing event with a combustion tube is triggered, the corresponding speaker that substitutes the combustion tube is excited with the normalized voltage signal shown in Fig. 2. This excitation of the loudspeaker defines a so-called firing event. The trigger signal, the aforementioned 1 in the discrete-time control vector, will be referred to as a firing instant in this contribution, and we will call the totality of firing instants of all actuators within an actuation period a firing pattern.

As the actual amplitudes of a detonation event through a PDC cannot be emulated fully by this setup, the substitute signal was used, as defined in [13]. The signal's definition ensures that the frequency range of the pressure inside the plenum excited by a sequential firing pattern matches the range shown in [15]. To characterize the pressure field in the plenum, measurements are required. The sensors are mounted in the outer cylinder of the plenum, as the positioning of pressure sensors within the plenum is not a viable option, due to the more complex necessary cooling for the sensors in a real setup. In the test rig, the pressure is measured using $n_y = 10$ microphones (G.R.A.S. 40BP 1/4 in. externally polarized pressure microphone) with respective preamplifiers (G.R.A.S. 26AC 1/4 in. standard preamplifier). The microphones are mounted at $z_0 = l = 0.1$ m and $\varphi \in \{0°, 36°, 72°, 108°, \ldots, 324°\}$. The measurements are taken at discrete time steps of $\Delta t = 0.1$ ms. Measurement and control routines are implemented on a digital signal processor by dSpace (DS1006 processor board with DS2103 DA-board and DS2004 AD-board). The system is operated so that every mockup combustion tube has to fire once in every actuation period. This is done to ensure a fixed, feasible actuation frequency for every tube. Therefore, no changes of the system response are a result of changes in the actuation frequency. Further, a firing pattern is applied for several cycles before the periodic steady-state of the system is reached. Only then are the measurements evaluated for the next control step. The actuation period is defined as the inverse of the firing frequency. This frequency is set to $f = 20$ Hz in all experiments.

2.1 Extension to a Nonlinear System

For some of the experiments using the model-free ILC approach, an output nonlinearity is applied to the actual signals measured at the test rig. Given the true measurement of the i-th sensor $\underline{y}_i(k)$ for a time step k, the transformed output is calculated by

$$\underline{y}_{i,nl} = 5 \cdot sgn(y_i(k)) \cdot y_i(k)^2 \quad , \tag{1}$$

with $sgn(\cdot)$ representing the signum function. This is chosen to preserve the sign orientation of the signal. The same output nonlinearity was used in [8] to examine the behavior of a binary ILC control to output nonlinearities that the model does not describe.

3 Binary Iterative Learning Control

In the following section, the mathematical background for the applied control methods is explained. We will indicate matrices in bold style, all vectors are underlined, and upper- and lower-case letters distinguish super vectors from ordinary vectors.

By applying a defined control input repeatedly to an asymptotically stable, linear system, we assume that the converged periodic steady state can be described using a linear input–output model

$$\underline{Y} = \mathbf{\Phi} \underline{U} \quad . \tag{2}$$

Here, $\underline{Y} \in \mathbb{R}^{n_y \cdot p}$ and $\underline{U} \in \mathbb{B}^{n_u \cdot p}$ represent the so-called super vectors of measurement and actuation, respectively, that are generated by stacking the output vectors $\underline{y}(k)$ and control inputs $\underline{u}(k)$ of a period of discrete length p, as follows:

$$\underline{Y} = [\underline{y}(1)^T, \underline{y}(2)^T, \underline{y}(3)^T, \ldots, \underline{y}(p)^T]^T \quad , \tag{3}$$

$$\underline{U} = [\underline{u}(1)^T, \underline{u}(2)^T, \underline{u}(3)^T, \ldots, \underline{u}(p)^T]^T \quad . \tag{4}$$

Notice that $p = 500$ discrete-time instants are stacked in this contribution due to the arbitrarily chosen firing frequency of $f = 20$ Hz and the maximum sampling period of the used hardware of $\Delta t = 0.1$ ms used. The transition matrix $\mathbf{\Phi}$ can be generated by different methods, such as by identifying a linear time-invariant model [7], or through impulse experiments [13]. The latter approach is used in this contribution.

Within the control input, as stated above, an entry $u_i(k) = 1$ indicates that a firing event of the i-th loudspeaker is triggered at time instant k. The applied voltage progression to the loudspeaker is assumed to be part of the dynamic to be described by the model. This means that the time series of the actuation, as shown in Fig. 2, that is generated while $\underline{U}$ contains just a 1, needs to be captured by Φ. Based on this model, an optimal open-loop control that minimizes the tracking error of a reference $\underline{R}$ can be calculated using the cost function

$$J(\underline{U}) = (\underline{R} - \underline{Y}(\underline{U}))^T \mathbf{W}_E (\underline{R} - \underline{Y}(\underline{U})) \ , \tag{5}$$

with a positive definite weighting matrix $\mathbf{W}_E$. Because such a control approach using open-loop control alone would not be robust against disturbances and because the linear model is only an approximation of the real input–output behavior, a complementary closed-loop control shall ensure that the control objectives are achieved. One possible closed-loop control approach is an iterative calculation of the open-loop control of the following iteration $\underline{U}_{j+1}$ using an iteratively adapted cost function

$$J_{j+1}(\underline{U}_{j+1}) = \underline{E}_{j+1}^T \mathbf{W}_E \underline{E}_{j+1} + \Delta\underline{U}_{j+1}^T \mathbf{W}_{\Delta U} \Delta\underline{U}_{j+1} \ , \tag{6}$$

with the corresponding tracking error $\underline{E}_{j+1} = \underline{R} - \underline{Y}_{j+1}$. In addition to Eq. (5), a term to penalize the change in the control trajectory $\Delta\underline{U}_{j+1} = \underline{U}_{j+1} - \underline{U}_j$ is added using a positive symmetric weighting matrix $\mathbf{W}_{\Delta U}$, as shown in Eq. (6). By assuming a cyclic invariant disturbance, e.g., the same disturbances in every iteration respectively at each time step, and applying the model Eq. (2), the cost function can be reformulated to the following expression [8], which has to be minimized in each iteration:

$$J_{j+1} = \underline{U}_{j+1}^T \mathbf{Q} \underline{U}_{j+1} + \underline{q}^T \underline{U}_{j+1} + c \tag{7}$$

with

$$\begin{aligned}
\mathbf{Q} &= \mathbf{\Phi}^T \mathbf{W}_E \mathbf{\Phi} + \mathbf{W}_{\Delta U} \\
\underline{q}^T &= -2(\underline{U}_j^T (\mathbf{\Phi}^T \mathbf{W}_E \mathbf{\Phi} + \mathbf{W}_{\Delta U}) + \underline{E}_j^T \mathbf{W}_E \mathbf{\Phi}) \\
c &= \underline{E}_j^T \mathbf{W}_E (\underline{E}_j + 2\mathbf{\Phi}\underline{U}_j) + \underline{U}_j^T (\mathbf{\Phi}^T \mathbf{W}_E \mathbf{\Phi} + \mathbf{W}_{\Delta U}) \underline{U}_j \ .
\end{aligned}$$

While this could be solved easily in a real-valued domain of the control by equating the derivative of the cost function with respect to the control input of the $\underline{U}_{j+1}$ to zero, the binary domain given here does not enable such a solution. Instead, the binary QP

$$\begin{aligned}
\underline{U}_{j+1}^* = \arg\min \ & (\underline{U}_{j+1}^T \mathbf{Q} \underline{U}_{j+1} + \underline{q}^T \underline{U}_{j+1}) \\
\text{subject to } & \underline{U}_{j+1} \in \mathcal{S}_{j+1} \\
\text{with } & \mathbf{Q} = \mathbf{\Phi}^T \mathbf{W}_E \mathbf{\Phi} \\
& \underline{q}^T = -2(\underline{U}_j^T (\mathbf{\Phi}^T \mathbf{W}_E \mathbf{\Phi}) + \underline{E}_j^T \mathbf{W}_E \mathbf{\Phi})
\end{aligned} \tag{8}$$

has to be solved in each iteration [8] given the set of all feasible solutions $\mathcal{S}_{j+1}$, which replaces the penalization of the change of the control input in Eq. (6). Due to the high computational cost of solving this problem based on numerical methods, such an ILC approach is not suitable for real-time application if the dimension of $\underline{U}_{j+1}$ is high and no solutions from $\mathcal{S}_{j+1}$ can be eliminated beforehand. A potential approach that reduces the complexity of the problem significantly and scales linearly with the number of firing instances was introduced and applied in [8]. In [8], a real-time capable solution was achieved by reducing the set of possible solutions $\mathcal{S}_{j+1}$ in each iteration. The full set of possible control inputs was reduced in a way that only the control input of the last iteration, $\underline{U}_j$, and control inputs in a neighborhood of defined size around the last control input were part of the solution space. This neighborhood could be expressed by a surrogate control input that was chosen so that additional constraints, such as the requirement that every actuator should only be used once in an actuation period, were met. Therefore, the size of the set of possible control inputs was defined using a tuning parameter a as $|\mathcal{S}_{j+1}| = n_u \cdot a$. For details, see [8]. By using the defined solution space, a transformation matrix T_{j+1} could be formulated that could be applied to transform the QP starting from Eq. (8). By setting

$$\tilde{\underline{U}}_{j+1} = \mathbf{T}_{j+1}\underline{U}_{j+1} \tag{9}$$

the QP could be expressed as

$$\begin{aligned} \tilde{\underline{U}}^*_{j+1} &= \underset{\tilde{\underline{U}} \in \mathbb{B}^m}{\arg\min}\ (\tilde{\underline{U}}^T_{j+1}\tilde{\mathbf{Q}}\tilde{\underline{U}}_{j+1} + \tilde{\underline{q}}^T\tilde{\underline{U}}_{j+1}) \\ \text{with } \tilde{\mathbf{Q}} &= \mathbf{T}_{j+1}\mathbf{Q}\mathbf{T}^T_{j+1} \\ \tilde{\underline{q}}^T &= \underline{q}^T\mathbf{T}^T_{j+1} \ . \end{aligned} \tag{10}$$

The transformation effectively reduces the dimension of the original $\mathbf{Q}$ by cutting out columns that belong to control inputs that are not part of the set $\mathcal{S}_{j+1}$. After a solution has been calculated, it is then re-transformed into the original domain to be applied to the system.

3.1 Eigenvector-Based Reduction of Solution Space

In [8], the set $\mathcal{S}_{j+1}$ was chosen in such a way that only the last applied control input and control inputs in the vicinity were elements of the set of possible solutions. The vicinity was defined using the parameter a. Only control inputs were allowed, where the trigger instants for each actuator were at maximum a discrete time steps away from the last position. This was done for the sake of simplicity. In [8], it was speculated that a better choice could possibly be evaluating the eigenspace of $\mathbf{Q}$, which is investigated in this contribution.

Given the original problem formulation starting with Eq. (8), a possible method to determine the solution set could be the evaluation of the orientation of $\mathbf{Q}$ by means of its eigenvectors. Assuming that the directions of eigenvectors and the eigenvalues characterize the direction and steepness of the descent in

Algorithm 1. Eigenvector-based determination of binary solution space

Initialize $\mathcal{S}_{j+1} := \{\underline{U}_j\}$
$\{\underline{V}_1, \ldots, \underline{V}_p\} :=$ Eigenvectors($\mathbf{Q}$)
$\{D_1, \ldots, D_p\} :=$ Eigenvalues($\mathbf{Q}$)
$\mathcal{P} := \{\underline{P}_1, \ldots, \underline{P}_{2p}\} = \{D_1 \cdot \underline{V}_1, \ldots, D_p \cdot \underline{V}_p, -D_1 \cdot \underline{V}_1, \ldots, -D_p \cdot \underline{V}_p\}$
$counter := 0$
while $counter < a$ **do**
    Initialize $\Delta\underline{U} := \underline{0}$
    $k_- :=$ time instant at which the tube was fired in cycle j
    $\Delta\underline{U}[k_-] = -1$
    $\underline{\tilde{P}} :=$ scaled eigenvector in $\mathcal{P}$ with minimal value in row k_-
    $k_+ :=$ row index of $\underline{\tilde{P}}$ with largest value
    $\Delta\underline{U}[k_+] = 1$
    $\underline{\tilde{U}}_{j+1} := \underline{U}_j + \Delta\underline{U}$
    if $\underline{\tilde{U}}_{j+1} \notin \mathcal{S}_{j+1}$ **then**
        Add $\underline{\tilde{U}}_{j+1}$ to $\mathcal{S}_{j+1}$
        $counter = counter + 1$
    end if
    Remove $\underline{\tilde{P}}$ from $\mathcal{P}$
end while

the QP, respectively, it is anticipated that such information can be exploited to increase the speed of convergence of the ILC.

As usually the eigenvectors of the QP cannot be expected to be of a binary nature, a direct adaptation of the binary control input in the direction of the eigenvectors is in most cases not feasible. Instead of such a direct approach, a feasible binary control vector has to be determined that is most similar to a respective eigenvector of the QP. Therefore, a method to determine the similarity has to be defined that allows for a comparison of eigenvectors and possible control inputs. While a multitude of different methods can be defined, such as the 2-norm of the difference of an eigenvector and a feasible binary control vector, here we choose a naive similarity comparison method that directly enables an efficient numerical determination of the most suitable control vector. Assuming that for a single tube the control vector $\underline{U}$ is zero in every entry but one, which indicates the firing instant, this vector should be changed in such a way that its updated version yields a better control output. To update $\underline{U}$, a second vector $\Delta\underline{U}$ has to be added that contains a -1 at the entry where $\underline{U}$ contains a 1 and a 1 somewhere else to ensure a different control input. This vector is determined using the eigenvectors of the QP and the defined method. In this method, an eigenvector is defined as the most similar to a $\Delta\underline{U}$ vector if the smallest entry of the eigenvector and the smallest entry of $\Delta\underline{U}$, that is, -1, lie within the same row and the biggest entry of this eigenvector and $\Delta\underline{U}$, that is 1, lie within the same row. This definition allows for an efficient calculation of $\Delta\underline{U}$ by comparing the entries of all the calculated eigenvectors. The required solution set is determined by Algorithm 1 in this contribution. For simplicity and readability, the algorithm shows how the set is determined for a setup with only one combustion tube. However, an extension to the multi-tube case is straightforward.

4 ILC Based on Iterative Model Identification

An ILC approach that requires no model of the system before its application is introduced here. It is based on a two-step regime performed in each iteration. In the first step, a substitute model is identified that describes the change of the measurement signals with respect to a change of the control input. In the second step, a new open-loop control is calculated and applied in the next iteration.

4.1 Iterative Model Identification

While the model described by Eq. (2) represents the systems output behavior as a function of the binary input actuation, the proposed iterative model tries to avoid the complex calculations resulting from the input constraints of such a model by using a real-valued control input. Here, the output behavior of the system in the $(j+1)$-th iteration is described by

$$\underline{Y}_{j+1} = \underline{Y}_j + \mathbf{M}_j \Delta\underline{u}_{j+1} \; . \tag{11}$$

This deviation model contains two parts, a constant super vector $\underline{Y}_j$ that represents the super vector of measurements, which is obtained by applying the open-loop control of the j-th iteration, and a linear part $\mathbf{M}_j \Delta\underline{u}_{j+1}$ representing the actuation change for the next iteration. As a firing event with each of the n_u firing tubes is performed once within a period, an alternative representation of a firing pattern defined through $\underline{U}$ could be generated by a vector $\underline{u} \in \mathbb{R}^{n_u}$. Each element of $\underline{u}$ indicates the instant in time within an actuation period at which the corresponding tube is triggered. Therefore, the entries of $\underline{u}$ are real-valued. Similarly, a deviation of such an applied pattern could be defined by a vector $\Delta\underline{u} \in \mathbb{R}^{n_u}$. By using this representation, a matrix $\mathbf{M}_j \in \mathbb{R}^{n_y \cdot p \times n_u}$ can be generated through the subsequent application of different firing patterns that are defined through the deviation of an initial pattern $\underline{u}_{j,0}$ by different $\Delta\underline{u}$. As will be shown, using additional $2 \cdot n_u$ executions with altered firing patterns in each iteration, $\mathbf{M}_j$ can be generated to formulate a model that can be used to predict $\underline{Y}_{j+1}$. The altered patterns are defined by

$$\underline{u}_{j,\pm x} = \underline{u}_{j,0} + \Delta\underline{u}_{j,\pm x} \; , \tag{12}$$

where $x \in \{1, 2, \ldots, n_u\}$ indicates which firing instant of the pattern is altered by a value of d. The sign indicates in which direction it is altered. All other entries but the x-th are set to zero. The corresponding measured outputs $\underline{Y}_{j,\pm x}$ can then be used to define $\mathbf{M}_j$ by applying a central differencing scheme such as

$$\mathbf{M}_j = \frac{1}{2 \cdot d}[\underline{Y}_{j,+1} - \underline{Y}_{j,-1}, \underline{Y}_{j,+2} - \underline{Y}_{j,-2}, \ldots, \underline{Y}_{j,+n_u} - \underline{Y}_{j,-n_u}] \; . \tag{13}$$

A perturbation value of $d = 0.1$ ms was used in all experiments in this contribution.

4.2 Iterative Control Calculation Using the Deviation Model

As the control objective is to minimize the pressure fluctuation, one way to define the objective mathematically is through the minimization of the control error $||\underline{E}||_2^2$ with a reference of $\underline{R} = \underline{0}$. Using the model with the real-valued control input of Eq. (11) and the control error in super-vector representation for the $(j+1)$-th iteration

$$\underline{E}_{j+1} = \underline{R} - \underline{Y}_{j+1} = \underline{0} - \underline{Y}_j - \mathbf{M}_j \Delta \underline{u}_{j+1} \ , \tag{14}$$

the cost function formulated in Eq. (6) transforms into

$$J_{j+1} = \underline{E}_{j+1}^T \mathbf{W}_E \underline{E}_{j+1} + \Delta \underline{u}_{j+1}^T \mathbf{W}_{\Delta u} \Delta \underline{u}_{j+1} \tag{15}$$

$$= (-\underline{Y}_j - \mathbf{M}_j \Delta \underline{u}_{j+1})^T \mathbf{W}_E (-\underline{Y}_j - \mathbf{M}_j \Delta \underline{u}_{j+1}) + \Delta \underline{u}_{j+1}^T \mathbf{W}_{\Delta u} \Delta \underline{u}_{j+1} \ . \tag{16}$$

Notice that the primary variable in the model is defined as the change of the control input compared to the last iteration $\Delta \underline{u}_{j+1}$. The cost function derivative with respect to $\Delta \underline{u}_{j+1}$ is calculated, equated to zero, and rearranged to determine how the control input should be changed, resulting in

$$\Delta \underline{u}_{j+1} = -(\mathbf{M}_j^T \mathbf{W}_E \mathbf{M}_j + \mathbf{W}_{\Delta u})^{-1} \mathbf{M}_j \mathbf{W}_E \underline{Y}_j \ . \tag{17}$$

The calculated change of the control input $\Delta \underline{u}_{j+1}$ is then added to the firing pattern of the previous iteration to define the firing pattern for the next iteration

$$\underline{u}_{j+1} = \underline{u}_j + \Delta \underline{u}_{j+1} \ . \tag{18}$$

It should be noted that the weight $\mathbf{W}_{\Delta u}$ can be used to ensure that the change in control input is limited to a surrounding that is described well by the model. In addition, no drastic changes of the control input are avoided during the calculation. As the influence of this weight is not part of this contribution, the weight is chosen as $\mathbf{W}_{\Delta u} = \mathbf{0}$. Further, $\mathbf{W}_E = \mathbf{I}$ is set in all experiments.

5 Results

We perform multiple experiments using both proposed ILC approaches. The experiments differ in the initial firing pattern and are all performed sequentially, starting with the experiments using the model-free ILC. The procedure results in a short-term aging effect in the loudspeaker's membranes, which affects the results slightly. This can be seen by comparing the results obtained by applying the optimal open-loop control performed at the end of each line of experiments with either one of the control approaches. It is shown in the respective Figs. 3 and 4.

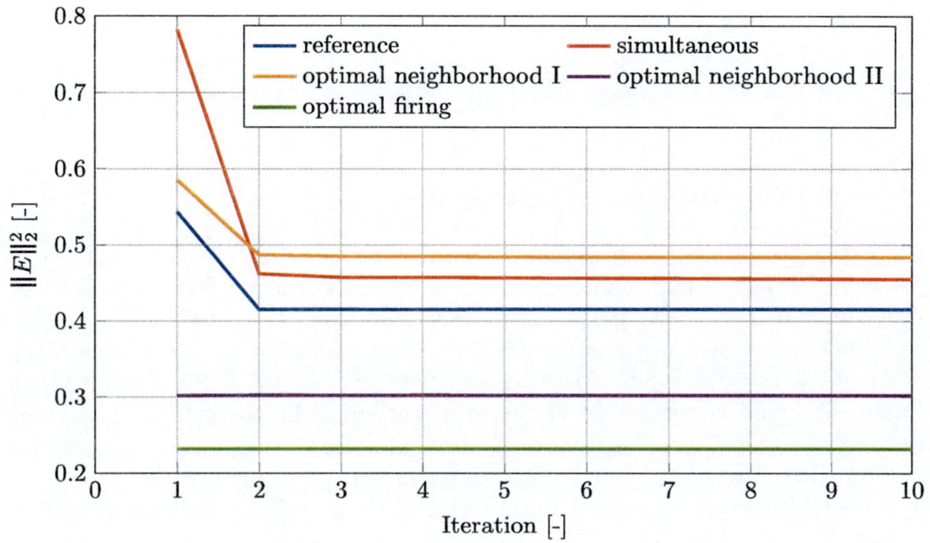

Fig. 3. Development of the error $||E||_2^2$ over iterations using the binary ILC

5.1 ILC Using Eigenvector-Based Binary Solution Sets

Four different experiments are performed using a neighborhood of $a = 4$ and the weighting matrices $\mathbf{W}_{\Delta u} = \mathbf{0}$ and $\mathbf{W}_E = \mathbf{I}$. In all experiments, a reference $\underline{R} = \underline{0}$ is chosen, similar to [8]. Accordingly, the sum of the squared errors, that is $||E||_2^2$, is minimized. Because only the periodic steady state of the system is of interest, a control input is repeated 20 times to ensure that the effect of the system's initial conditions subsides before an evaluation of the measurements is performed. Figure 3 shows the error development of the four experiments over the iterations and the minimum possible error $||E||_2^2$ achievable by a pre-calculated optimal control as reference, as defined in [13]. In the first experiment, a firing pattern is chosen where each firing event is performed in sequential order with an equally distant delay between firing triggers in the actuation period of 0.01 s (reference pattern). Starting from about 0.55, the error $||E||_2^2$ of the binary ILC converges in one iteration and decreases to about 0.41, that is by 25%. An even larger reduction can be seen in an experiment starting with a firing pattern where all speakers actuate simultaneously (simultaneous pattern). This results in an error $||E||_2^2$ of 0.78 to 0.45, equating to a reduction of about 42%. In an experiment starting from a control input close in value to the optimal pattern (optimal neighborhood pattern I), a reduction of about 17% of error $||E||_2^2$ from 0.58 to 0.48 is reached. In an experiment starting with a firing pattern that lies even closer to the optimal pattern (optimal neighborhood pattern II), no significant changes to the control input are made. The system response has a high sensitivity to the control input, and a multitude of local minima with respect to the objective function exist. Therefore, even slight changes in the initial firing patterns result in different final firing patterns, as can be seen in all experiments, similar to the results in [8]. However, as a major improvement to the method in [8], this approach converges very fast, and all significant changes are reached within one iteration.

5.2 ILC Using Iterative Model Identification

Four experiments using the same initial firing patterns and steady-state evaluation as for the experiments with the binary ILC are performed using 100 iterations and $2 \cdot n_u$ perturbations in each iteration, respectively. The corresponding development of the error $||E||_2^2$ can be seen in Fig. 4. Starting with the reference pattern, the model-free ILC reduced the error $||E||_2^2$ from 0.55 to 0.30 (about 45%) and converges in about 55 iterations. The spikes and deviations after the 55th iteration arise from measurement noise that is not particularly filtered out here, resulting in slight readjustments of the firing pattern. By applying the simultaneous pattern as an initial pattern, a reduction of about 38% is reached (from 0.79 to 0.49). The control input converges after 15 iterations on a first plateau, which is held until iteration 83. In the following iterations, the error converges further to the final plateau. We assume that this occurs due to the influence of some external acoustic disturbance, which changed the function value significantly enough to calculate a more favorable control input allowing for a further decrease of error $||E||_2^2$. By starting from the optimal neighborhood pattern I, the system converges into the optimal pattern, corresponding to an error reduction of about 55% (from 0.6 to 0.27) in 30 iterations. The subsequent changes in the control input and therefore the error $||E||_2^2$ result from small changes in the control input through the influence of measurement noise. A similar result is achieved using the optimal neighborhood pattern II. An 18% decrease in error $||E||_2^2$ (0.33 to 0.27) is achieved in about 10 iterations.

An additional experiment is performed using the output nonlinearity defined in Eq. (1). As described above, the nonlinearity is applied artificially to the measurements before the evaluation within the control to check whether the model-free ILC approach is suitable for systems with nonlinear behavior.

Figure 5 shows the error evolution $||E||_2^2$ over the iterations and the error $||E||_2^2$ of an experiment starting from the optimal pattern. The ILC successfully minimizes the error $||E||_2^2$ and converges in around 47 iterations into a local optimum but does not reach the precalculated optimal pattern. Additionally, an experiment starting with the optimal pattern is performed to exemplify that the control approach would not diverge from the found optimum.

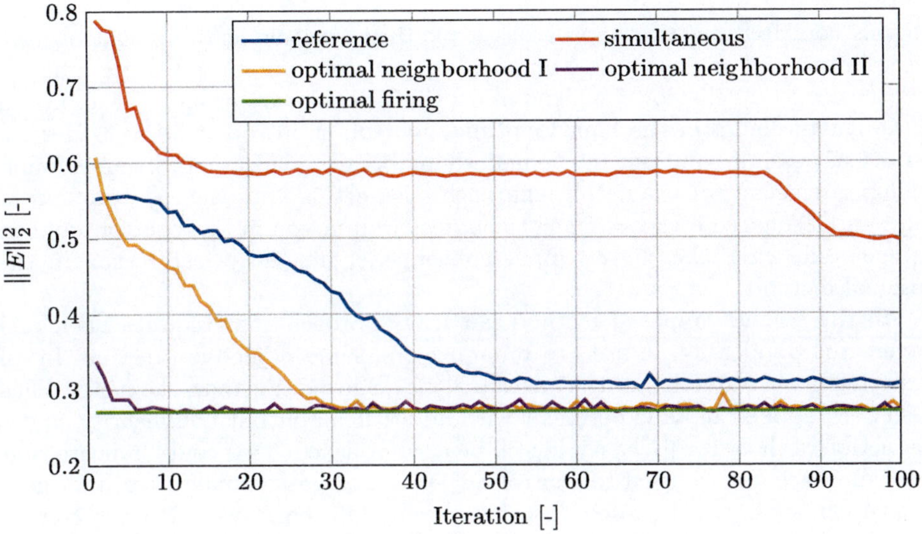

Fig. 4. Development of the error $||E||_2^2$ over iterations using the model-free ILC

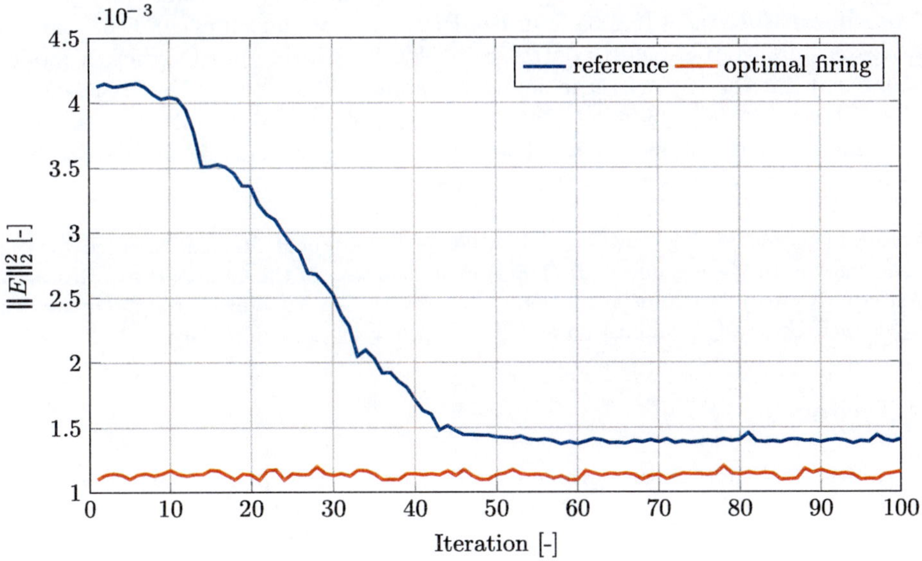

Fig. 5. Development of the error $||E||_2^2$ of a system with artificially added output nonlinearity over iterations using the model-free ILC.

6 Conclusion and Outlook

In this contribution, two approaches exploiting iterative learning control have been introduced and applied to an annular combustion mockup test rig mimicking the acoustic behavior of a plenum in a PDC-based gas turbine. One approach is an enhancement of the binary optimal control, as introduced in [8]. Here, a subset of possible solutions to the underlying binary quadratic problem is found using an eigenvector evaluating approach. Using this approach, a reduction of the sum of squared errors defining the objective function in this control problem is achieved, while the control input converges significantly faster than in the original method.

In the second proposed method, an ILC approach that requires no model before the application is used to minimize the same objective function. In all experiments, a reduction of the error $||E||_2^2$ is achieved. Further, the method has also been applied in an experiment with an artificial output nonlinearity, and a reduction of the error $||E||_2^2$ could still be accomplished. This could indicate that this approach can be used to control a PDC-based gas turbine, even if a linear model cannot be used to describe the system sufficiently well. For the control of an actual PDC-based gas turbine, a combination of both approaches seems reasonable. A fast-reacting binary eigenvector ILC could be cascaded with the model-free ILC in operation points where the system behavior deviates from the used model until an operation point is reached again, which is described sufficiently by the model. Such a combined master control for PDC-based gas turbines will be the focus of future research. Additional aspects, such as rejection of possible disturbances (e.g., a failure of PDC combustion tubes) and the startup process using an ILC-based control, will also be considered. Further, a model update rule for the ILC with the deviation model will be focus of future research, which allows for the avoidance of any additional, expensive sub-iterations, and a combination of the shown binary ILC with sparsity-promoting methods to find a sparse control eigenvector will be conducted.

Acknowledgment. The authors acknowledge the support for this research by the Deutsche Forschungsgemeinschaft (DFG) in the context of the Collaborative Research Center CRC 1029 'Substantial Efficiency Increase in Gas Turbines through Direct Use of Coupled Unsteady Combustion and Flow' through sub-project A05.

References

1. Gülen, S.: Étude on gas turbine combined cycle power plant - next 20 years. J. Eng. Gas Turbines Power **138**(5), 051701–1–051701–10 (2015)
2. Bussing, T., Pappas, G.: An introduction to pulse detonation engines. In: 32nd Aerospace Sciences Meeting and Exhibit (1994)
3. Li, J.M., Teo, C.J., Khoo, B.C., Wang, J.P., Wang, C.: Detonation Control for Propulsion: Pulse Detonation and Rotating Detonation Engines. Springer, Cham (2018)

4. Bobusch, B.C., Berndt, P., Paschereit, C.O., Klein, R.: Shockless explosion combustion: an innovative way of efficient constant volume combustion in gas turbine. Combust. Sci. Technol. **186**(10–11), 1680–1689 (2014)
5. Nicholls, J.A., Wilkinson, H.R., Morrison, R.B.: Intermittent detonation as a thrust-producing mechanism. J. Jet Propul. **27**(5), 534–541 (1957)
6. Gray, J., et al.: Thermodynamic evaluation of pulse detonation combustion for gas turbine power cycles. In: Proceedings Turbo Expo 2016 ASME, GT2016-57813 (2016)
7. Topalovic, D., Wolff, S., King, R., Heinrich, A., Peitsch, D.: Minimization of pressure fluctuations in an axial turbine stage under periodic inflow conditions. In: AIAA 2019-4213 (2019)
8. Arnold, F., Topalovic, D., King, R.: Efficient implementation of a binary iterative learning control. In: IFAC-PapersOnLine, vol. 53, no. 2 (2020)
9. Ahn, H.S., Chen, Y.Q., Moore, K.L.: Iterative learning control: brief survey and categorization. IEEE Trans. Syst., Man Cyber. Part C: Appl. Rev. **37**(6), 1099–1121 (2007)
10. Bristow, D.A., Tharayil, M., Alleyne, A.G.: A survey of iterative learning control. IEEE Control Syst. Mag. **26**(3), 96–114 (2006)
11. Wolff, S., Schäpel, J.S., King, R.: Application of artificial neural networks for misfiring detection in an annular pulsed detonation combustor mockup. J. Eng. Gas Turbines Power **139**(4), 041510-1–1-041510-7 (2016)
12. Wolff, S., King, R.: Model-based detection of misfirings in an annular burner mockup. In: King, R. (ed.) Active Flow and Combustion Control 2014. NNFMMD, vol. 127, pp. 229–244. Springer, Cham (2015). https://doi.org/10.1007/978-3-319-11967-0_15
13. Wolff, S., King, R.: Optimal control for firing synchronization in an annular pulsed detonation combustor mockup by mixed-integer programming. In: AIAA 2019-1742 (2019)
14. Schobeiri, M.: Turbomachinery Flow Physics and Dynamic Performance. Springer, Berlin (2005)
15. Rasheed, A., Furman, A.H., Dean, A.J.: Pressure measurements and attenuation in a hybrid multitube pulse detonation turbine system. J. Propuls. Power **25**(1), 148–161 (2009)

Part III: Combustor-Turbine-Integration: Cooling and Unsteady Performance

Dynamic Forced Impingement Cooling: Latest Experimental Results Regarding Variations in Flow Guidance and Pulse Parameters

Frank Haucke(✉), Arne Berthold, and Niklas Meyners

Department of Aeronautics and Astronautics, Chair of Aerodynamics, Technische Universität Berlin, Marchstr. 12–14, 10587 Berlin, Germany
{frank.haucke.1,arne.berthold}@tu-berlin.de,
niklas.meyners@campus.tu-berlin.de
https://www.sfb1029.tu-berlin.de/menue/sfb_1029/

Abstract. Convective heat transfer of a hot surface in a regime of a periodically pulsed 7 by 7 cooling air impingement jet array is experimentally investigated. In the category of geometrical parameters wall curvature was added. Based on a flat target plate configuration two additional wall curvatures for impingement jet flow guidance were focused on to enhance geometrical boundary conditions in style of real turbine blades. Thereby, internal structure of pressure and suction side is represented by a convex and a concave shaped impingement jet flow space with the same constant radius. For all three experimental configurations nozzle and impingement distance were kept constant at three nozzle diameters. Convective heat transfer behaviour achieved for flat plate setup is transferable for both additional investigated curvatures. The dynamic parameters, frequency, phase shift and duty cycle, separately as well as combined have a significant influence on convective heat transfer improvements compared to the corresponding steady blowing case.

Keywords: Heat transfer · Experimental · Impingement cooling · Periodic forcing · Pulsed blowing · Phase shift · Duty cycle · Crossflow · Wall curvature

1 Introduction

The efficiency of modern gas turbines has increased over the last few decades constantly, which has led to a turbine inlet temperature of more than 1600 K [1]. The increased turbine inlet temperature is one main method for the efficiency gain, while the other one is a raised turbine pressure ratio. In order to reach such high temperatures, several improvements to the maximum permitted thermal material load are needed, which are achieved by the use of thermal barrier coatings, the development of super temperature resistant alloys as well as the convective cooling of turbine blades via internal and external mechanisms.

R. King and D. Peitsch (Eds.): AFCC 2021, NNFM 152, pp. 135–152, 2022.
https://doi.org/10.1007/978-3-030-90727-3_9

In order to further increase the efficiency of gas turbines, either the maximum allowed temperature of the turbine blade material needs to be increased or the cooling mechanisms of the blades need to be improved.

Internal steady blowing impingement cooling is an already implemented concept, which features a higher local heat transfer coefficient in comparison to classical convective cooling inside of the turbine blade. The improved cooling efficiency achieved by this concept is applied on the inner surface, where the cooling air mass flow impinges as jets before being channelled through the blade and exiting it at the trailing edge. Due to this flow, a crossflow interacting with downstream jets is created. To improve this cooling concept, several geometric parameters have been determined as important by Florschuetz et al. [2,3], Weigand and Spring [4] and Xing et al. [5]. These parameters include the nozzle diameter, nozzle arrangement, nozzle distance, impingement distance and the Reynolds number of the cooling air mass flow. An improvement to this concept is dynamically forced impingement cooling, which features the generation of strong vortex structures leading to an increased local convective heat transfer in comparison to the steady blowing method. It was shown by Liu and Vejrazka [6,7] using a single forced impingement jet that the heat transfer in the wall jet region is affected by a dynamically forced impingement jet without significantly influencing the stagnation zone. The distance to the impingement plate H/D has been determined as a driving parameter. Since this improvement is achieved by the vortex structures, the interaction of these vortices in an array of impinging jets is of particular interest for an increased gain of cooling efficiency.

These results were confirmed by Hofmann et al. [8], who showed that the main reason for the change in heat transfer depending on the nozzle to impingement plate distance H/D is the mixing effect of the jets with the environment due to their interaction. Depending on the impingement distance, the mean jet velocity and therefore the heat transfer on the target plate is reduced by the mixing effects. This effect is larger the longer the impingement distance is. A possible increase of the heat transfer due to the turbulence magnitude and the Strouhal number of the dynamically forced jets being of the same order was found by Hoffman, which led to a threshold Strouhal number of $Sr_D = 0.2$ [8]. A formation number $t^* = \frac{(u_p \cdot t)}{D}$ describing the generation of high-energy ring vortices in dependency of the nozzle diameter D, the nozzle exit velocity u_p and the process time t was defined by Gharib et al. from 1998 [9]. This formation number and the threshold Strouhal number were linked by Janetzke [10] which in turn links the works of Herwig et al. [11], Middleberg et al. [12] and Janetzke et al. [10] to the limits for the generation of ring vortices with maximized vorticity, size and amplitude found by Gharib. These works describe the use of velocity square wave signals to produce strong vortices by which strong temporal and local velocity gradients are created which lead to a maximized local convective heat transfer. The local Nusselt numbers increase and the combination of amplitude and frequency are dependent on each other while also being influenced by system resonance and other actuator characteristics which further underlines the importance of an detailed investigation of the connection of local heat transfer in the wall jet and stagnation zone and the dynamically forced impinging jets.

To further increase the cooling ratio the additional parameters duty cycle and phase shift between adjacent nozzle should be considered. In the field of flow control it was shown that flow stability is enhanced while the control mass flow is reduced due to a reduction of the duty cycle at constant maximum jet exit velocity [13]. It was shown by Scholz et al. [14] that duty cycles as low as $DC = 25\%$ are increasing normal force coefficient very effectively. Furthermore it was stated by the authors that a high potential for flow conditioning lies in the optimization of the duty cycle. Even though both studies were focused on jets for separation control, the findings are also applicable for impingement cooling. The momentum coefficient c_μ of a jet in a pulsed flow regime with a constant mass flow rate and a square wave form is proportional to $\frac{1}{DC}$ [15]. This correlation together with the increased temporal velocity leads to the generation of strong vortex structures. In the application of impingement cooling the wall shear stress is increased by the increased local and temporal wall gradients induced by the vortices generated by the jets. Since convective heat transfer is directly linked to the wall shear stress, the duty cycle is most likely a driving parameter for further optimization of the impingement cooling concept. The phase shift between adjacent nozzles can be used to optimize the vortex ring interaction and therefore potentially increasing the cooling ratio even further.

The present experimental investigation is focused on the local convective heat transfer under a 7 by 7 dynamically forced impinging jet array with side-wall induced crossflow. In comparison to previous investigations the nozzle spacing is reduced to $S/D = 3$, which causes stronger jet interactions. As a new geometrical parameter wall curvature was included in this experimental investigation. Based on a flat target plate configuration two additional wall curvatures for impingement jet flow guidance were focused on enhancing the geometrical boundary conditions in style of real turbine blades. Thereby, internal structure of pressure and suction side is represented by a convex and a concave shaped impingement jet flow space. In detail, the study focuses on the local convective heat transfer for several combinations of the excitation frequency, the phase shift between adjacent rows of nozzles, the duty cycle and the nozzle Reynolds number. Thereby, the focus is on maximization of the local convective heat transfer.

2 Experimental Setup

The following section contains information regarding the experimental setup. The basic testing configuration was already utilized by previously conducted work [16–21]. Convective heat transfer between an electrically heated wall and a set of interacting dynamically forced impingement jets is investigated. Figure 1 illustrates the current experimental setup of this work. The geometric impinging jet pattern is produced by jet nozzles arranged in a standard inline 7 (N_X) by 7 (N_Y) array with a reduced dimensionless spacing of $S/D = 3$ in the main directions of the X-Y plane. The nozzle X-locations are defined as follows:

$$\frac{X}{D}(n_X) = \left(n_X - \frac{1}{2} \cdot (N_X + 1)\right) \cdot \frac{S_X}{D} \tag{1}$$

This nozzle plate is positioned above the target plate in various distances ($Z/D = H/D = 2, 3, 5$). Three corresponding side walls, one upstream at $X_{wall}/D = -3.5 \cdot S_X/D$ and two at $Y_{wall}/D = \pm 3.5 \cdot S_Y/D$, border the flow space laterally and define the resulting crossflow in positive x direction. The coordinate origin is located at the intersection of the central longitudinal nozzle axis with the target plate. The inner nozzle geometry is equivalent as utilized in previous work [16–21] and is based on the geometry introduced by Janetzke [10]. Thus, the average crossflow Reynolds number Re_{CF} is increased with every row of impingement nozzles until a maximum is generated shortly upstream of the last row. This kind of crossflow generation is derived from a turbine blade, where the internal structure is channelling the exploited cooling mass flow to the trailing edge of the turbine blade. With every nozzle row the mean Reynolds number ratio of crossflow (CF) and nozzle jet (D) increases and follows the simplified equation:

$$\frac{Re_{CF}}{Re_D}(n_X) = \frac{\overline{U}_{CF}(n_X) \cdot H}{\overline{U}_D \cdot D} \cdot \frac{\nu_D}{\nu_{CF}} \approx \frac{\pi}{4} \cdot \frac{n_X - 1}{S_Y/D} \tag{2}$$

Furthermore, wall curvature was considered in this work as a new geometrical parameter taken into account for enhancing experimental boundary conditions towards more realistic flow topology. Therefore, in addition to the standard flat target plate setup two more experimental setups were investigated, one with a convex and another one with a concave curved impingement plate flow space (see Fig. 1). These 2D curvatures represent structural conditions of internally cooled parts in style of turbine blades. The wall curvature radius was derived from NASA E^3 profile shape [22,23]. The realistic radius was upscaled and matched to the generic nozzle diameter employed in this work. Both curved setups have the same radius but with convex and concave shaped target plates representing inner walls, one on the pressure side and the other one on the suction side of turbine blade. Thereby, the experimental focus lies on the curvature driven influence on wall jet propagation and the resulting convective wall heat transfer. With introducing 2D wall curvature nozzle axes of adjacent rows are inclined due to the perpendicular alignment to the target plate. Within every row nozzle axes remain parallel. Apart from this all test setups are basically equivalent.

The coolant air mass flow is provided by an in house compressed air system. Based on the experimental configuration seven individual mass flow control units (MFCU) were utilized. With an accuracy of 0.1–0.5% each of them feeds a set of seven nozzle jets in each row, located at a constant crossflow position, see Fig. 2. This connection technique combined with a symmetric air flow divider (not displayed) enables the cooling air mass flow to be distributed as evenly as possible among the nozzles in each row, particularly in the case of non-coherent actuation.

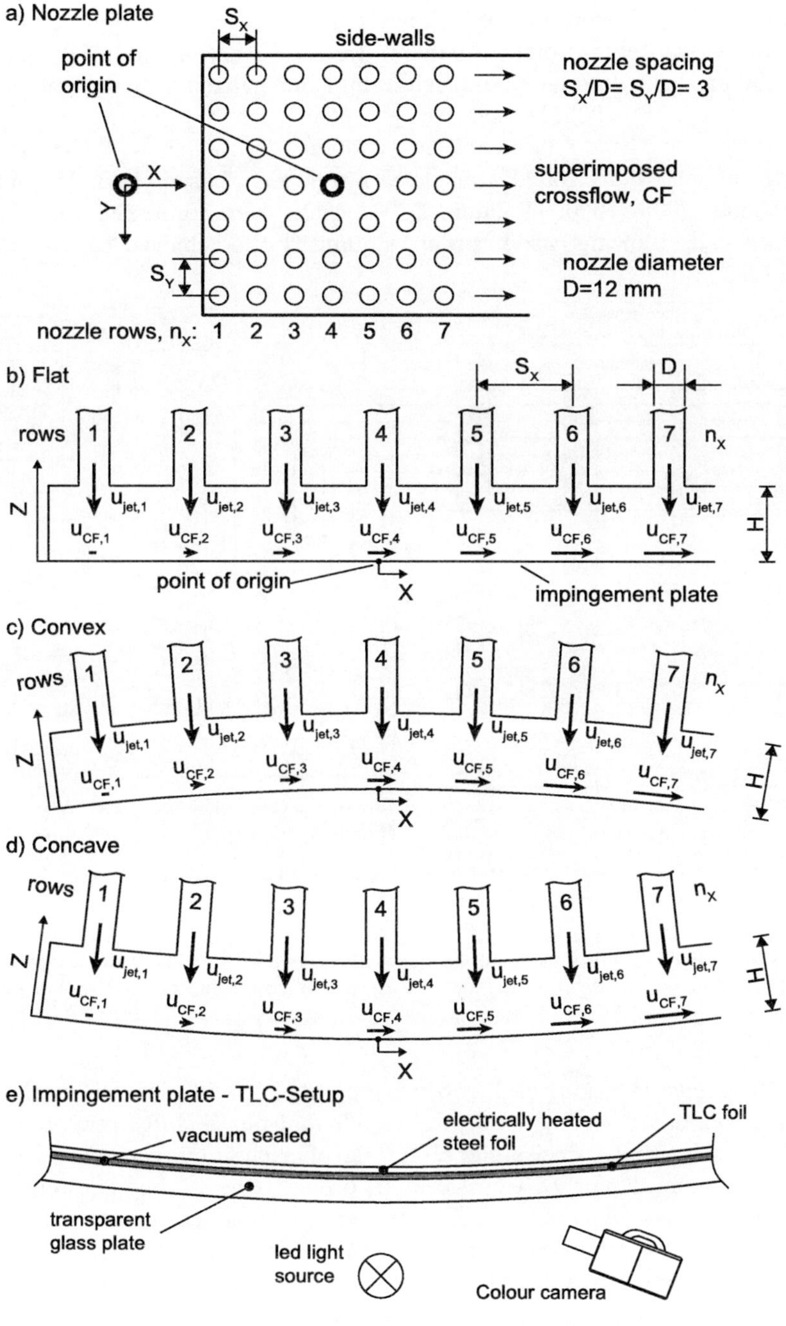

Fig. 1. a) Bottom view of nozzle plate with schematic nozzle arrangement, X, Y plane at Z = H, S/D = H/D = 3. b–d) Side views of crossflow mid sections, X, Z plane at Y = 0, Z > 0: b) Flat configuration, c) Convex configuration, d) Concave configuration. e) Impingement plate with TLC-setup.

The dynamic deformation of impinging jets was enabled by employing an individual fast switching solenoid valve in front of every single nozzle. Thereby, the maximum pulse frequency and maximum normalized volume flow rate are limited at $f_{max} = 1000\,\text{Hz}$ and $V_N \leq 160\,\text{l}_\text{N}/\text{min}$ by the utilized solenoid valves. Based on these specifications, switching frequency was varied within the range of $f = 0$ to $1000\,\text{Hz}$ at three nozzle exit based Reynolds numbers $Re_D = 3000, 5000, 7000$. Dynamic forced cases were compared to the corresponding steady blowing case in order to quantify the achieved benefit.

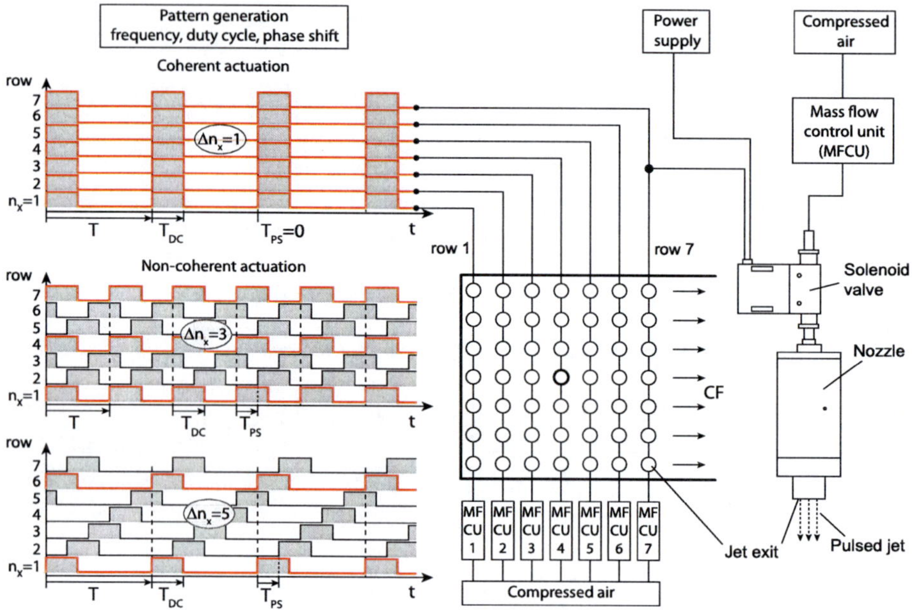

Fig. 2. Pulse pattern after start-up process, parameters and schematic presentation for dynamic control of impinging jets. Examples for coherent and non-coherent pulse patterns.

For these investigations each row was operated simultaneously with parameter variations based on frequency, duty cycle and phase-shift between adjacent rows. In this context, precise signal timing of each impingement jet pulse is a very fundamental part of this work. In order to meet the requirements for valve control most efficiently an Arduino microcontroller board was modified and programmed for multiple signal generation with a reference input square wave signal provided by a function generator. Thereby, pulse parameters can be set independently and are defined as followed.

Pulse frequency f (Hz) defines the temporal repetition of pulses of one jet and focuses primarily on the dynamic development of wall boundary layers in comparison to steady blowing case.

$$f = \frac{1}{T}, \ T \text{ - periodic time in (s)} \tag{3}$$

In Fig. 2 different signal patterns are displayed exemplarily. For coherent actuation seven square wave signals are generated simultaneously or in phase. In case of non-coherent pulsation, pulse time shift defines the time delay between control signals of adjacent nozzle rows. It impacts temporal and spatial propagation of wall jets and boundary layer development compared to the coherent pulsed flow case. Due to the periodic property of each control signal defined by pulse frequency the period duration can be expressed by the following expressions:

$$PS = \frac{\Delta T_{PS}}{T} = \frac{1}{\Delta n_X}, \ \Delta T_{PS} \text{ - pulse time shift in (s)} \tag{4}$$

$$T = \Delta n_X \cdot \Delta T_{PS}, \ \Delta n_X \text{ - repetition number (-)} \tag{5}$$

The repetition number Δn_X is calculated by the reciprocal value of phase shift PS. If Δn_X, as a real number, takes an even value, it can be interpreted as a spatial interval between simultaneously switching valves in X direction, see Fig. 2. According to Eq. 4 a phase shift of $PS = 0\%$ is mathematically not allowed but it is equivalent to $PS = 100\%$ after the start-up process when all valves are switching. In this case Δn_X equals one, which means adjacent jets are pulsed coherently.

Duty cycle DC (%) defines the duration of jet pulse within one period, which follows in good approximation the square wave control signal for the valves.

$$DC = \frac{T_{DC}}{T}, \ T_{DC} \text{ - pulse duration time in (s)} \tag{6}$$

The transfer behaviour of a square wave signal to a jet does not have to be one-to-one because it may be influenced by flow and geometry properties and have to be considered if necessary. Depending on the amount of mean cooling air mass flow or mean nozzle Reynolds number combined with geometry of air supply the jet properties such as exit velocity, amplitude, kinetic energy and momentum are defined. Thus, at a constant nozzle mass flow rate or nozzle Reynolds number a reduction of pulse duration ideally leads to a inversely proportional increase of jet momentum [15] (Eq. 7).

$$\dot{I}_D \approx \frac{1}{DC_D} \cdot \overline{\dot{m}}_D \cdot \overline{U}_D \tag{7}$$

Due to the valves switched on and off a pulsated jet flow is generated. Thus, the mean momentum ratio between crossflow and impinging jet is driven by the pulse parameters discussed above and can be expressed by the simplified Eq. 8. Due to the incoming pulsed impinging jets the resulting crossflow underlies fluctuations as well. Since the jets impinge on the target plate directional fluctuations are dampened and deflected partly in crossflow direction but they are reduced to a minimum due to strong mixing interactions. Thus, the cross flow duty cycle DC_{CF} can be set to one in good approximation.

$$\frac{\dot{I}_{CF}}{\dot{I}_D}(n_X) \approx (n_X - 1)^2 \cdot \frac{\pi}{4} \cdot \frac{1}{S_Y/D \cdot H/D} \cdot \frac{DC_D}{DC_{CF}}, DC_{CF} \approx 1 \tag{8}$$

The structure of the heated impingement plate is illustrated in Fig. 1e). This version contains a 0.05 mm thin steel foil with the self-adhesive commercial thermochromic liquid crystal-film (TLC) attached to the reverse side. The film combination was vacuum sealed on a glass plate. Electrical power was used to generate wall heat flux using a power supply unit with $P_{max} = 3700\,\text{W}$. This interacts directly with the attached TLC-foil resulting in a temperature distribution visible by LED illumination from the bottom side through the glass plate. Liquid crystals reflect temperature specific wave lengths of the illuminating light source and were captured by a colour camera. The electrical current through the steel foil was continuously controlled until temperature range defined by TLC-foil and a thermal equilibrium inside the test rig chamber was achieved for each operation point. Furthermore, each colour picture was de-warped and combined with the corresponding colour-calibration-matrix which results in the time averaged temperature distribution on the wall. More detailed information regarding the whole processing chain are contained in the work of Berthold et al. [20].

In order to evaluate convective heat transfer on the target zone induced by impinging jets some underlying assumptions need to be mentioned. It is postulated that the 3D heat flux into the structure is negligibly small compared to the heat transfer into the cooling air flow. Operating points are measured at a state of thermal equilibrium inside the whole test rig chamber. Thereby, the temperature gradient driven structural heat flux as well as radiation effects are minimized and are almost constant. In this context it is assumed that the thermal energy supplied by electrical power was transferred completely into the cooling air mass flow of the impinging jets which is expressed by the following Eq. 9 for Nusselt number calculation.

Based on the assumptions mentioned above Nusselt number can be expressed by the electrical Power P, the heated area A_{heat}, the balance between local wall and the nozzle jet temperature $(T_W - T_D)$, the thermal conductivity of air λ_{air} and the nozzle diameter D. Due to the low Mach numbers ($Ma = 0.03$) the ratio between total and static temperature is nearly one ($T_{D0}/T_D \approx 1$).

$$Nu = \frac{\alpha \cdot D}{\lambda_{air}} = \frac{\dot{q}}{T_W - T_D} \cdot \frac{D}{\lambda_{air}} = \frac{P}{A_{heat} \cdot (T_W - T_D)} \cdot \frac{D}{\lambda_{air}} \tag{9}$$

Considering all possible measurement uncertainties, the overall theoretical uncertainty of the Nusselt number was determined as $\delta Nu/Nu = 3$–8%. Nevertheless, reproducibility studies revealed an maximal experimental random uncertainty of 3%.

3 Experimental Results

In the following section results of experimental investigations are presented and discussed.

Flat Impingement Plate: In Fig. 3a) the reference 2D Nusselt number distribution on the flat plate configuration with a reduced nozzle spacing of $S/D = 3$ and an impingement distance of $H/D = 3$ at a steady blowing ($DC_D = 100\%$) nozzle Reynolds number of $Re_D = 7000$ is illustrated. The depicted Y/D range represents the mid section of the target zone in crossflow direction. Here, it is visible that the convective heat transfer is sufficiently symmetric to the centre line represented by $Y/D = 0$.

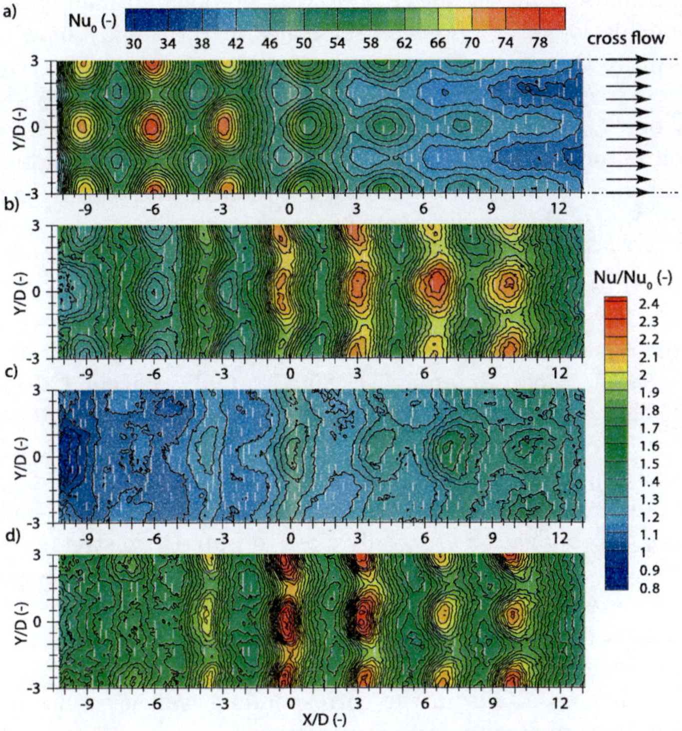

Fig. 3. Local convective heat transfer distributions for flat plate configuration with $S/D = 3$ and $H/D = 3$ at $Re_D = 7000$. a) Local absolute Nusselt numbers, steady blowing $DC_D = 100\%$. b–d) Local Nusselt number ratios, coherent pulsed blowing vs. steady blowing: b) $f = 300\,\text{Hz}, DC_D = 50\%, PS = 0\%$, c) $f = 500\,\text{Hz}, DC_D = 50\%, PS = 0\%$, d) $f = 800\,\text{Hz}, DC_D = 50\%, PS = 0\%$.

Due to the blocked discharge upstream at $X/D = -10.5$ crossflow is generated in positive X/D direction. Thereby, the local crossflow component is accumulated by the discharge of upstream nozzle rows, see Eq. 2 and overlays the local impinging jets. Due to the increasing crossflow component in positive X/D direction the deflection of downstream impinging jets increases with every row which is visible by the increasing downstream shift of impinging zones with maximum local Nusselt numbers and the simultaneous deformation of the circular jet shape. The contour plots b), c) and d) of Fig. 3 illustrate the distributed enhancement in convective heat transfer expressed by the local Nusselt numbers related to the steady blowing case Nu_0 for selected pulse frequencies. For $f = 300$ Hz and $f = 800$ Hz the local Nusselt number ratios are consequently larger than one ($Nu/Nu_0 > 1$) which indicates positive enhancements in convective heat transfer induced by pulsed blowing. Due to the larger temporal jet momentum in case of pulsed actuation the impinging jets can better penetrate the local crossflow which leads to significantly reduced deflection jet axes. Therefore, particularly high values of Nu/Nu_0 are located mainly in secondary flow areas outside local jet impingement zones of the steady blowing case. In picture c) of Fig. 3 the effect of coherent pulsed blowing on local convective heat transfer using a frequency of $f = 500$ Hz is illustrated. Thereby, Nusselt number ratios below one $Nu/Nu_0 \approx 0.88$ can be detected for $X/D < -8.5$ which corresponds to a reduction in heat transfer compared to the steady blowing case. In downstream direction $X/D > -8.5$ local enhancements can be achieved but they undercut by far the achievements of case b) and d).

In Fig. 4 extracted data from contour plots discussed above are displayed. Data curves represent the corresponding centre lines at $Y/D = 0$. In case of steady blowing local stagnation zones upstream of the centre nozzle match very good with their geometrical X/D positions ($n_X = 1, 2, 3$). This is also valid for pulsed cases $f = 500$ Hz and $f = 800$ Hz. Only marginal deviations are visible. In 300 Hz pulsed case stagnation zones for $n_X = 2$ and $n_X = 3$ are shifted even upstream from their geometrical position against local crossflow component more significantly. This indicates different outflow conditions for the local pulsed impinging jets and the accompanying generation of enclosing ring vortices and their propagation especially close to wall. Downstream of the centre nozzle ($n_X > 4$, $X/D > 0$) stagnation zone shift increases with every nozzle row and is most deviated for steady blowing while it is minimized for 800 Hz pulsed blowing case. Nusselt number ratios indicate that periodic actuation is capable of increasing convective heat transfer on the wall significantly almost everywhere on target plate but especially in the corresponding wall jet zones upstream of the stagnation areas of steady impingement jets.

Based on contour plots discussed above (Fig. 3) spatially averaged Nusselt number ratios were calculated and plotted versus pulse frequency. These results are displayed in Fig. 5. Additionally, experimental data of two more geometric flat plate configurations represented by $H/D = 2$ and $H/D = 5$ are inserted for equivalent pulse parameters. All Nusselt numbers are related to the corresponding steady blowing case. Pulse frequency with duty cycle of $DC_D = 50\%$ can be considered as an important actuation parameter. In all three depicted geometric cases the mean averaged Nusselt number ratio is increased significantly for each investigated pulse frequency. Thereby, two significant maxima in Nusselt number enhancement for each function graph can be detected. One maximum is localized at $f = 300$ Hz independently from the depicted impingement distances. This can be explained by geometric resonance effects induced mainly by the cooling air supply lines used [24]. These additional influences of cooling air supply

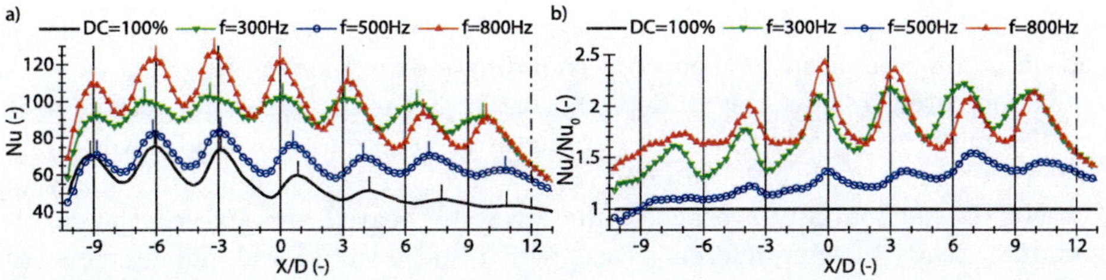

Fig. 4. Steady vs. coherent pulsed blowing ($DC_D = 50\%, PS = 0\%$) local convective heat transfer for flat plate configuration with $S/D = 3$ and $H/D = 3$ at $Re_D = 7000$. a) Absolute local Nusselt numbers. b) Local Nusselt number enhancements.

infrastructure are not discussed in this work, but this resonance frequency can be varied by changing the tube length before and after solenoid valves or changing the cooling air mass flow.

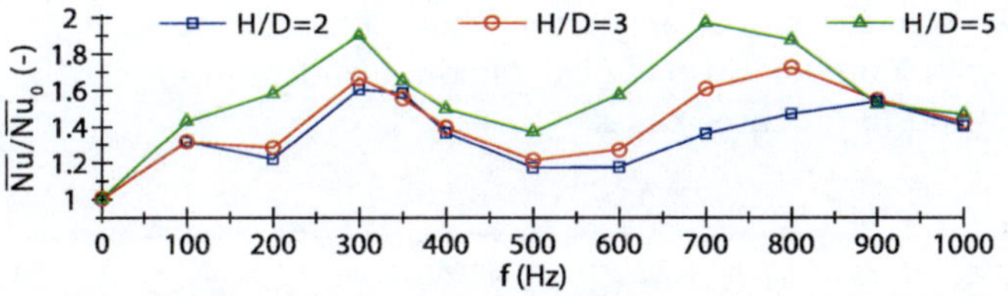

Fig. 5. Global Nusselt number ratio vs. pulse frequency with $DC_D = 50\%$ and $PS = 0\%$ for flat plate configuration with $S/D = 3$ at $Re_D = 7000$. Various impingement distances $H/D = 2$, $H/D = 3$ and $H/D = 5$.

$H/D = 5$ is the largest impingement distance investigated in this work. Thus, outflow conditions are mostly similar to free stream conditions at least close to the jet exit. The second maximum in Nusselt number enhancement for $H/D = 5$ is located at a pulse frequency of $f = 700$ Hz which can be traced back on nozzle geometry based resonance. The nozzle geometry itself has a resonance frequency at about $f = 700$ Hz which can be detected under free stream conditions without impingement plate as well [24]. Similar effects were also observed by Janetzke et al. [10,25].

For decreasing impingement distance to $H/D = 3$ and $H/D = 2$ the second maximum in Nusselt number ratio is shifted to higher pulse frequency $f = 800$ Hz and $f = 900$ Hz under otherwise identical conditions. This can be interpreted as influence of wall distance. Based on the results presented here it is concluded that the smaller the impingement distance the higher the pulse frequency of the second maximum of Nusselt number enhancement. Furthermore, the level of Nusselt number enhancement increases with increasing impingement distance, because pulsed jets at an equivalent mean jet exit velocity are capable of transporting their kinetic energy and momentum over a wider range before they break down and dissipate compared to their corresponding steady blowing jet.

As indicated by flow case c) depicted in Fig. 3 a reduction in convective heat transfer must be expected if pulse parameter are chosen accordingly. Although, local reductions in Nusselt number ratio was observed for pulse frequency of $f = 500\,\text{Hz}$ the spatially averaged value over the heated area exceed the corresponding steady blowing reference case, which is also valid for all depicted curves in Fig. 5 representing impingement distances $H/D = 2$ and $H/D = 5$.

Wall Curvature Variation: Plate curvature as a new geometric parameter was considered, which is a vital step towards investigating more realistic internal flow guidance in style of turbine blades. The flat plate configuration initially investigated serves as reference flow case.

In Fig. 6 the mean Nusselt number ratio is plotted against coherent pulse frequency with a duty cycle of $DC_D = 50\%$. In general all three graphs follow the course as it has been established for the flat plate configuration. Apart from minor deviations in convective heat transfer for almost every single pulse frequency except for $f = 200\,\text{Hz}$ in the convex flow case the applied radius plays a subordinate role for global actuation characteristics at least for the surface averaged Nusselt number ratio. Local maxima and minima can be detected at the same location in all three cases.

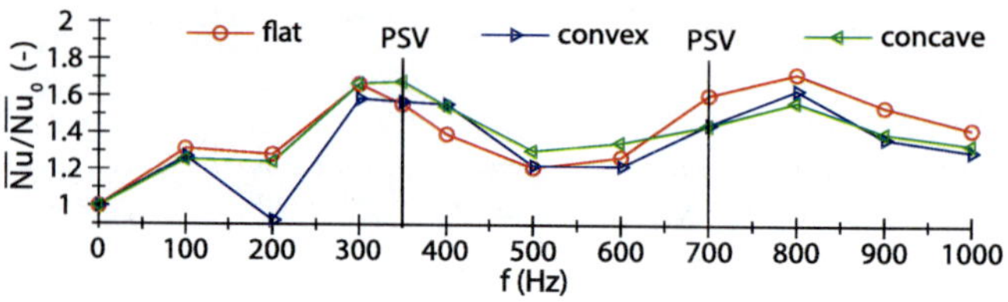

Fig. 6. Global Nusselt number ratio vs. pulse frequency for $S/D = 3$, $H/D = 3$, $DC_D = 50\%$ and $PS = 0\%$ at $Re_D = 7000$. Various target plate curvatures: flat, convex and concave.

Pulse phase shift between adjacent impingement jets has crystallised as additional time parameter for periodic heat transfer optimizations. Therefore, in Fig. 6 two frequencies were marked for phase shift variations (PSV). Figure 7a) illustrates additional significant increases in Nusselt number ratio for $f = 350\,\text{Hz}$ and $DC_D = 50\%$ compared to the corresponding coherently pulsed flow case at $PS = 0\%$. For the flat configuration a maximum increase of $\Delta\overline{Nu}/\overline{Nu}_0 \approx 89\%$ is achieved located at $PS = 15\%$ with a constantly high level up to $PS = 75\%$ with $\Delta\overline{Nu}/\overline{Nu}_0 \approx 87\%$. 10% and 85% pulse time shift lead to maximum Nusselt numbers for the convex case whereas maximum heat transfer is enhanced by up to $\Delta\overline{Nu}/\overline{Nu}_0 \approx 87\%$ with a constant high level in between. In case of the concave wall shape a phase shift between $30\% \leq PS \leq 70\%$ leads to a evenly high increase in heat transfer of about $\Delta\overline{Nu}/\overline{Nu}_0 \approx 80\%$ compared to the steady blowing case.

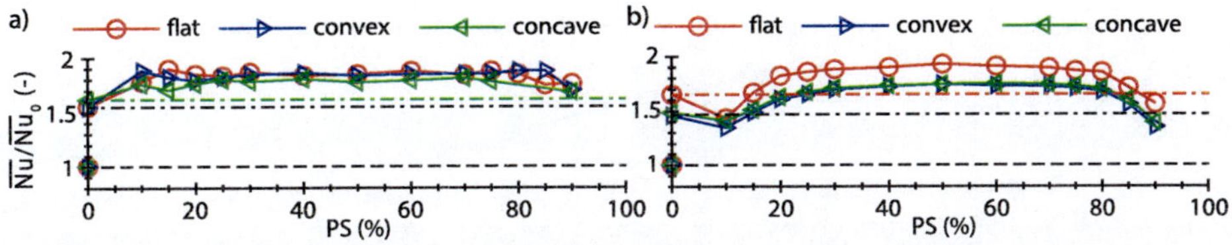

Fig. 7. Global Nusselt number ratio vs. pulse phase shift for $S/D = 3$ and $H/D = 3$ at $Re_D = 7000$. Various target plate curvatures: flat, convex and concave. a) Pulse frequency $f = 350\,\text{Hz}$, $DC_D = 50\%$. b) Pulse frequency $f = 700\,\text{Hz}$, $DC_D = 50\%$.

Equivalent phase shift variations were conducted for 700 Hz pulse frequency, see Fig. 7b). Between $PS = 15\%$ and $PS = 85\%$ additional significant increases in Nusselt number ratio compared to the corresponding coherently pulsed flow case at $PS = 0\%$ can be achieved. In all three geometric cases the maximum increase in convective heat transfer is located almost symmetrically at 50% phase shift which represents perfectly alternating pulsed impingement jets of adjacent nozzle rows or simultaneously pulsed jets of every second down stream nozzle row. Thereby, a Nusselt number increase of $\Delta\overline{Nu}/\overline{Nu}_0 \approx 92\%$ was achieved for the flat target plate and $\Delta\overline{Nu}/\overline{Nu}_0 \approx 74\%$ for both curved configurations.

Duty Cycle Variation: Experimental results for wall curvature variations presented in the previous paragraph have shown very similar characteristics of convective heat transfer for pulse frequency and pulse phase shift variations. It can be assumed that pulse duty cycle characteristics are also equivalent at least between the considered wall curvatures. Thus, the following discussed experimental results are focused on the convex shaped configuration.

In Fig. 8 Nusselt number ratio versus duty cycle is illustrated for a pulse frequency of $f = 350\,\text{Hz}$ and various nozzle Reynolds numbers up to $Re_D = 7000$. In theory, convective heat transfer is connected to temperature and velocity wall boundary layers and their development. Apart from local temperature gradients which are fundamental for heat transfer local velocity gradients on the wall are essential as well. Due to pulsations velocity wall boundary layers and their generally higher transient wall gradients are subjected to periodic renewals before they would reach steady state properties with lower wall gradients and thus lowered convective heat transfer rates. Thereby, jet momentum is a driving parameter for wall boundary layer development. Thus, it could be presumed that heat transfer enhancements are inversely proportional to duty cycle as it is theoretically valid for jet momentum expressed in Eq. 7. But an inverse, almost linear connection between Nusselt number enhancement and duty cycle is discovered in the range of $50\% \leq DC_D \leq 100\%$ and is very similar with minor deviations for all three depicted Reynolds numbers.

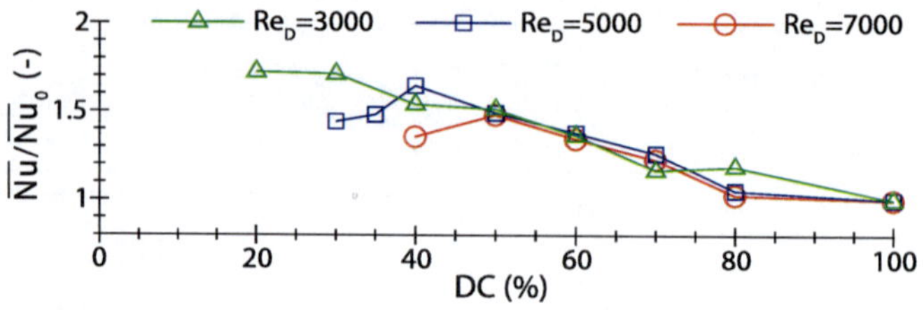

Fig. 8. Global Nusselt number ratio vs. pulse duty cycle at $f = 350\,\text{Hz}$ for $S/D = 3$, $H/D = 3$ and convex curved target plate. Various nozzle Reynolds numbers.

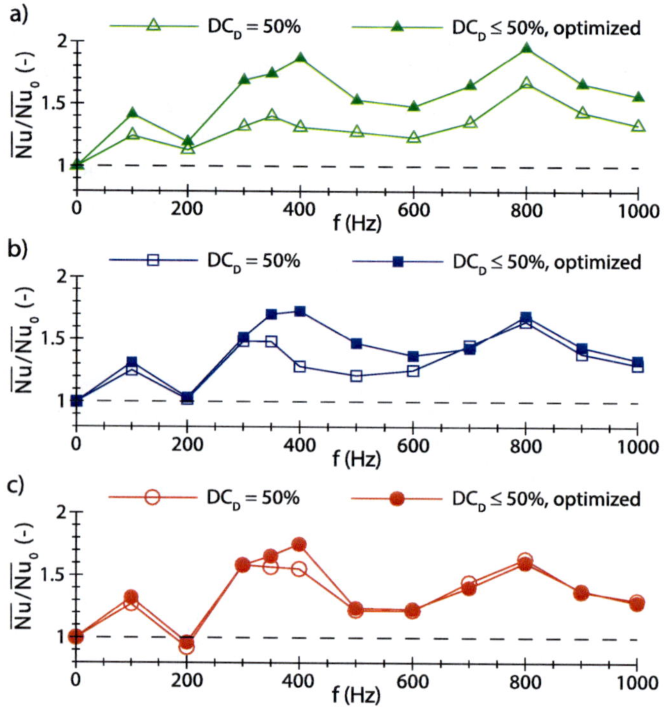

Fig. 9. Global Nusselt number ratio for coherent actuation and optimized pulse duty cycle for $S/D = 3$, $H/D = 3$ and convex curved target plate. Various nozzle Reynolds numbers: a) $Re_D = 3000$, b) $Re_D = 5000$, c) $Re_D = 7000$.

Lowering the pulse duration further to 40% or 20% of time period the inverse linear correlation is extended to higher convective heat transfer values for $Re_D = 5000$ and $Re_D = 3000$. Below specified duty cycles the increases collapse. The deviation of the temporal jet exit velocity profile from the perfect square wave signal is a result of the transmission behaviour of the pneumatic valve-nozzle actuation system. More detailed results regarding this topic are documented in [19].

In the further course optimizations regarding heat transfer enhancements based on duty cycle variations were extended. Results depicted in Fig. 8 exemplarily point out that only a reduction below $DC_D \leq 50\%$ is capable of further increasing the Nusselt number ratio. Figure 9 illustrates the achievements for the nozzle Reynolds numbers already discussed. The duty cycle of 50% is defined as reference value because it is the design point of the valves delivering a constant symmetric switching performance within the depicted frequency range. Thus, the corresponding Nusselt number curves are plotted for every Reynolds number as lower target value curve. Thereby, the largest potential in heat transfer enhancements due to duty cycle optimizations was found for the lowest investigated nozzle Reynolds number $Re_D = 3000$ within the whole frequency range of $0 < f \leq 1000\,\text{Hz}$ (case a)). Increasing the Reynolds number to $Re_D = 5000$ the frequency range is significantly reduced to $300\,\text{Hz} < f < 700\,\text{Hz}$ (case b)). Finally, the discussed frequency range is minimized to $300\,\text{Hz} < f < 500\,\text{Hz}$ in case c) at $Re_D = 7000$.

4 Summary and Conclusions

This work presents latest experimental results from the sub project B03 as part of the Collaborative Research Centre 1029 (SFB 1029). Thereby, the influence of periodically pulsed impingement jets arranged in a 7 by 7 inline pattern on the local and global convective heat transfer on a target plate was investigated experimentally. Regarding this topic, several investigations were performed by the authors in the recent past. This study extends the already existing database by considering primarily curvature of flow guidance as important geometrical parameter to comply step by step with realistic boundary conditions. Starting from a flat impingement plate configuration with reduced nozzle spacing of $S/D = 3$ and a fixed target plate distance of $H/D = 3$ two more curvatures, one in a convex and the other one in a concave design, were integrated in the same global experimental test setup. Convex and concave wall shapes represent internal structure of pressure side as well as of suction side of real turbine blades. Based on liquid crystal thermography measurement technique applied on a electrically heated impingement plate the local wall temperature distributions were determined. In post processing local and global Nusselt numbers were calculated representing convective heat transfer between target plate and impinging jets.

The flat target plate configuration serves as geometrical reference for variations in wall curvature. Contour plots for local Nusselt number and their ratio to corresponding steady blowing case illustrate the local convective heat transfer absolute and relative in stagnation and wall jet zones exemplarily for the flat plate configuration. Thereby, pulse frequency was defined as one main driving parameter for heat transfer enhancements. For the presented flat target plate setup a pulse frequency of $f = 300\,\text{Hz}$ increases the Nusselt number by up to 66% for a target plate distance of $H/D = 3$. Changing only the distance to $H/D = 5$ convective heat transfer is increased by about 90% compared to corresponding steady blowing case. A second maximum in Nusselt number ratio was found at $f = 800\,\text{Hz}$ for the reference plate distance of $H/D = 3$ and was calculated to $\Delta\overline{Nu}/\overline{Nu}_0 \approx 72\%$ above steady blowing case. Furthermore, a dependency between target plate distance and pulse frequency for the second maximum Nusselt number ratio was discovered. Lowering the target plate distance to $H/D = 2$ results in a Nusselt number maximized pulse frequency of $f = 900\,\text{Hz}$ with an increase of $\Delta\overline{Nu}/\overline{Nu}_0 \approx 53\%$. Increasing impingement distance to $H/D = 5$ results in a maximum located at $f = 700\,\text{Hz}$ with an increase of $\Delta\overline{Nu}/\overline{Nu}_0 \approx 96\%$. Thereby, the local heat transfer maximisations take place mainly in the wall jet zones between adjacent impinging jets upstream the stagnation areas of steady blowing case.

For all three investigated curvatures dynamic actuation of impinging jets was performed including frequency and phase shift variations. In general the qualitative comparison of frequency depending Nusselt number ratios show equivalent results. The discussed local extreme value points are located at the same frequencies. Except for one data point for convex configuration similar enhancements in convective heat transfer can be achieved for all curvatures with satisfactorily small deviations. Results for non-coherent actuation, which is expressed by pulse time shift between adjacent jet rows, discovers further space for heat transfer optimisations. Depending on curvature additional maximum Nusselt number increases lie between $54\% \leq \Delta\overline{Nu}/\overline{Nu}_0 \leq 61\%$ coherently pulsed with $f = 350\,\text{Hz}$. Phase shift optimized values are further maximized to $68\% \leq \Delta\overline{Nu}/\overline{Nu}_0 \leq 89\%$. Coherent actuation with $f = 700\,\text{Hz}$ leads to similar increased values in the range of $45\% \leq \Delta\overline{Nu}/\overline{Nu}_0 \leq 64\%$ which can be extended by phase shift optimisations to $74\% \leq \Delta\overline{Nu}/\overline{Nu}_0 \leq 92\%$. Thereby, pulsations with 50% phase shift is capable of delivering very high Nusselt number ratios. This phase shift represents perfectly alternating pulsations between adjacent nozzle rows which is an interesting fact for actuator development. For example, self switching actuators such as fluidic oscillators could be designed for specific operating frequencies to reduce mechanical complexity. Furthermore, based on advanced manufacturing methods this kind of actuators could be integrated in components during manufacturing process resulting in only one part.

Finally the effect of pulse duration was analysed. This parameter defines the momentum rate of the impinging jets. This interacts directly with velocity wall boundary layer and influences convective heat transfer. Thereby, the transformation behaviour between switching characteristic of the employed valves and the pulsed jets have a significant impact on resulting vortex flow structures in proximity of the jets. In general a reduction of pulse duty cycle within permissible working limits is capable of significantly increasing jet momentum at constant cooling air mass flow rate. Depending on nozzle based Reynolds number and pulse frequency the maximum Nusselt number increase can amount up to $\Delta\overline{Nu}/\overline{Nu}_0 \leq 95\%$.

The combination of the three pulse parameters frequency, duty cycle and time shift of adjacent jets creates a huge parameter space with enormous potential for optimisations in any direction but not only for heat transfer matters.

Acknowledgement. "Funded by the Deutsche Forschungsgemeinschaft (DFG, German Research Foundation) - Projektnummer 200291049 - SFB 1029". Additionally the authors thankfully acknowledge the support of the student research assistants B.Sc. Burcu Ataseven, B.Sc. Lennart Rohlfs and B.Sc. Melik Keller during the measurement process.

References

1. Bräunling, W.J.G.: Flugzeugtriebwerke, vol. 3. Springer, Heidelberg (2009). https://doi.org/10.1007/978-3-540-76370-3
2. Florschuetz, L.W., Metzger, D.E., Takeuchi, D., Berry, R.: Multiple jet impingement heat transfer characteristic - experimental investigation of in-line and staggered arrays with crossflow. NASA-CR-3217. Department of Mechanicle Engineering, Arizona State University, Tempe, January 1980
3. Florschuetz, L.W., Truman, C.R., Metzger, D.E.: Streamwise flow and heat transfer distributions for jet array impingement with crossflow. J. Heat Transf. **103**, 337–342 (1981)
4. Weigand, B., Spring, S.: Multiple jet impingement - a review. Heat Transf. Res. **42**(2), 101–142 (2010)
5. Xing, Y., Spring, S., Weigand, B.: Experimental and numerical investigation of heat transfer characteristics of inline and staggered arrays of impinging jets. J. Heat Transf. **132**, 092201/1–11 (2010)
6. Liu, T., Sullivan, J.P.: Heat transfer and flow structures in an excited circular impingement jet. Int. J. Heat Mass Transf. **39**, 3695–3706 (1996)
7. Vejrazka, J., Tihon, J., Marty, P., Sobolik, V.: Effect of an external excitation on the flow structure in a circular impinging jet. Phys. Fluids **17**, 105102 (2005)
8. Hofmann, H.M., Movileanu, D.L., Kind, M., Martin, H.: Influence of a pulsation on heat transfer and flow structure in submerged impinging jets. Int. J. Heat Mass Transf. **50**, 3638–3648 (2007)
9. Gharib, M., Rambod, E., Shariff, K.: A universal time scale for vortex ring formation. J. Fluid Mech. **360**, 121–140 (1998)
10. Janetzke, T.: Experimental investigations of flow field and heat transfer characteristics due to periodically pulsating impinging air jets. Heat Mass Transf. **45**, 193–206 (2008). https://doi.org/10.1007/s00231-008-0410-8
11. Herwig, H., Middelberg, G.: The physics of unsteady jet impingement and its heat transfer performance. Acta Mech. **201**, 171–184 (2008). https://doi.org/10.1007/s00707-008-0080-0
12. Middelberg, G., Herwig, H.: Convective heat transfer under unsteady impinging jets: the effect of the shape of the unsteadiness. J. Heat Mass Transf. **45**, 1519–1532 (2009). https://doi.org/10.1007/s00231-009-0527-4
13. Bons, J., Sondergaard, R., Rivir, R.: The fluid dynamics of LPT blade separation control using pulsed jets. J. Turbomach. **124**(1), 77–85 (2002)
14. Scholz, P., Ortmanns, J., Kaehler, C., Radespiel, R.: Leading edge separation control by means of pulsed jet actuators. In: 3rd AIAA Flow Control Conference, AIAA AVIATION Forum, no. AIAA 2006-2850 (2006)
15. Haucke, F.: Aktive Stroemungskontrolle zur Effizienzsteigerung von Hoch\-auf\-triebs\-kon\-fi\-gu\-ra\-tion\-en. Ph.D. thesis, Technische Universitaet Berlin (2016)

16. Haucke, F., Nitsche, W., Peitsch, D.: Enhanced convective heat transfer due to dynamically forced impingement jet array. In: Proceedings of ASME Turbo Expo 2016, no. GT2016-57360 (2016)
17. Berthold, A., Haucke, F.: Experimental investigation of dynamically forced impingement cooling. In: Proceedings of ASME Turbo Expo 2017: Heat Transfer, ASME Turbo Expo 2017: Turbomachinery Technical Conference and Exposition, vol. 5A, no. GT2017-63140 (2017)
18. Berthold, A., Haucke, F.: Experimental study on the alteration of cooling effectivity through excitation-frequency variation within an impingement jet array with side-wall induced crossflow. In: King, R. (ed.) Active Flow and Combustion Control 2018. NNFMMD, vol. 141, pp. 339–353. Springer, Cham (2019). https://doi.org/10.1007/978-3-319-98177-2_21
19. Berthold, A., Haucke, F., Weiss, J.: Flow field analysis of a dynamically forced impingement jet array. In: AIAA Scitech 2020 Forum, no. AIAA 2020-2081 (2020)
20. Berthold, A., Haucke, F.: Influence of excitation frequency, phase shift and duty cycle on cooling ratio in a dynamically forced impingement jet array. J. Turbomach. **142**(5), 051001 (2020)
21. Neumann, N., Berthold, A., Haucke, F., Peitsch, D., Stathopoulos, P.: Pulsed impingement turbine cooling and its effect on the efficiency of gas turbines with pressure gain combustion. J. Turbomach. **143**(7), 071016 (2021)
22. Thulin, R.D., Howe, D.C., Singer, I.D.: Energiy efficient engine high pressure turbine detailed design report. Technical report, NASA CR-165608, NASA - Lewis Research Center (1982)
23. Timko, L.P.: Energy efficient engine high pressure turbine component test performance report. Technical report, NASA CR-168289, NASA - Lewis Research Center (1984)
24. Ataseven, B., Berthold, A., Haucke, F.: Influence of system resonance on a dynamically forced impinging jet. In: Proceedings of ASME Turbo Expo 2019, Poster GT2019-92266 (2019)
25. Janetzke, T.: Experimentelle Untersuchungen zur Effiziensteigerung von Prallkühlkonfigurationen durch dynamische Ringwirbel hoher Amplitude. Ph.D. thesis, Technische Universität Berlin (2010)

Time-Resolved Analysis of Film Cooling Effects Under Pulsating Inflow Conditions

Alexander Heinrich(✉), Markus Herbig, and Dieter Peitsch

Chair for Aero Engines, Institute for Aeronautics and Astronautics, Technische Universität Berlin, Berlin, Germany
alexander.heinrich@tu-berlin.de

Abstract. The development of modern gas turbines requires higher turbine inlet temperatures for an increase in thermal efficiency. With a change to a pressure gain combustion concept to increase the efficiency significantly, more challenges for the cooling of the first turbine stages must be overcome. For this purpose an array of 777 fan-shaped cooling holes on a flat plate are exposed to a series of different pulsating inflow conditions. Varying the amplitude up to 100% to the mean differential pressure, the film cooling performance is analyzed and evaluated. Adjusting the pulsating frequencies from 1 Hz–5 Hz further allows to gain a comprehensive understanding of the influence of the main parameters affecting the cooling film development. The experimental data recorded with an infrared thermography system reveals a strong impact of the pulsating inflow conditions on the adiabatic film cooling effectiveness.

Keywords: Film cooling · Pressure gain combustion · Turbine

η	Adiabatic film cooling effectiveness
ρ_∞	Main flow air density
ρ_c	Secondary flow air density
CRC	Collaborative Research Centre
DR	Density ratio
f_p	Pulse frequency
l_{ax}	Axial position of cooling hole
M	Blowing ratio
NGV	Nozzle guide vane
Re	Reynolds number
T_∞	Main flow air temperature
T_{aw}	Adiabatic surface temperature
T_c	Secondary flow air temperature
V_∞	Main flow velocity
V_c	Secondary flow air density

R. King and D. Peitsch (Eds.): AFCC 2021, NNFM 152, pp. 153–168, 2022.
https://doi.org/10.1007/978-3-030-90727-3_10

1 Introduction

In order to significantly increase overall gas turbine efficiency by up to 10%, a change from a classical constant pressure to a pressure gain combustion might be the necessary adjustment to realise the aforementioned goal [10,19]. Recent research efforts by the Collaborative Research Centre (CRC) 1029 have thoroughly examined the feasibility of such an approach and also highlighted the challenges for the turbomachinery components in terms of unsteady combustion and flow dynamics. The influence of these boundary conditions on the integrity and performance of the turbine have been subject to numerous experimental investigations.

A change from a constant pressure to a pressure gain combustion has a direct impact on the cooling concepts of the downstream turbine. In contrast to the classical combustion method, the new combustion concept imposes strong temporal variations in the total pressure, in temperature and in an incidence variation of the incoming flow. Especially the Nozzle Guide Vane (NGV) of the first turbine stage is poised to withstand those new boundary conditions. In order to gain a comprehensive understanding of the impact on the cooling concepts the turbine research efforts have been clustered. Whereas this paper concentrates on the impact on the film cooling formation, accompanying research by Heinrich et al. [7] and Topalovic et al. [20] has looked at the turbine performance with varying incidence angle due to the pulsating inflow conditions and means to homogenize these distortions.

The basic idea of film cooling is to act as a heat sink and to provide a thin cooling boundary layer around a body to protect it from the hot main gas flow. For this purpose cooling air is fed through holes from the inside of a blade to the outside, where it is supposed to unite and build an uniform cooling film, shielding the blade material from the hot main gas flow. Recent developments have seen a shift from classical cylindrical holes towards so-called fan-shaped or laidback cooling holes. Cylindrical holes have the disadvantage of producing a lift-off of the cooling jet from the blade surface at higher blowing ratios. The resulting jets rather resemble the desired characteristics for flow control applications and do not favor the formation of a uniform cooling film [1,3]. Further research has then analysed the aforementioned fan-shaped or laidback coolings holes. As Auf dem Kampe et al. have been reporting, the benefit of these configurations is the expanded exit which decreases the jet entrainment velocity thus resulting in a laterally wide spread cooling film [8]. Various researchers have analysed the effect of adjusting the outlet geometry for their desired applications and made recommendations which, however, cannot easily be transferred to general applications. Schroeder and Thole have researched the various available geometries and published a condensed geometry which serves as a baseline for adopting different experimental and numerical investigations [14–18]. This thoroughly examined open-access geometry, called 777 geometry, is intended to be used by other researchers and also serves as a baseline for these research efforts.

Little research was carried out, investigating the direct influence of the pulsating inflow conditions on the film cooling effects. Some conclusions can be drawn from the work of Heidmann et al. [6] and Womack et al. [21]. Heidmann looked at the impact of wake passing on the film cooling effectiveness and observed a decrease in effectiveness for all analyzed blowing ratios. Womack followed the same question, using a flat plate geometry and noted that for blowing ratios which experienced a slight lift-off, the wake can actually promote reattachment in the near hole region. However a strong disturbance of the film cooling jets was the overall perception. Bakhtiari and Schiffer [2] have, as one of the few researchers, numerically analyzed the influence of transient inflow conditions originating from a pressure gain combustion on the film cooling effects. For their numerical investigations they used cylindrical holes which were exposed to a sinusoidal oscillation of the pressure amplitude. They report that in the case of pulsating inflow, the film cooling effectiveness decreases and the heat transfer coefficient rises.

2 Experimental Facility

The experimental film cooling measurements are conducted in a low-speed wind tunnel at the Chair of Aero Engines at the Technische Universität Berlin. Operated by a radial compressor, that is sucking the air through the measurement section, a main flow velocity (V_∞) up to 40 m/s can be realized. Through the periodic blocking of the flow path a periodic unsteady main flow can be produced. This concept of generating flow instabilities was proved by several investigations before [4,5,9]. An integrated air heater for the main flow as well as a heat exchanger for the secondary air flow provides the necessary density ratio to determine the resulting film cooling effectiveness. A full overview of the setup is given in Fig. 1.

For a detailed acquisition of a surface temperature field an infrared thermography system is used. Optical access is provided by a 5 mm thick Germanium window with a diameter of 75 mm which is placed above the flat plate. The camera lens is positioned almost flush mounted on the Germanium window to reduce any backlight interference. In addition, a special anti-reflection coating is chosen to allow for sufficient transmission for thermal imaging applications in the wavelength range of 8.000–12.000 nm.

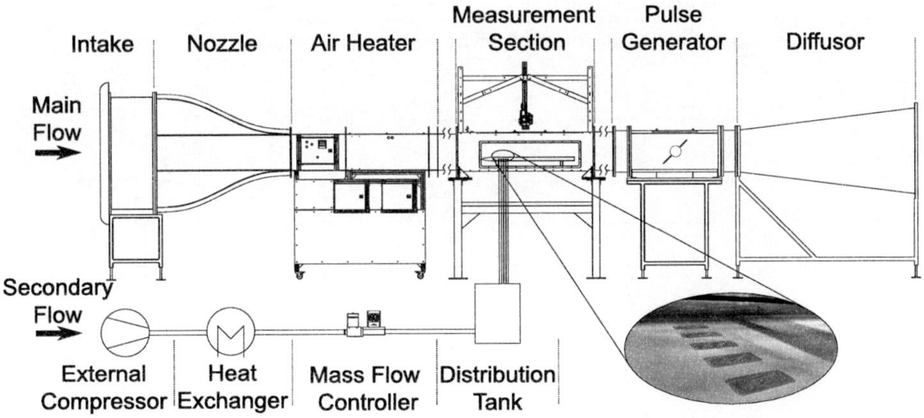

Fig. 1. Scheme of the low-speed wind tunnel

The main flow is sucked into the nozzle from ambient. Downstream, a 40 kW air heater enables maximum main flow temperatures of 323 K. From an aerodynamic point of view the air heater causes unwanted disturbances in the velocity and temperature distribution. However, a homogeneous temperature field can be reached using a turbulence grid that improves the temperature mixing process at the outlet of the air heater. To reduce the generated turbulence intensity a honey-comb flow straightener is positioned upstream the measurement section. Due to the limited available space, the turbulence grid as well as the flow straightener are not shown in Fig. 1. Positioned 1.6 m behind the air heater, the square shaped inlet of the measurement section has a side length of 0.4 m and houses a flat plate made of *NECURON* 1007 with an axial chord length of 1m. The flat plate is designed with an elliptic nose as well as an elliptic trailing edge. In case of backflow, as a result of the high amplitude pulsating pressure fluctuations, a straight trailing edge would be unfavorable. A set of five 777-shaped film cooling holes with a lateral spacing of 45 mm is integrated into the plate. The design is related to experimental investigations by Schroeder and Thole [15], who developed a fan-shaped outlet for film cooling holes. The lateral spacing of $c_{lat}/D = 6$ is chosen to focus on periodicity rather than on forcing a uniting of the discrete film cooling flows.

The hole geometry parameters are listed and visualized in the Table 1 and Fig. 2 respectively.

Table 1. 777-hole geometry

Name	Parameter	Value
Injection angle	α	30°
Laidback angle	β_{fwd}	7°
Lateral angle	β_{lat}	7°
Hole diameter	D	7, 5 mm
Lateral hole spacing	c_{lat}	45 mm
Axial position of cooling hole	l_{ax}	38 mm

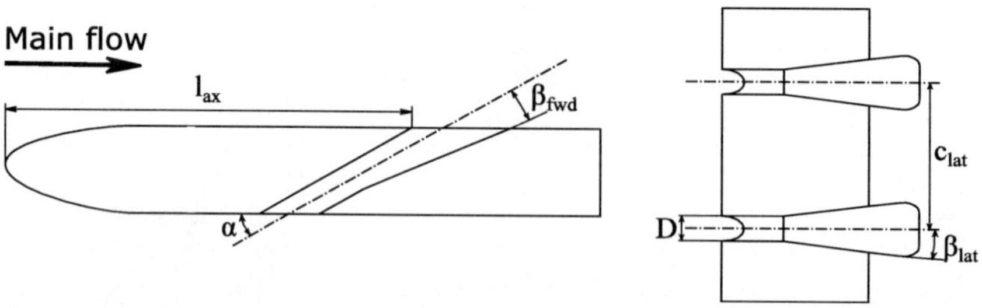

Fig. 2. 777-hole geometry (adapted from Schroeder and Thole [15])

Cooling air for the experiments is provided by an external compressor. The air is dried and cooled before it is stored in a styrofoam-isolated distribution tank from which every cooling hole is fed separately. Besides the flow distribution the tank serves as a damping volume for experiments under pulsating flow conditions. Without this volume the pressure fluctuations would disturb the mass flow controller, which is used to set the desired blowing ratio between the main and secondary air flow. For investigations under periodic unsteady flow conditions a rotating metal disk driven by an electric motor is used. Since the speed of sound at given flow conditions is an order of magnitude larger than the actual main flow velocity the position of the pulse generator will not affect the results of the experimental investigations.

3 Methods

All experiments were performed with a main flow velocity V_∞ of 10 m/s, which corresponds to a mean Reynolds number based on l_{ax} of $Re = 2.2 \times 10^5$ and with a freestream turbulence intensity of $Tu = 5.27\%$. For the determination of the Reynolds number the distance from the leading edge to the cooling hole (l_{ax}) is used (see Fig. 2). The velocity measurement is done via a Prandtl probe with an integrated temperature sensor mounted upstream the flat plate. Due to high blockage under unsteady boundary conditions backflow may occur. Therefore, a second Prandtl probe, directing in the opposite direction of the main flow, is installed downstream the flat plate. In addition, further temperature and total pressure probes are installed inside the nose and the trailing edge of the plate. Pressure data is recorded using *First Sensor* HDO pressure sensors with a 5 mbar pressure range. The length of the pressure tubing is kept to a very minimum and thermocouples are chosen over pt100 temperature sensors. In this way the main flow parameters, e.g. the density ρ_∞, can be characterized even under highly unsteady boundary conditions.

As described in Sect. 2 the periodic pressure fluctuations are produced with a rotating metal disk. With different sized disks pressure amplitudes from 2% to 100% of the mean differential pressure are achievable. The investigated pulse frequencies f_p range from 1 Hz to 5 Hz. For example Fig. 3 shows the main velocity and the total pressure fluctuations during five periods at a pulse frequency of 1 Hz.

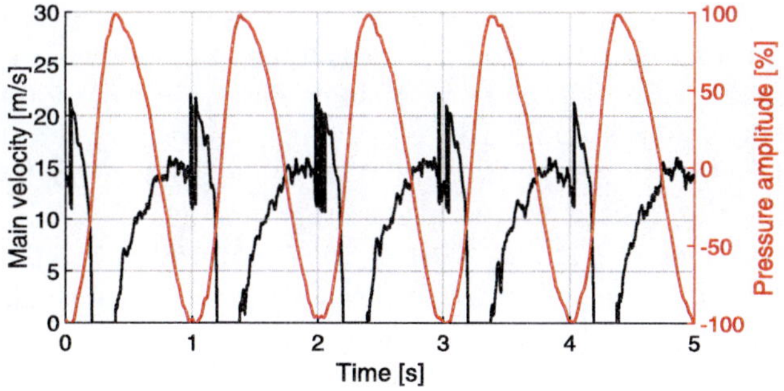

Fig. 3. Velocity and pressure fluctuations during unsteady experiments at 1 Hz

The secondary airflow velocity is controlled by a *Omega FMA 2621A* mass flow controller. For air temperature measurement a PT100-probe inside the distribution tank as well as two Type-K thermocouples inside the outer film cooling holes are used. The mass flow controller, which is connected with the measurement computer, provides necessary data to calculate the density ρ_c and the flow velocity V_c of the cooling flow. Further Type-K thermocouples are flush mounted to the flat plate surface and inserted into the flat plate body to ensure a homogeneous heating of the test specimen prior to each measurement and to also verify the measured temperature of the infrared camera. To characterize the cooling mechanism in a gas turbine, the density ratio between the hot gas from the combustion chamber and the coolant flow extracted from the compressor is used. In modern gas turbines the density ratio ranges up to 1.8 at a pressure ratio between coolant and main flow of 1.02 [12]. The low pressure ratio means, that the density ratio is only driven by the temperature of the cooling flow. Whereas other researchers used different gases to generate a density ratio without a high temperature ratio, the authors in this publication used ambient air for the main and secondary flow due to a long operation time of the wind tunnel to reach a thermal steady state. Therefore only a low density ratio (DR) of 1.1 can be investigated. Nevertheless a comparison to other experimental studies [13,15] is still possible for a smaller range of blowing ratios. As mentioned above the blowing ratio has a high impact on the flow topology. In order to keep the pressure ratio low, as in real gas turbines, only blowing ratios (M) from 0.5 to 1.5 are investigated. The variation of the blowing ratio in that range does not affect the density ratio which was intended by the authors.

The density ratio and the blowing ratio are defined as:

$$DR = \frac{\rho_c}{\rho_\infty} \qquad M = \frac{\rho_c V_c}{\rho_\infty V_\infty}.$$

As mentioned before an infrared thermography system (VarioCam HD 600) from *Infratec* with a resolution of 640×480 pixels at a sampling rate of 30 Hz is used for adiabatic film cooling measurements. This allows to record the desired surface on the flat plate in a single shot with a resolution of 0.3 mm/Px. A lens correction is applied to correct non-uniformity due to the lens curvature. The temperature resolution of 0.05 K is achieved through a manufacturer calibration for the desired temperature range and checked with two flush mounted thermocouples. The *NECURON 1007* material of the flat plate is chosen due to its mat and non-reflecting surface. A logical trigger chain synchronizes the camera sampling rate with the sampling rate of the measurement system and it sets the measurement starting point depending on the position of the rotating metal disk.

In order to set the boundary conditions for the measurement campaign, the temperature of the main flow, the secondary air mass flow and the rotating frequency of the metal disk are adjusted in a closed-loop manner to meet the desired conditions. A *Dewetron 50 PCI* measurement system is used to record the data with a sampling rate of 10.000 Hz.

Table 2. Experimental conditions

Name	Parameter	Value
Main flow velocity	V_∞	10 m/s
Main flow temperature	T_∞	298–323 K
Cooling flow temperature	T_c	290 K
Pulse frequencies	f_p	1–5 Hz
Jet Reynolds number	Re_D	2200–7200

Each measurement point is recorded over five seconds, after the surface temperature of the flat plate reaches a steady state. A full overview of the experimental conditions is given in Table 2.

4 Results

This chapter shows the results of the adiabatic film cooling measurements under steady and unsteady boundary conditions. The measured temperatures will be used to calculate the adiabatic film cooling effectiveness η, which is defined as:

$$\eta = \frac{T_\infty - T_{aw}}{T_\infty - T_c}.$$

Due to the periodicity of the surface visualization, the line plots refer to the middle of the five cooling holes.

Steady Measurements

The investigations under steady conditions serve as a baseline for the unsteady experiments. Furthermore the results allow a comparison to the results from Schroeder and Thole [15]. The surface visualization of the adiabatic film cooling effectiveness η for different blowing ratios M can be seen in Fig. 4. The displayed effectiveness upstream the cooling holes results from heat conduction inside the film cooling holes. Due to a low density ratio DR of 1.1 and a relative high freestream turbulence intensity of 5.27%, the effectiveness distribution downstream of the cooling holes behaves similar for all investigated blowing ratios. The highest film cooling effectiveness with the highest spread in lateral and axial direction can be observed at a blowing ratio of 0.75. For blowing ratios above 1.0 a smaller lateral spread of the cooling flow can be detected. This effect results of strong counter-rotating vortex pairs as described by Ostermann et al. [11].

There is a small skewness in lateral distribution for low blowing ratios detectable. The authors believe, that the air heater, which is located upstream the measurement section, creates an inhomogeneous flow field. The chosen spacing between the turbulence grid and the flow straightener seems not sufficient for a complete mixing process. In addition a little difference in the supply for

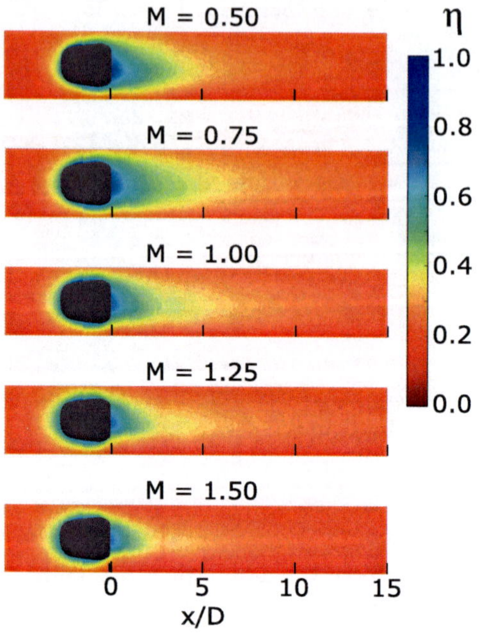

Fig. 4. Adiabatic film cooling effectiveness under steady inflow conditions

the film cooling holes can also result in an uneven lateral distribution of film cooling effectiveness.

Figure 5 shows the adiabatic effectiveness along the centre line of the middle film cooling hole for a normalized axial distance x/D of 15. Additional experimental data from Schroeder and Thole [15] is included. The centre line data from Schroeder and Thole was measured with a density ratio of 1.5 and a freestream turbulence intensity of 0.5. These setup differences are the reason for the lower cooling performance within the presented study, especially the turbulence intensity is a major factor as observed by Saumweber et al. [13].

Nevertheless the results for low blowing ratios are similar to those for low blowing ratios from Schroeder and Thole. The lower the blowing ratio, the better is the adiabatic effectiveness - except for a ratio of $M = 0.5$. It shows best results in the near field, but the mass flow is just not sufficient to maintain the protective shield for values $x/D > 5$. All blowing ratios indicate a steep negative gradient over an axial distance from 0 to 4 x/D with. For higher blowing ratios > 1.00 a saddle point between 5 and 6 x/D can be detected. This behavior can be explained with a reattached cooling flow.

The lateral distribution at a fixed axial position ($x/D = 5$) is illustrated in Fig. 6. As seen in Fig. 4 the lateral spread of the cooling film for blowing ratios above 1.0 is smaller than for low blowing ratios. The lateral distribution for a blowing ratio of 1.0 at a high density ratio from Schroeder and Thole is displayed as well. The maximum film cooling effectiveness in the centre is significantly

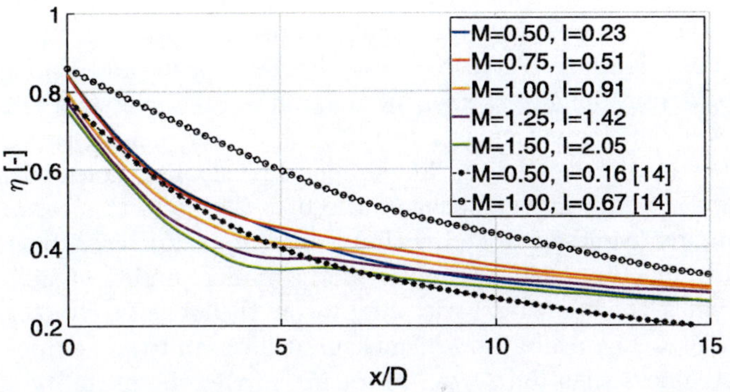

Fig. 5. Adiabatic film cooling effectiveness along the centre line of a film cooling hole

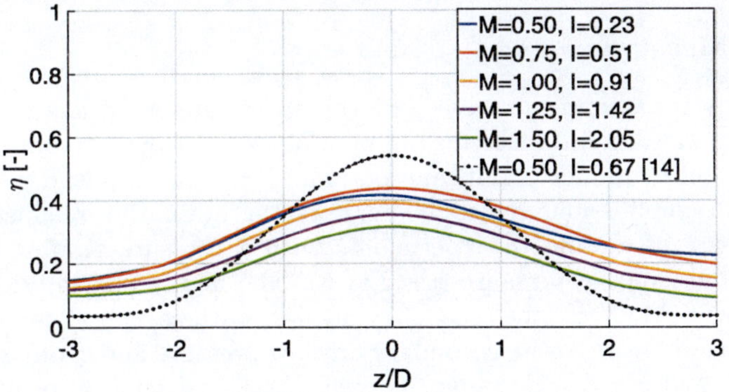

Fig. 6. Lateral distribution of film cooling effectiveness at x/D = 5

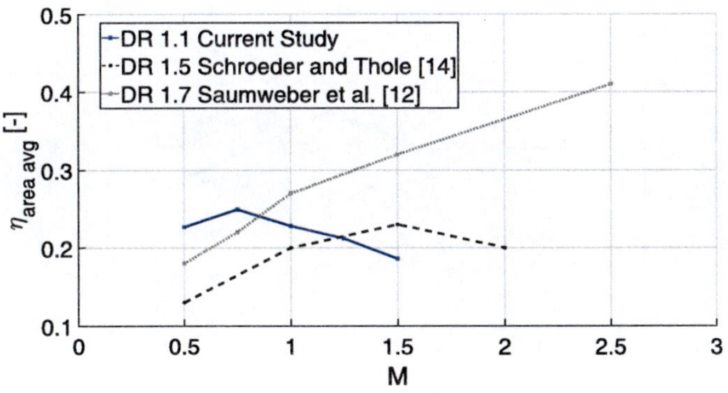

Fig. 7. Area-averaged film cooling effectiveness over x/D = 2–22

larger compared to the evaluated blowing ratios in this study. However, the lateral spread for the blowing ratio of 1.0 is similar in both studies.

In Fig. 7 the area-averaged film cooling effectiveness is compared to other film cooling investigations. As also observed by Schroeder and Thole, who investigated two different density ratios, the maximum area-averaged effectiveness is shifted to lower blowing ratios as a result of the lower density ratios [15].

Saumweber et al. investigated fan-shaped film cooling holes with a reduced injection angle α and a higher density ratio of 1.7. Due to this setup there is no maximum in effectiveness reached within a blowing ratio range from 0.5 to 2.5.

In conclusion, the overall cooling performance under steady conditions depends, besides the blowing ratio, highly on the density ratio and the turbulence intensity. The evaluation of measured data and the comparison of film cooling effectiveness with data from literature proves the usability of the setup. Referring to the steady measurements, the results of the experiments under unsteady boundary conditions will be presented in the following chapter.

Unsteady Measurements

The unsteady measurements were performed for different pressure amplitudes with different sized metal disks for periodic blockage. However, only the largest metal disk could produce significant periodic pressure amplitudes to mimic a pressure gain combustion concept. Therefore the evaluation of unsteady measurement data refers only to experiments with a pressure amplitude of 100% to the mean differential pressure (see Fig. 3). The highly unsteady character of the main flow field affects the secondary air flow path as well. The main control parameter, e.g. blowing ratio, secondary air flow pressure and cooling air density are fluctuating also in a wide range. Therefore, only blowing ratios up to 1.0 are evaluable for a density ratio of 1.1.

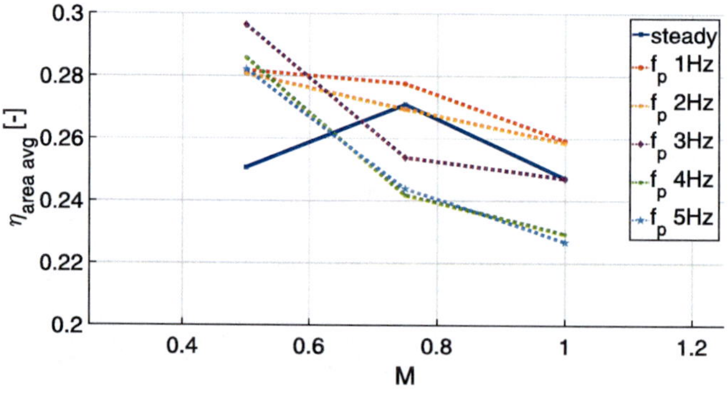

Fig. 8. Area- and time-averaged film cooling effectiveness over x/D = 0–20 for different pulse frequencies under periodic pressure amplitude of 100%

Figure 8 shows the area- and time-averaged film cooling effectiveness for different pulse frequencies f_p and blowing ratios M. The steady measurement data is also included for a better comparison of the different boundary conditions. The trend of the steady measurement with a peak effectiveness at a blowing ratio of 0.75 does not apply for a main flow with periodic pressure fluctuations. For all investigated frequencies a low blowing ratio seems beneficial for a higher cooling effectiveness compared to the steady case. With an increasing blowing ratio the area-averaged effectiveness decreases for all investigated pulse frequencies. Compared to the steady case, low frequencies from 1 Hz to 2 Hz produce better or equal time-averaged cooling effectivenesses for all investigated blowing ratios. The evaluation of higher pulse frequencies results in a steeper negative gradient with increased blowing ratio. Under the influence of higher frequencies (>3 Hz) of pressure fluctuations the area-averaged cooling effectiveness is reduced by a maximum of 11% for the blowing ratios 0.75 and 1.00.

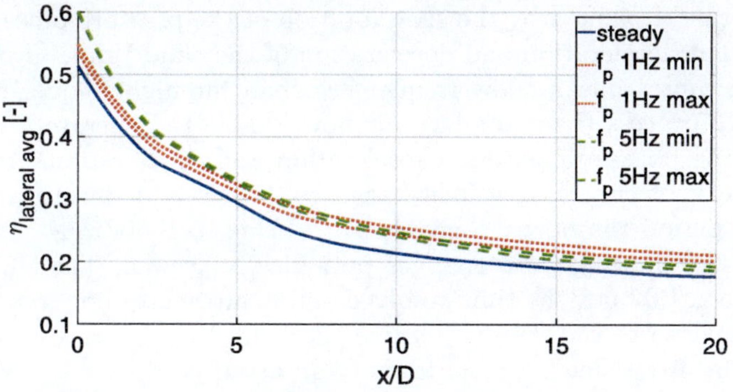

(a) Blowing ratio 0.5

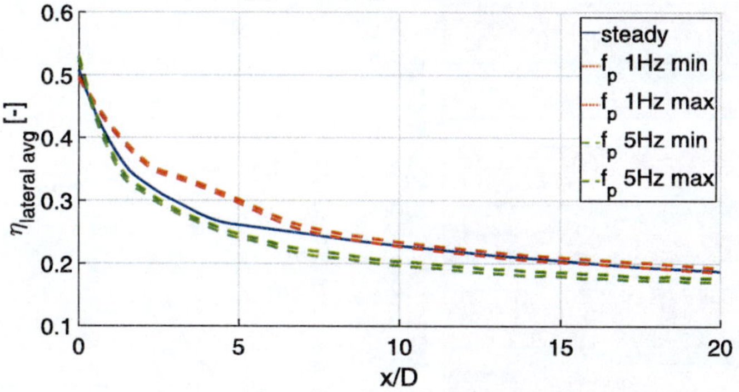

(b) Blowing ratio 1.0

Fig. 9. Minimum and maximum lateral-averaged film cooling effectiveness for pulse frequencies 1 Hz and 5 Hz

Figure 9 shows the lateral-averaged film cooling effectiveness over a distance x/D of 20. In both figures the steady baseline and the pulse frequencies 1 Hz and 5 Hz are compared. In addition, for each frequency the minimum and maximum lateral-averaged film cooling effectiveness over one pulse period is shown. For a blowing ratio of 0.5 (see Fig. 9a) the lateral-averaged effectiveness is increased for both unsteady cases. At the first half of the investigated distance a higher pulse frequency seems beneficial, whereas a pulse frequency of 1 Hz produce an increased lateral averaged cooling effectiveness above a distance x/D of 10. In both Figs. 9a and 9b the spread between minimum and maximum film cooling effectiveness increases along the axial distance for all experiments under the influence of periodic pressure fluctuations. This effect will cause fluctuating high thermal loads downstream the cooling holes which can influence the lifetime of the turbine negatively. Therefore a detailed investigation of the downstream area x/D above 20 is highly recommended for further experiments.

For a blowing ratio of 1.0 only a low pulse frequency results in a higher lateral-averaged cooling effectiveness, which can be seen in Fig. 9b. Especially over a distance x/D from 1 to 6 a significant rise in effectiveness is observable. This behavior can be explained by the slow rotation of the pressure pulse generating metal disk. The acceleration and deceleration of the wind tunnel over one pulse period takes more time for low frequencies than for high frequencies. During this unsteady process the secondary air flow is able to generate a higher heat transfer during phases of periodic acceleration and deceleration. According to Fig. 3 the peak of the main velocity reaches 20 m/s and during a short time of the pulse period the flow velocity is 0 m/s. Due to the strong main velocity fluctuations the blowing ratio changes periodically as well. The time-averaged blowing ratio is 1.02 but the time-resolved values alternates between 0.5 and 6.5 mainly. Compared to the steady baseline there are time steps during one pulse period, in which the blowing ratio leads to an attached flow.

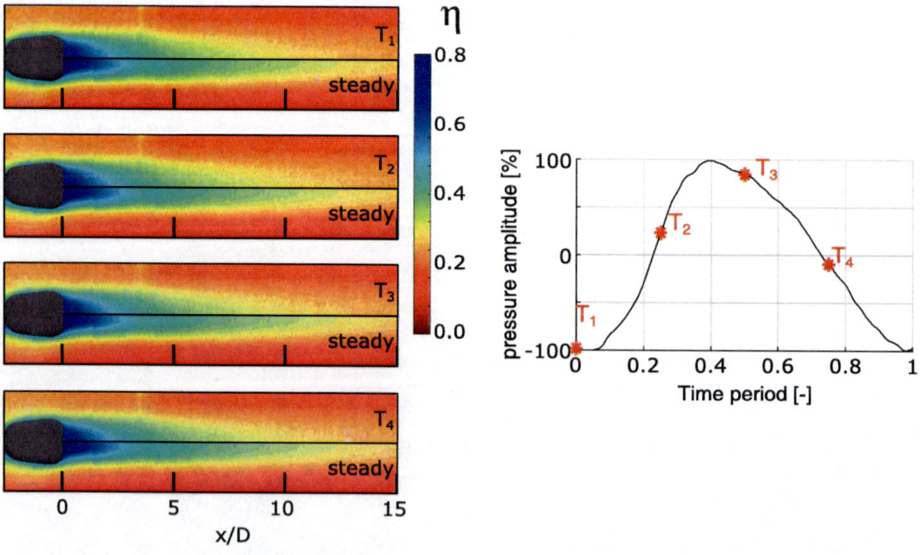

Fig. 10. Time-resolved field of adiabatic film cooling effectiveness with a blowing ratio of 1.0 and a pulse frequency of 1 Hz

A time-resolved field of the adiabatic film cooling effectiveness at specific points of time over an entire period is presented in Fig. 10. The steady measurement is inserted in the lower half of the effectiveness field to allow for comparison. The time steps are equally quartered over a full pulse period T. Compared to the lower half of each time step the lateral spread is larger under periodic pressure fluctuations, as seen in Fig. 9b. However, the axial expansion of the cooling flow under steady conditions is larger by almost 33%. This effect has to take into account for an arrangement of cooling holes under the influence of pressure fluctuations caused by a pressure gain combustion. All four time frames show that there is no constriction of cooling effectiveness at the outlet of the fan-shaped hole. The periodic fluctuating main flow seems to be the reason for a better distribution of the cooling flow at higher blowing ratios.

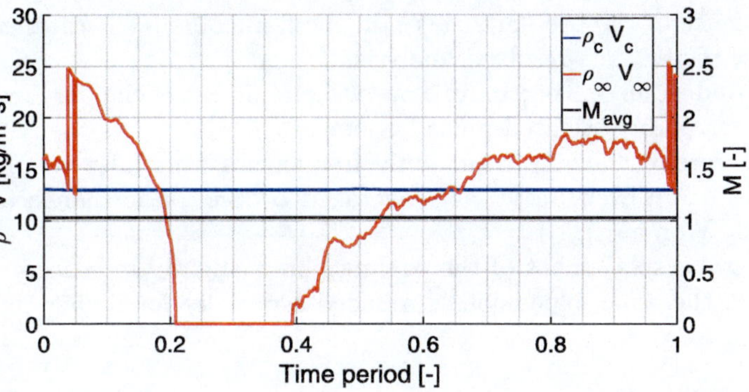

Fig. 11. Time-resolved blowing ratio for a pulse frequency of 1 Hz

Referring to the evaluation in Fig. 10, Fig. 11 shows the time-resolved blowing ratio which is phase-averaged over five pulse periods. The time-averaged blowing ratio is displayed as well. The used coolant mass flow is equal for steady and unsteady investigations as long as the time-averaged blowing ratio and the density ratio is kept constant during the investigations. As a result of the high blockage and the pressure wave, the main flow velocity experiences its minimum which results in the peak in blowing ratio at time point 0.4 s (due to $M = \rho_c V_c / \rho_\infty V_\infty$). A detached flow during this short period of time can be assumed. This effect and the impact on film cooling effectiveness needs to be investigated in further experiments. Future experiments will additionally be carried out at the Hot-Acoustic-Testrig (HAT) which allows for engine realistic conditions. Also the development of control algorithms for an adapted blowing ratio depending on the pulse frequency and pressure amplitude is conceivable.

5 Conclusion

In this contribution the adiabatic film cooling effectiveness for a set of 777 fan-shaped cooling holes under the influence of pulsating inflow conditions is analyzed. First, fundamental steady measurements were performed and compared to results from literature. Due to a low density ratio of 1.1 and a high turbulence intensity of 5.27% a peak effectiveness at a blowing ratio of 0.75 could be evaluated.

In a further step, the influence of periodic pressure fluctuations on the area-averaged film cooling effectiveness is analyzed. Although this investigation has only considered isolated cooling holes on a flat plate, neglecting induced pressure gradients resulting from real geometry blade curvatures, a few conclusions can be carefully drawn:

- An influence of the pressure waves on the film cooling effectiveness field was shown with the time-resolved analysis.
- The periodic fluctuating main flow velocity is beneficial for lower blowing ratios over all investigated pulse frequencies f_p.
- With increasing blowing ratios only frequencies from 1 Hz to 2 Hz result in a similar or an even better area-averaged cooling performance (as was also stated by Womack [21]).
- The lateral displacement of the cooling film is slightly increased.
- However, the axial expansion is reduced especially for higher pulsating frequencies.

Figure 12 summarizes the findings of this publication in a heat map, highlighting the beneficial areas. Following the question for the overall cooling requirement the authors conclude an increased demand in cooling air. The decreased axial expansion demands a new array of cooling holes to be positioned further upstream thus increasing the number of cooling holes which increases the cooling air demand. Further, higher frequencies of the pulsating inflow conditions actually induce a new steady state. In this condition the cooling film has no time to recover. In order to ensure a sufficient blade protection, the cooling air supply has to be adjusted to the new back pressure resulting in a higher demand. Last but not least, the investigations using the TU-Pulse testrig [7] have shown a movement of the stagnation line on the NGV. As a consequence, the vulnerable temperature peak area is increased which calls for an increase of the film cooling area thus again demand more cooling mass flow.

In summary, these results represent a basic set of measurement data for film cooling effectiveness of fan-shaped under the influence of high amplitude inflow conditions. Further investigations with higher density ratios and a wider range of blowing ratios are recommended. Especially a quantification of the increased cooling air mass flow is necessary in order to derive possible counter measures.

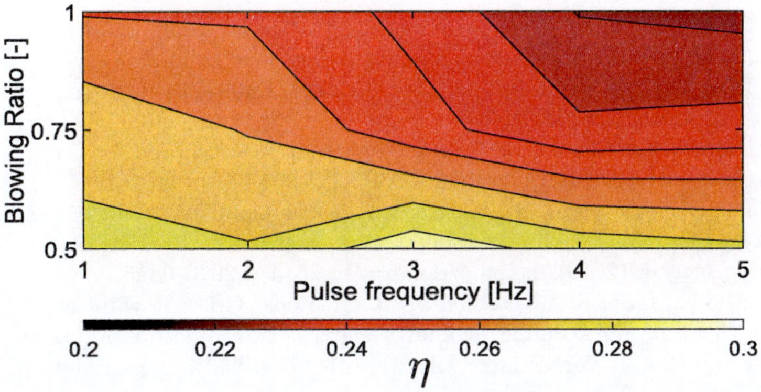

Fig. 12. Surface plot of adiabatic film cooling effectiveness compared to investigated parameters for unsteady main flow

Acknowledgements. The authors acknowledge the support for this research by the Deutsche Forschungsgemeinschaft (DFG) in the context of the Collaborative Research Center CRC1029 'Substantial Efficiency Increase in Gas Turbines through Direct Use of Coupled Unsteady Combustion and Flow' through sub-project B05.

References

1. Baheri, S., Tabrizi, S.P.A., Jubran, B.A.: Film cooling effectiveness from trenched shaped and compound holes. Heat Mass Transf. **44**(8), 989–998 (2008). https://doi.org/10.1007/s00231-007-0341-9
2. Bakhtiari, F., Schiffer, H.P.: Numerical investigation of film cooling effects under transient inflow. In: 24th International Symposium on Air Breathing Engines (2019)
3. Bunker, R.S.: A review of shaped hole turbine film-cooling technology. J. Heat Transf. **127**(4), 441 (2005). https://doi.org/10.1115/1.1860562
4. Dähnert, J., Lyko, C., Peitsch, D.: Transition mechanisms in laminar separated flow under simulated low pressure turbine aerofoil conditions. J. Turbomach. **135**(1), 011007 (2012). https://doi.org/10.1115/1.4006393
5. Deinert, M., Hourmouziadis, J.: Film cooling in unsteady flow with separation bubble. In: Volume 3: Turbo Expo 2004, pp. 55–66. ASMEDC (2004). https://doi.org/10.1115/GT2004-53075
6. Heidmann, J.D., Lucci, B.L., Reshotko, E.: An experimental study of the effect of wake passing on turbine blade film cooling. J. Turbomach. **123**(2), 214 (2001). https://doi.org/10.1115/1.1354621
7. Heinrich, A., Herbig, M., Peitsch, D., Topalovic, D., King, R.: A testrig to evaluate turbine performance and operational strategies under pulsating inflow conditions. In: AIAA Propulsion and Energy 2019 Forum. American Institute of Aeronautics and Astronautics (2019). https://doi.org/10.2514/6.2019-4039
8. auf dem Kampe, T., et al.: Experimental and numerical investigation of flow field and downstream surface temperatures of cylindrical and diffuser shaped film cooling holes. In: Volume 5: Heat Transfer, Parts A and B, pp. 21–34. ASMEDC (2011). https://doi.org/10.1115/GT2011-45106

9. Lyko, C., Dähnert, J., Peitsch, D.: Forcing of separation bubbles by main flow unsteadiness or pulsed vortex generating jets - a comparison. J. Turbomach. **136**(5), 051016 (2013). https://doi.org/10.1115/1.4025214
10. Neumann, N., Peitsch, D.: A comparison of steady-state models for pressure gain combustion in gas turbine performance simulation (accepted for publication). In: Proceedings of GPPS Beijing 2019. GPPS-BJ-2019-0198 (2019)
11. Ostermann, F., Woszidlo, R., Nayeri, C., Paschereit, C.O.: The time-resolved flow field of a jet emitted by a fluidic oscillator into a crossflow. In: 54th AIAA Aerospace Sciences Meeting (2016). https://doi.org/10.2514/6.2016-0345
12. Sargison, J.E., Guo, S.M., Oldfield, M.L.G., Lock, G.D., Rawlinson, A.J.: A converging slot-hole film-cooling geometry–part 2: transonic nozzle guide vane heat transfer and loss. J. Turbomach. **124**(3), 461–471 (2002). https://doi.org/10.1115/1.1459736
13. Saumweber, C., Schulz, A., Wittig, S.: Free-stream turbulence effects on film cooling with shaped holes. J. Turbomach. **125**(1), 65–73 (2003). https://doi.org/10.1115/1.1515336
14. Schroeder, R.P.: Influence of in-hole roughness and high freestream turbulence on film cooling from a shaped hole. Ph.D. thesis, The Pennsylvania State University (2015)
15. Schroeder, R.P., Thole, K.A.: Adiabatic effectiveness measurements for a baseline shaped film cooling hole. In: Proceedings of ASME Turbo Expo 2014: Turbomachinery Technical Conference and Exposition, Düsseldorf, Germany, p. V05BT13A036 (2014). https://doi.org/10.1115/GT2014-25992
16. Schroeder, R.P., Thole, K.A.: Effect of high freestream turbulence on flowfields of shaped film cooling holes. J. Turbomach. **138**(9), 091,001-1–091,001-10 (2016). https://doi.org/10.1115/1.4032736
17. Schroeder, R.P., Thole, K.A.: Effect of in-hole roughness on film cooling from a shaped hole. J. Turbomach. **139**(3), 031,004-1–031,004-9 (2016). https://doi.org/10.1115/1.4034847
18. Schroeder, R.P., Thole, K.A.: Thermal field measurements for a shaped hole at low and high freestream turbulence intensity. J. Turbomach. **139**(2), 021,012-1–021,012-9 (2016). https://doi.org/10.1115/1.4034798
19. Stathopoulos, P.: Comprehensive thermodynamic analysis of the humphrey cycle for gas turbines with pressure gain combustion. Energies **11**(12), 3521 (2018). https://doi.org/10.3390/en11123521
20. Topalovic, D., Wolff, S., Heinrich, A., Peitsch, D., King, R.: Minimization of pressure fluctuations in an axial turbine stage under periodic inflow conditions. In: AIAA Propulsion and Energy 2019 Forum. American Institute of Aeronautics and Astronautics (2019). https://doi.org/10.2514/6.2019-4213
21. Womack, K.M., Volino, R.J., Schultz, M.P.: Measurements in film cooling flows with periodic wakes. J. Turbomach. **130**(4), 041008 (2008). https://doi.org/10.1115/1.2812334

Rotating Detonation Combustor Downstream Transition Passage Design Considerations

James Braun[1(✉)], Guillermo Paniagua[1], and Donald Ferguson[2]

[1] Purdue University, West Lafayette, IN, USA
[2] NETL, Morgantown, WV, USA

Abstract. A key enabler to integrate turbines downstream of rotating detonation combustors is the design of an optimal combustor-turbine passage. Precise estimates of fluctuations, losses, and heat loads are required for the turbine design as rotating detonation combustors feature transonic flow with rotating shocks moving at few kilohertz. This paper analyzes fluctuations and heat loads of the Purdue Turbine Integrated high-Pressure RDE through reactive unsteady Reynolds Averaged Navier-Stokes (URANS) simulations. CFD++, a commercial CFD software package from Metacomp, is employed to solve the unsteady RANS equations through a one-step reaction mechanism. The inlet of the combustor is fed with a hydrogen-air mixture at mass flows of ~1 kg/s with two different back pressures to obtain supersonic and subsonic outlet flows. The mesh featured around 36 million grid points to ensure the resolving of the boundary layer. Finally, a methodology to lower computational time tenfold for the supersonic and subsonic passage is presented based on non-reacting unsteady RANS simulations.

Keywords: Rotating detonation combustors · Supersonic passage · Subsonic passage

1 Introduction

Rotating detonation combustors (RDCs) are a promising technology for power generation [1]. However, the gains of RDCs are accompanied by supersonic rotating shock waves with large fluctuations in Mach number, flow angle, pressure, and temperature [2]. In the past decade, significant efforts have been attributed to the understanding of combustors; literature from Anand et al. [3] and Ma et al. [4] provide a review of the operability of RDEs. Early on, Euler solvers were employed to understand their flow field [5], followed by three-dimensional Reynolds-averaged Navier-Stokes solvers [6] with induction parameter models, one-step and multi-step reaction mechanisms [7], and more recently Large Eddy Simulations through Converge CFD [8] with 21 species, 38 reactions reaction mechanisms for ethylene air are employed or through opensource OpenFoam flow solvers [9]. Experimental validation of pressure gain is performed through total pressure measurements [6] or equivalent available pressure with choked back-pressured RDCs [10].

R. King and D. Peitsch (Eds.): AFCC 2021, NNFM 152, pp. 169–183, 2022.
https://doi.org/10.1007/978-3-030-90727-3_11

While some researchers have focused on nozzle guide vanes on the combustor (Asli et al. [11]) other teams have focused on optimizing the turbine downstream of the RDE. Power extraction from RDCs can be achieved through axial supersonic [12], bladeless [13], or radial outflow turbines [14] and require supersonic flow, or transonic turbines with modified end wall contouring which require subsonic inflow [15]. However, there is still a dearth of information on the optimal design of the passage between combustor and turbine.

Two objectives are outlined in this manuscript. The first objective is to assess heat load, outlet fluctuations, and losses within the coupled combustor and downstream passage. Two different transition passages are analyzed; the first is a supersonic passage that expands flow to ~Mach 1.6. The second is a subsonic passage suitable for subsonic turbines [16]. The second objective is to determine a strategy to investigate those supersonic and subsonic passages downstream of the combustion region at a reduced computational cost for future optimization of downstream transition elements.

2 Methodology

2.1 Solver Description

CFD++ from Metacomp [17] is employed to solve the URANS equations with a one-step reaction mechanism for stoichiometric H_2-air [6]. Limitations on the one-step reaction mechanism include the reduced performance at handling for rich mixtures [18]. The solver is a finite-volume density-based solver. Convective fluxes were solved through the Harten-Lax-Van Leer contact approximate Riemann Scheme, and a second Order Total Variation Diminishing (TVD) polynomial interpolation was selected with a continuous limiter. Implicit time-integration with fixed a global timestep of 0.1 μs was established with an internal iteration termination criterion of 0.1, an essential parameter for unsteady flows [19]. Time step and grid spacing were chosen according to [20] to achieve grid independence. The turbulence closure is provided by the k-omega SST model and based on previous validation in a high-speed environment [13]. Validation of the solver is presented in the appendix for a supersonic shock boundary layer interaction [13] and against coherent anti-Stokes Raman scattering (CARS) thermometry experiments at the exhaust of the Turbine High-pressure Optical RDC [21].

2.2 Investigated Passages Geometries

The two different downstream passages are depicted in Fig. 1 and share the same injection geometry as the experimental counterpart described in [22]. This diverging geometry (Fig. 1a) was inspired by previous work to optimize nozzles for supersonic flows [23] and installed in the Purdue Turbine High-pressure Optical RDC [22]. The diverging geometry has an outlet-to-inlet area ratio of 1.6, measured downstream of the backward-facing step. Figure 1b shows a converging-diverging passage, with a throat suitable for a mass flow averaged Mach number of 1.2 upstream of the throat and decelerate the flow to Mach 0.6 downstream of the throat (outlet-to-inlet area = 1.2). Simulation time for one geometry was three weeks on five nodes with two 10-core Intel Xeon-E5 processors.

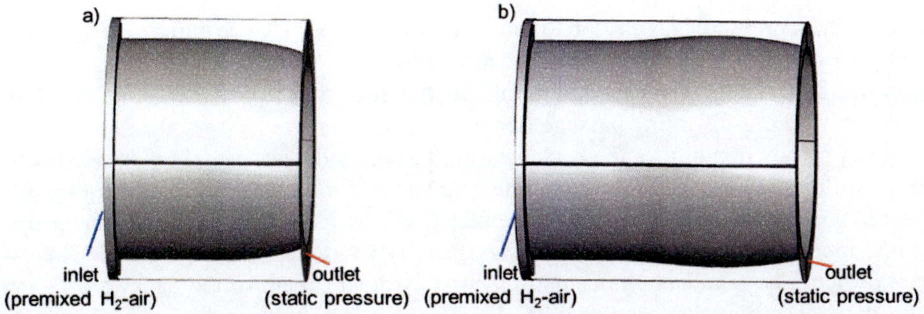

Fig. 1. Selected rotating detonation combustor with two different passages: a) diverging geometry, b) converging-diverging geometry

3 Aerothermal Characterization

3.1 Aerothermal Characterization Within the Transition Passage

This section describes the phenomena occurring within the coupled combustor (region defined by the detonation wave) and downstream passage (through which one or more oblique shocks travel). The pressure contour of the combustor and passage is depicted in Fig. 2a.

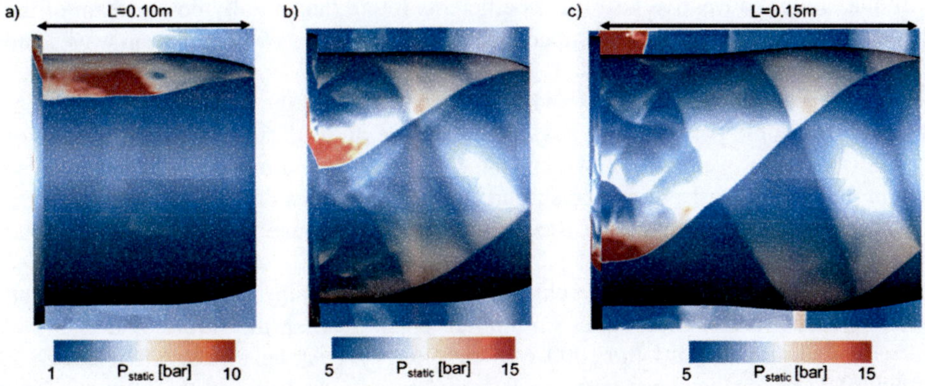

Fig. 2. Static pressure flow field of a) low back pressured RDC with diverging nozzle, b) high back pressured RDC with diverging geometry, c) high back pressured RDC with converging-diverging geometry

A low back pressure (1 bar) and an inlet total pressure of 10 bar and 290K for the premixed reactants were imposed to achieve the required pressure ratio for supersonic exit conditions. Exit conditions are expected to be supersonic and required for supersonic axial, radial, or bladeless turbine designs and will be highlighted in the next section. The same combustor and passage geometry is highlighted in Fig. 2b, but with the outlet pressure increased to 7 bar to model the downstream turbine's effect/blockage. The selected back pressure was chosen to account for a maximum Mach number of 0.6 at the passage's outlet, suitable for accommodating optimized stator end walls such as those proposed by Liu et al. [16]. Plotted in Fig. 2c, the third one depicts the combustor with a long converging-diverging passage (length 0.15 m).

The pressure contour for the back pressured RDCs indicate several reflective shocks occur within the passage with three reflective waves for the converging-diverging diffuser in contrast to the low back pressured RDC in which continuous expansion occurs. Furthermore, the detonation region is significantly shortened for the higher backpressure device. The combustion occurs immediately downstream of the backward-facing step, with all combustion occurring roughly 0.02 m downstream of the backward-facing step and extending to up to 0.05 m for the supersonic passage. Besides, the converging-diverging geometry features three detonation waves at this condition, while the diverging passage supports four detonation waves at increased backpressure. Local separation occurred tangentially upstream of the shock in the low momentum region.

For cooling estimates, convective heat fluxes are calculated through a simulation with isothermal wall boundary conditions (the walls' temperature was set at 800K). The convective heat flux allows for the scaling of heat flux estimates for a range of wall temperatures. Figure 3 plots the instantaneous convective heat flux ($h = \frac{Q}{T_g - T_{wall}}$) of the supersonic nozzle coefficient, more specifically the unwrapped shroud end wall (a) and hub end wall (b) with a wall temperature of 800K and a gas total temperature of 2300K, corresponding to the mass flow averaged total temperature at the exit of the passage). The detonation wave travels across both end walls in the supersonic passage, with maximum convective heat flux coefficients found tangentially downstream of the detonation wave. The diverging supersonic passage features two detonation waves and is visualized by two high heat flux regions.

Figure 3c,d depict the convective heat flux coefficient of the unwrapped converging-diverging subsonic passage with a wall temperature of 800K and a gas temperature of 2200K (mass flow averaged outlet total temperature). Maximum heat fluxes are observed within the detonation front for the shroud (Fig. 3c), whereas the hub (Fig. 3d) features a more constant convective heat flux as the detonation wave rides across the shroud (Fig. 3c).

The spanwise-averaged convective heat flux for the supersonic nozzle is around 2000 W/K/m^2 with higher values on the hub end wall than the shroud end wall and decreases towards the outlet to 1000 W/K/m^2 (Fig. 4). Peak heat fluxes are retrieved at around 0.04 m for the supersonic combustor passage, which lies in the aft part of the combustion region. The spanwise integrated heat flux coefficient indicates that the hub and shroud have similar heat loads, with 94 kW for the hub and 100 kW for the shroud. The required coolant heat load to keep the end walls at 800K is 200 kW.

In contrast to the supersonic passage, the convective heat flux coefficient for the converging-diverging downstream of the combustion region remains constant at about 2000 W/K/m^2 before decreasing downstream of the throat. Higher heat fluxes are noted in the combustion region compared to the supersonic case due to the higher operating static pressure. Additional cooling is required for longer diffusing passages.

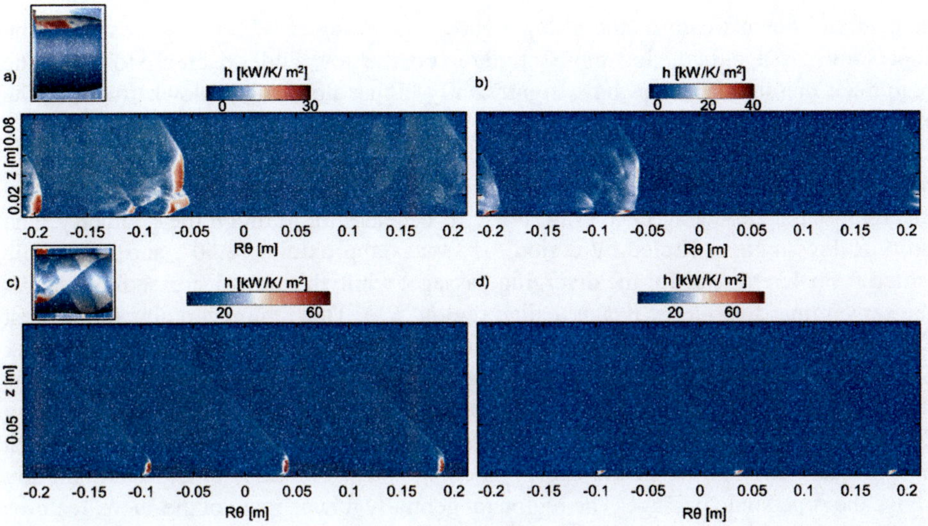

Fig. 3. Instantaneous convective heat flux coefficient for the supersonic passage on the shroud (a), hub (b), and for the subsonic passage with converging-diverging passage shroud (c) and hub (d)

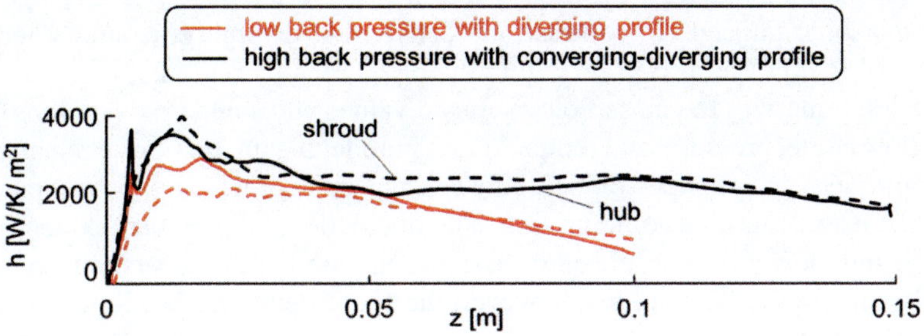

Fig. 4. Average convective heat flux for the diverging geometry and converging-diverging geometry

3.2 Spatio-Temporal Evolution at the Outlet of the Passage

Figure 5a plots the radial mass flow averaged outlet Mach number across the spanwise location (θ). The results demonstrate that the desired supersonic exit conditions for the diverging geometry (solid lines) are indeed achieved for high mass flows, with a mass flow averaged Mach number at the exit of around 1.65. In contrast, the mass flow averaged Mach number for the high back pressured RDC is 0.29 for the diverging passage and 0.36 for the converging-diverging passage. The converging-diverging passage features higher peak Mach numbers of 0.6 due to a lower area passage ratio compared to the diverging passage. Mach numbers of only 0.4 are reached for the diverging geometry at increased backpressure, owing to the more extensive diffusion for the diverging geometry. Mach number is below 0.1 tangentially upstream of the shock, caused by mainly stagnant swirling flow moving at the speed of the oblique shock. In terms of flow angle (Fig. 5b), tangentially downstream of the oblique shock, flow angles of 20° are present in the supersonic passage. In the low momentum region, the flow angles decrease to –20°. The high back pressured devices have significantly higher flow angles downstream of the shock (~40°) than the supersonic passage. Although those maximum flow angles are similar and dictated by the oblique shock, the flow angle decay differs between the two passages, with higher flow angle decay for the diverging passage than the converging-diverging passage, owing to the difference in deceleration through the geometry. The drop in flow angle encircled by region "1", spans approximately 50°, and flow angle varied from 40° to –30° for the diverging passage, while this only decreased to ~–5 for the converging-diverging passage (called region "2"). This region contains the highest energy and is computed by the local total enthalpy ($\dot{m}H_{total}$), visualized in Fig. 5c. Peak enthalpy resides tangentially downstream of the rotating shock. The red line depicts the mass-flow averaged total enthalpy ($\dot{m}H_{total}$), and 50% of the flow's total energy is contained within 25% of the circumference for the supersonic passage. For both backpressure cases, 50% of the energy is contained within 33% of the span, slightly above the supersonic passage. The region tangentially downstream of the shock features low momentum/enthalpy with locally negative values tangentially upstream of the shock. The precise identification of regions with high enthalpy is critical for the nozzle guide vane inlet metal angle selection. The radially mass flow averaged total pressure (Fig. 5d) indicates lower local total pressures due to expansion at the exit of the supersonic passage. The total pressure fluctuations are 160% (min-to-max) of the mass flow averaged value for the supersonic passage. In contrast, the converging-diverging geometry features a fluctuation of ~100% (min-to-max) within the high enthalpy region.

Table 1 summarizes the mass flow averaged values and standard deviation of Mach number, flow angle, pressure, and temperature to model the fluctuations emanating from a supersonic passage to a supersonic turbine configuration or from a subsonic passage as an inlet to transonic turbine configuration. The fluctuations of pressure and temperature are significantly lower for the subsonic passage compared to the supersonic passage. To accurately capture the fluctuations, however, the spatio-temporal profile is required, as shown in Fig. 5.

3.3 Impact of the Back Pressure on Pressure Gain

The impact of the back pressure on the losses of the combustor-passage is plotted in Fig. 6 as a function of the mass-flow averaged Mach number and total pressure loss at the passage's outlet. For low back pressured devices (required for supersonic bladed and bladeless concepts), Mach numbers can reach up to Mach 1.65 for a diverging geometry at the expense of a total pressure loss of 55% for this injector geometry. A significant pressure loss occurs through the injection, with a drop of about 20% downstream of the backwards-facing step. For high back pressured devices, the mass flow averaged Mach number decreases with a consequent reduction in total pressure loss at the exit of the passage (around 20%). Kaemming and Paxson [24] observed similar phenomena in which pressure gain was increased for higher back pressured devices for a fixed injection geometry due to the lower flow speeds across the passage.

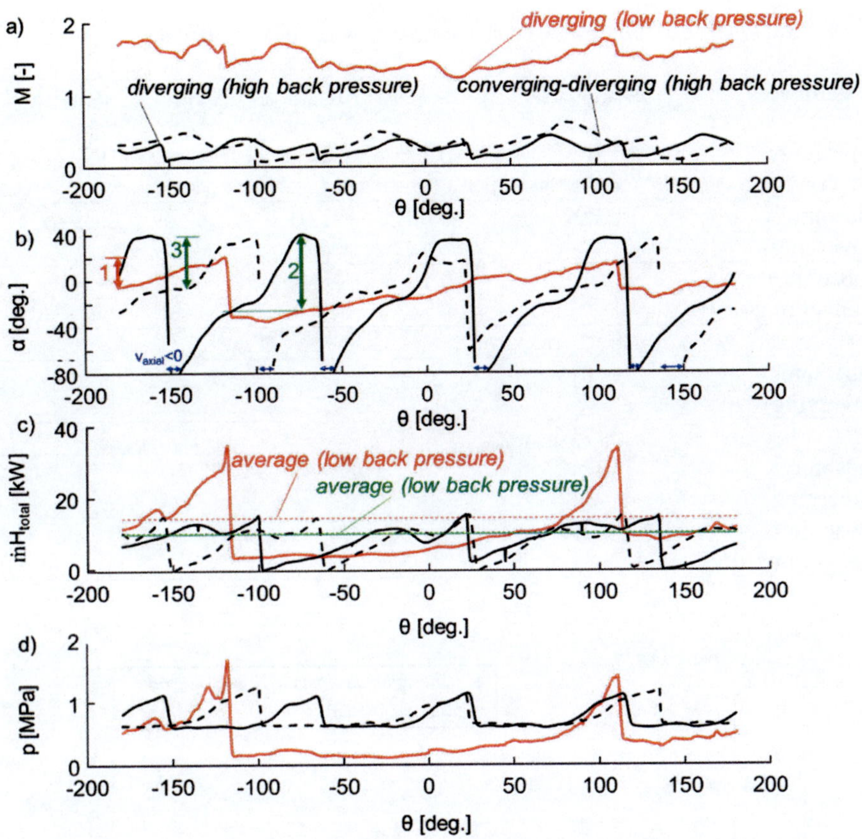

Fig. 5. Radially mass flow averaged quantities as a function of the span at the outlet of the passage: a) Mach number, b) flow angle, c) $\dot{m}H_{total}$ c) and d) total pressure

3.4 Pressure Loss Across the Transition Passage

The total pressure drop attributed to the supersonic or subsonic passage is investigated by isolating the passage downstream of the combustion zone. This is 0.025 m for the diverging subsonic passage, 0.03 m for the converging-diverging subsonic passage, and 0.065 m for the supersonic passage, as sketched in Fig. 7a. The total pressured drop across the supersonic passage is around 13%, and this pressure drop is dependent on the oblique shock strength and rotational speed. Figure 7b plots the pressure drop as a function of axial length for the isolated passage with similar total pressure drops for the two subsonic passages but with a different decay due to the difference in length and curvature. The pressure drop for the converging-diverging passage is around 25%. Interestingly, the pressure loss decay of all passages is similar. Table 2 summarizes the losses of the respective passages.

Table 1. Mass-flow averaged characteristics with their standard deviation

geometry	$M_{massflow\ av.}$	M_{STD}	$\alpha_{massflow\ av.}$ [deg]	α_{STD} [deg]
supersonic (diverging)	1.6	~0.17	-2	15
subsonic (diverging)	0.29	~0.08	-1.01	~39
Subsonic (converging-div)	0.36	~0.13	-1.44	~34
	$p_{s,\ std.}/p_{s,\ massflow\ av.}$		$T_{std.}/T_{s,,massflow\ av}$	
supersonic (diverging)	~50%		~13%	
	$p_{0,\ std.}/p_{massflow\ av.}$		$T_{0,\ std.}/T_{massflow\ av}$	
subsonic (diverging)	~22%		~4.7%	
subsonic (converging-div)	~22.4%		~4.8%	

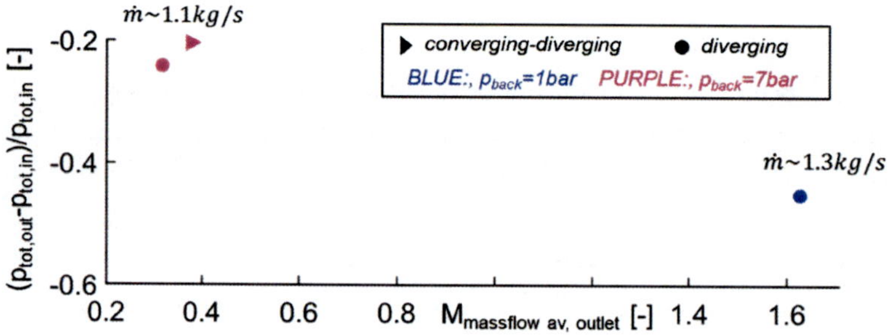

Fig. 6. Total pressure gain as a function of mass flow averaged Mach number for the investigated geometries

4 Assessment of the Chemistry Effects Across the Passage

4.1 Passage Inlet Profile Defined by the Combustor

The inlet of the transition elements is visualized in Fig. 8 as a function of the span to allow for its precise characterization. The location was determined based on the maximum mass-flow averaged total temperature, which occurred at an axial location of 0.025 m downstream of the combustor inlet for the diverging subsonic passage, 0.03 m for the converging-diverging subsonic passage, and 0.065 m for the diverging supersonic passage. Due to the higher expansion occurring in the supersonic passage, a lower total pressure plateau is reached; however, peak total pressures for the three passages are similar (around 20 bar).

Higher maximum total temperatures are achieved for the supersonic profile, compared to the two subsonic passages (Fig. 8b), which could be attributed to the difference

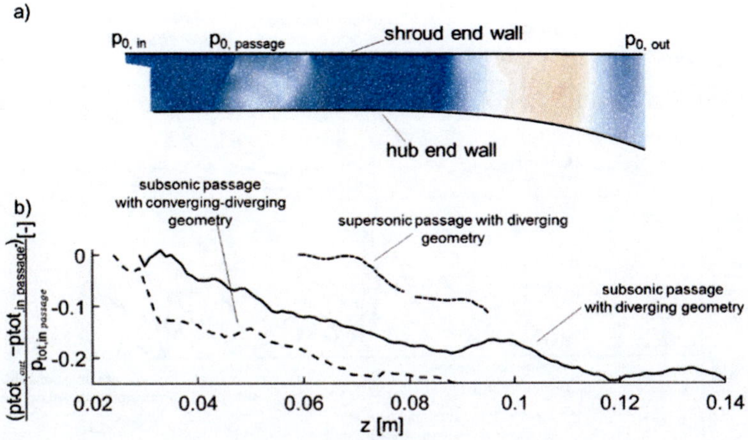

Fig. 7. a) Cross-section of the combustor-passage, and b) mass flow averaged total pressure loss across the isolated passage downstream of the combustion region

Table 2. Summary of the pressure losses for the isolated passages

	Subsonic passage (diverging geometry)	Subsonic passage (diverging geometry)	Subsonic passage (converging-diverging geometry)
Passage length without combustion	0.04 m	0.12 m	0.075 m
Pressure loss across the passage	12%	24.4%	25%

in static temperature upstream of the reactants ahead of the detonation with averaged injection speeds around Mach 1 and 240K of static temperature. By contrast, in the supersonic passage reactants travel at Mach 1.7 with static temperatures of 180K. The Mach number (Fig. 8c) is significantly higher for the supersonic passage, indicative of the combustion process occurring at high flow speeds. From the total enthalpy ($\dot{m}H_{total}$, Fig. 8e), 50% of the flow enthalpy lies within 31% of the span for the subsonic passage and is similar to the passage exhaust profile. In comparison, this is 22% for the supersonic passage. The maximum flow angle (Fig. 8d) for the subsonic passage is between 20° and 30°, while the supersonic passage is 40°. Consequently, flow angles decrease throughout the supersonic passage while they increase for the subsonic passage. Local regions with flow angles below –90° are found caused by areas of reversed flow.

Table 3 shows the mass flow averaged values sampled downstream of the combustion region. Compared to the subsonic passages, the supersonic passage features the highest total enthalpy (computed with an averaged specific heat, c_p, 1700 J/kgK).

Finally, this information is employed to model and isolate the transition passage between combustor and turbine without chemistry, as sketched in Fig. 9. The onset and boundary conditions of the non-reacting passage depend on the backpressure. Modeling

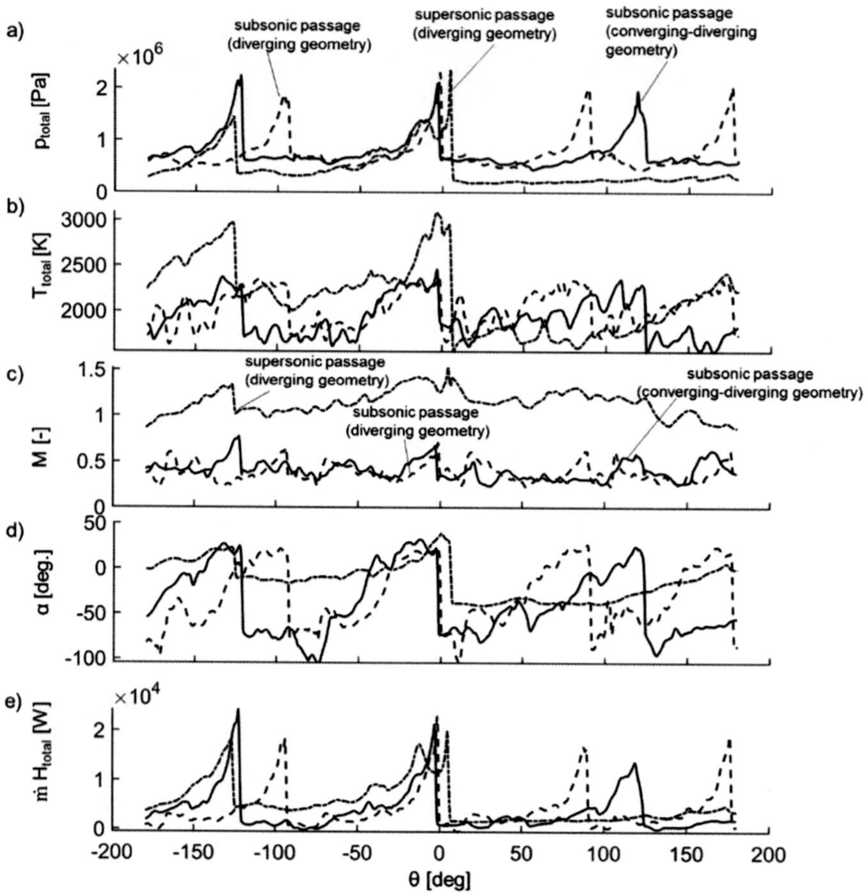

Fig. 8. Radially mass-flow averaged profiles downstream of the combustor and at the inlet of the transition passage: a) total pressure, b) total temperature, c) Mach number, d) flow angle, e) $\dot{m}H_{total}$ [28]

Table 3. Mass flow averaged inlet characteristics and total enthalpy at the onset of the non-reactive passage

	M [–]	α [deg]	p_0 [bar]	T_0[K]	$\dot{m}H_{total}$ [MW]
Supersonic passage (diverging)	1.17	–3.2	6.3	2300	5.13
Subsonic (diverging)	0.44	–6.9	10.4	2130	3.71
Subsonic (converging-diverging)	0.47	–9.14	10.6	2070	3.97

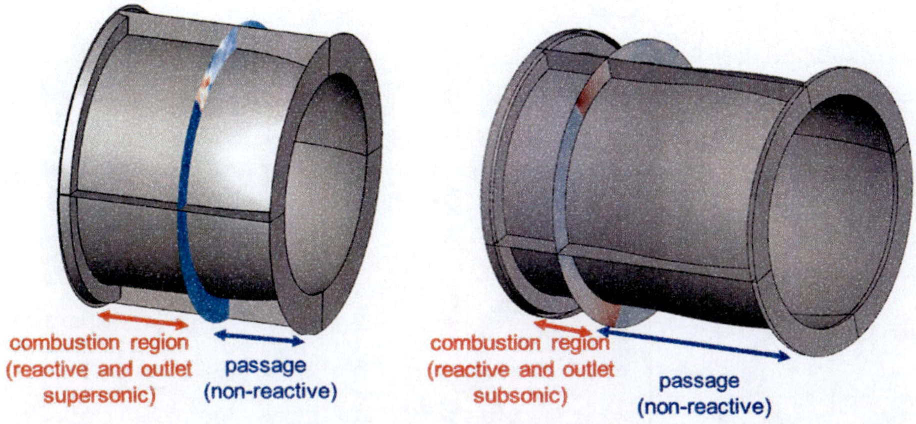

Fig. 9. Numerical domain for the supersonic passage (left) and the subsonic passage (right)

of isolated accelerating passages for RDCs has already been discussed in [23]. Schwer employed a similar method for modeling the plenum upstream of the RDC [25], and Rankin et al. modeled a converging-diverging nozzle to reduce the periodic fluctuations [26].

4.2 Simulation of the Supersonic Passage Without Chemistry

Figure 10a presents the boundary conditions to define the supersonic passage and consist of static pressure, a static temperature, an axial and tangential velocity which are functions of space and time. Those profiles are applied in a 'passage only' simulation (Fig. 10b); this approach was already successfully used [27]. In Fig. 10c the mass flow averaged total pressure loss within the passage from the three-dimensional URANS simulation without chemistry is compared to the reactive simulation of the combustor with comparable pressure loss decay rate over the axial distance.

4.3 Simulation of the Isolated Diffusing Passage Without Chemistry

For the subsonic passage, the total pressure, total temperature, and velocity direction (flow angles) downstream of the combustion region of the reactive simulations are imposed as a total pressure and temperature profile with flow angle profile (unsteady in time and space) in a non-reactive unsteady three-dimensional simulation. This non-reactive simulation (with ~6 million grid points) has a calculation time of around 48 h on two High-Performance nodes (Intel Xeon-E5 processors) compared to the coupled simulations, which require about 500 h on six High-Performance nodes. The contours of Fig. 11a,b,c represent the boundary conditions at the inlet interpolated on a 27 (radial) by 1200 grid. This fine interpolation allows an accurate rebuilt of the combustor outlet profile.

The mass flow averaged total pressure is extracted at each axial location. A 25% mass flow averaged total pressure decrease was measured throughout the passage, comparable

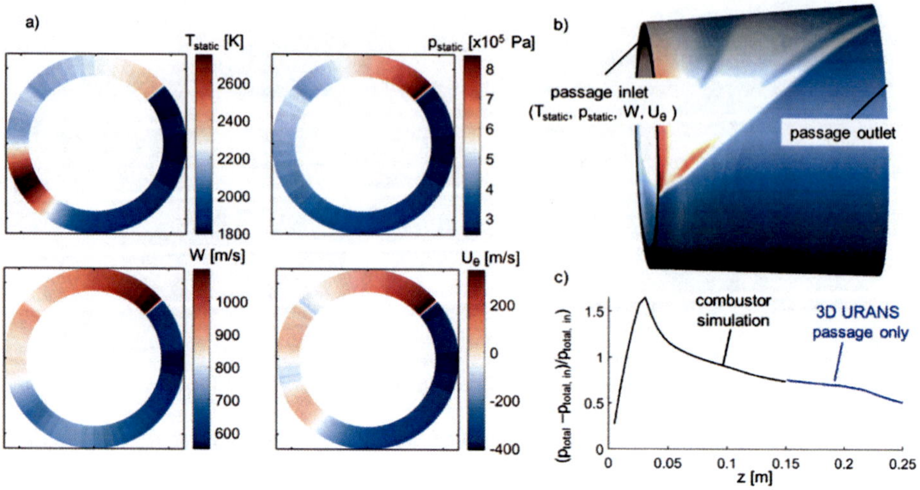

Fig. 10. Modeling of supersonic passage without chemistry: a) inlet boundary conditions, b) numerical domain with pressure signature, c) total pressure loss across the combustor via chemistry simulation, and the passage only simulation without chemistry

to the simulations with combustion (Fig. 11d). Additionally, the temporal evolution at three distinct axial locations from the 3D URANS with chemistry was added, indicating that the subsonic passage simulation without chemistry is within 2% of the combustor-diffuser simulations. The mass-flow averaged Mach number (Fig. 11e) at the subsonic passage outlet decreased from 0.37 to 0.32, and although local variations occur in the diffuser-only simulations, the outlet conditions were matched.

5 Conclusion

We propose design considerations through a multi-pronged approach to assess and model the high-speed transition elements downstream of the combustion region of a rotating detonation combustor. This requires the estimated pressure ratio across the combustor and transition element to assess flow characteristics, loss, and heat flux across combined combustor and transition elements. Second, we dissect the outlet conditions of the passage in terms of Mach number, flow angle, total pressure, and local enthalpy content, which are the four critical input parameters for the turbine design. The final step consists of modeling the transition passage through fast simulations without chemistry by imposing static quantities and velocities for a supersonic passage, or total conditions and flow angles for a subsonic passage based on one expensive reactive simulation.

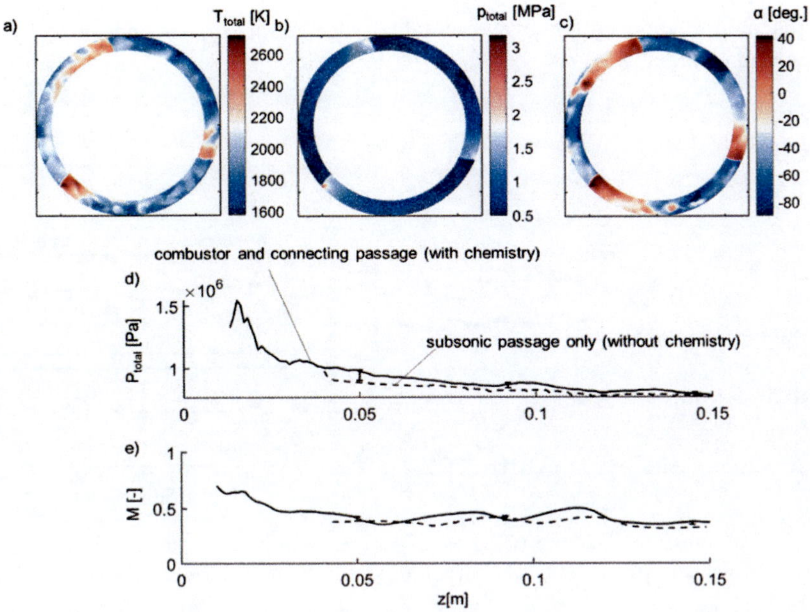

Fig. 11. Verification of the subsonic passage without chemistry vs. the 3D URANS with chemistry for the converging-diverging geometry: a) total inlet temperature, b) total inlet pressure, c) flow angle, d) mass-flow averaged pressure drop along the axial length, e) mass-flow averaged Mach number along the axial length [28].

Specifically, a supersonic passage with a target Mach number of 1.6 and two subsonic passages with an outlet Mach number of 0.6 are analyzed, both with the injector from Purdue's Turbine High-Pressure Optical RDC. The combustors are characterized through three-dimensional unsteady Reynolds Averaged Navier Stokes (URANS) simulations for a stoichiometric hydrogen-air mixture with a one-step reaction mechanism. We observed that the combustor pressure ratio significantly altered the combustion. Low back pressures resulted in a supersonic passage for diverging geometries, with the combustion zone covering 60% of the passage length. This resulted in complete supersonic flow across the span. The combustion zone was reduced to 20% for the subsonic passage with higher backpressure, and mass-flow averaged Mach numbers of around 0.32 were obtained. Significant differences were observed for the different passages concerning the peak Mach numbers and flow angle variation in the flow's high enthalpy region. The total pressure drop throughout the isolated subsonic passage downstream of the combustion region without chemistry was 25%, while this was around 12% for the nozzle, although decay rates were similar. The pressure drop across the injector and combustion zone was higher for the supersonic passage, resulting in an overall more considerable total pressure loss. The total pressure, total temperature, and flow angle profile at the exit of the combustor for the two subsonic passages with the same combustor inlet-to-outlet pressure ratio shared similar features. Those profiles were imposed for the isolated subsonic passage without chemistry. A similar mass flow averaged total pressure signature across the axial length was obtained, with a tenfold reduction in computational time

[28]. This method, independent of the actual passage geometry, can be used to optimize the transition element.

Appendix

See Fig. 12.

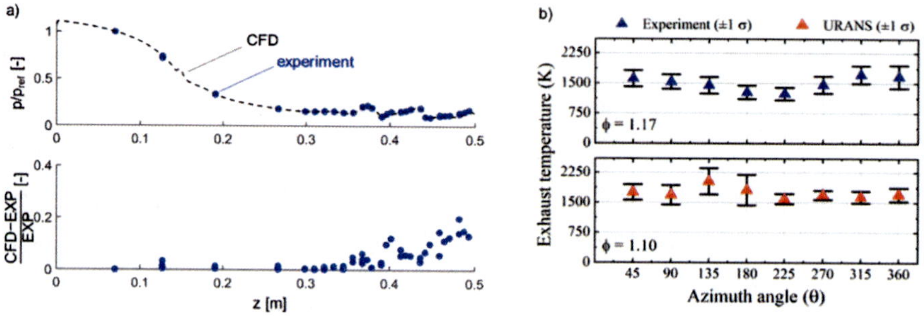

Fig. 12. a) Wall pressure signature (CFD vs. experiment) for Mach 2 supersonic wavy surface [13], b) experiments (top) vs. 3D URANS (bottom) temperature measurements at the exit of the THOR RDC [21]

References

1. Roy, A., Bedick, C.R., Ferguson, D.H., Sidwell, T., Strakey, P.A.: Investigating instabilities in a rotating detonation combustor operating with natural gas–hydrogen fuel blend—effect of air preheat and annulus width. J. Eng. Gas Turbines Power **141**(11) (2019)
2. Braun, J., Saracoglu, B.H., Paniagua, G.: Unsteady performance of rotating detonation engines with different exhaust nozzles. J. Propuls. Power **33**(1), 121–130 (2017)
3. Anand, V., Gutmark, E.: Rotating detonation combustors and their similarities to rocket instabilities. Prog. Energy Combust. Sci. **73**, 182–234 (2019)
4. John, Z.M., et al.: Recent progress, development trends, and consideration of continuous detonation engines. AIAA J. 1–59 (2020)
5. Schwer, D., Kailasanath, K.: Numerical investigation of the physics of rotating-detonation-engines. Proc. Combust. Inst. **33**(2), 2195–2202 (2011)
6. Frolov, S.M., Dubrovskii, A.V., Ivanov, V.S.: Three-dimensional numerical simulation of the operation of a rotating-detonation chamber with separate supply of fuel and oxidizer. Russ. J. Phys. Chem. B **7**(1), 35–43 (2013)
7. Cocks, P.A.T., Holley, A.T.: High Fidelity Simulations of a Non - Premixed Rotating Detonation Engine, pp. 1–18 (2016)
8. Pal, P., Xu, C., Kumar, G., Drennan, S.A., Rankin, B.A., Som, S.: Large-eddy simulations and mode analysis of ethylene/air combustion in a non-premixed rotating detonation engine. In: AIAA Propulsion and Energy 2020 Forum, pp. 1–12 (2020)
9. Sato, T., Chacon, F., White, L., Raman, V., Gamba, M.: Mixing and detonation structure in a rotating detonation engine with an axial air inlet. Proc. Combust. Inst. **000**, 1–8 (2020)
10. Bach, E., Stathopoulos, P., Paschereit, C.O., Bohon, M.D.: Performance analysis of a rotating detonation combustor based on stagnation pressure measurements. Combust. Flame **217**, 21–36 (2020)

11. Asli, M., Stathopoulos, P., Paschereit, C.O.: Aerodynamic investigation of guide vane configurations downstream a rotating detonation combustor. J. Eng. Gas Turbines Power (2020)
12. Liu, Z., Braun, J., Paniagua, G.: Characterization of a supersonic turbine downstream of a rotating detonation combustor. J. Eng. Gas Turbines Power **141**(3), 031501 (2018)
13. Braun, J., Paniagua, G., Falempin, F., Le Naour, B.: Design and experimental assessment of bladeless turbines for axial inlet supersonic flows. J. Eng. Gas Turbines Power **142**(4) (2020)
14. Inhestern, L.B., Braun, J., Paniagua, G., Serrano Cruz, J.R.: Design, optimization, and analysis of supersonic radial turbines. J. Eng. Gas Turbines Power **142**(3), 1–12 (2020)
15. Liu, Z., Braun, J., Paniagua, G.: Integration of a transonic high-pressure turbine with a rotating detonation combustor and a diffuser. Int. J. Turbo Jet-Engines (2020)
16. Liu, Z., Braun, J., Paniagua, G.: Thermal power plant upgrade via a rotating detonation combustor and retrofitted turbine with optimized endwalls. Int. J. Mech. Sci. **188**, 105918 (2020)
17. Chakravarthy, S., Peroomian, O., Goldberg, U., Palaniswamy, S.: The CFD++ computational fluid dynamics software suite. SAE Tech. Pap. Ser. **1** (2010)
18. Fernández-Galisteo, D., Sánchez, A.L., Liñán, A., Williams, F.A.: One-step reduced kinetics for lean hydrogen-air deflagration. Combust. Flame **156**(5), 985–996 (2009)
19. Saavedra, J., Paniagua, G., Lavagnoli, S.: On the transient response of the turbulent boundary layer inception in compressible flows. J. Fluid Mech. **850**, 1117–1141 (2018)
20. Braun, J., Sousa, J., Paniagua, G.: Numerical assessment of the convective heat transfer in rotating detonation combustors using a reduced-order model. Appl. Sci. **8**(6), 893 (2018)
21. Athmanathan, V., et al.: "Femtosecond/picosecond rotational coherent anti-Stokes Raman scattering thermometry in the exhaust of a rotating detonation combustor. Combust. Flame **231**, 111504 (2021)
22. Athmanathan, V., et al.: Turbine-integrated high-pressure optical RDE (THOR) for injection and detonation dynamics assessment. In: AIAA Propulsion and Energy 2019 Forum, pp. 1–15 (2019)
23. Braun, J., Saavedra Garcia, J., Paniagua, G.: Evaluation of the unsteadiness across nozzles downstream of rotating detonation combustors. In: 55th AIAA Aerospace Sciences Meeting, pp. 1–13 (2017)
24. Kaemming, T.A., Paxson, D.E.: Determining the pressure gain of pressure gain combustion. In: 2018 Joint Propulsion Conference, p. 4567 (2018)
25. Schwer, D.A., Kaemming, T.A., Kailasanath, K.: Pressure feedback in the diffuser of a ram-RDE propulsive device. In: 55th AIAA Aerospace Sciences Meeting (2017)
26. Rankin, B.A., Hoke, J., Schauer, F.: Periodic exhaust flow through a converging-diverging nozzle downstream of a rotating detonation engine. In: 52nd Aerospace Sciences Meeting, pp. 1–12 (2014)
27. Braun, J., et al.: Characterization of an integrated nozzle and supersonic axial turbine with a rotating detonation combustor. In: AIAA Propulsion and Energy 2019 Forum, pp. 1–11 (2019)
28. Braun, J., Paniagua, G., Ferguson, D.: Aero-thermal characterization of accelerating and diffusing passages downstream of rotating detonation combustors. In: ASME Turbo Expo 2021, GT2021–59111

Part IV: Numerical Methods

Linear Forcing of Compressible Isotropic Turbulence in Rectangular Domains with Adapted Locally Refined Grids

Mario Sroka(✉) and Julius Reiss

Technische Universitaet Berlin, Mueller-Breslau-Strasse 15, 10623 Berlin, Germany
Mario.Sroka@tnt.tu-berlin.de, Julius.Reiss@tnt.TU-Berlin.DE

Abstract. This study investigates the applicability of the linear forcing method at rectangular domains with an adapted grid via local refinement. The advantages of the linear forcing method, using in a physical space solver for combustion simulations, are discussed. We present test cases for the different modifications of the forcing term and the major drawbacks occurring when using non-cubic domains. The use of a filtered velocity field within the forcing term is investigated, first as a solution for the described problems with rectangular domains and second as an attractive method to control the integral length scale of the turbulent field. Finally, we present results for various DNS computations in preparation for future studies of turbulence-flame interactions, and a few statistical properties of the turbulence are discussed.

Keywords: Compressible Navier-Stokes · Isotropic turbulence · Linear forcing · Adaptive locally refined block-structured grid · DNS

1 Motivation and Introduction

Flames are in most technical applications turbulent. The interaction between flames and turbulence is decisive for most of the flame properties. Thereby the study of this phenomenon is a central aspect of reactive flow research.

Previous investigations of a pulsed detonation combustor (PDC) at the TU Berlin revealed turbulence-flame interactions in an interesting and little-studied regime. [2] The turbulence is determined due to inflow conditions and geometry of the combustion chamber. A propagating flame creates a shock wave, which triggers the desired detonation in a focal point, making the turbulent flame acceleration decisive for the whole process. The oxygen-enriched hydrogen/air mixture combustion takes place under elevated pressure, and we observe a highly accelerated turbulent flame front. The strength of experimentally observed shock waves allows us to estimate the main properties of this strongly accelerated and compressible flame under elevated pressure. Note that the turbulence is not used to initialize the detonation (DDT), which is forced due to shock focusing.

A generic turbulent field needs to be produced and maintained with adequate parameters to investigate such turbulence-flame interactions numerically. The

R. King and D. Peitsch (Eds.): AFCC 2021, NNFM 152, pp. 187–202, 2022.
https://doi.org/10.1007/978-3-030-90727-3_12

characteristic turbulence parameters depend on the inflow conditions and the geometry of the combustor and are not fully known from the experiment.

If we take a look at a specific turbulent premixed combustion in the Borghi diagram (Fig. 1), we can that the two most important turbulent parameters are the RMS-velocity u_{rms} and the integral length scale l_t. In previous experiments, we observed a turbulent flame speed up to $50\frac{m}{s}$, which corresponds to an at least equal-sized RMS-velocity. For the cases from the experiments, the turbulence-flame interaction occurs at a turbulent Mach number of $\mathrm{Ma}_t = 0.12$ (for hydrogen/air mixture with equivalence ratio $\Phi = 1$ at $p = 1$ bar). Due to quenching effects, the RMS-velocity could even be much larger and exceed $\mathrm{Ma}_t > 0.12$. Under these conditions, the combustion will occur in the thin or broken reaction zone regime. Both regimes are from a numerical perspective interesting research areas. At higher pressures, the laminar flame speed increases while the flame thickness becomes smaller. This leads to a shifted (to the lower right corner) area of investigation in Fig. 1, but the above assumptions about the turbulent Mach number should be valid even for those cases.

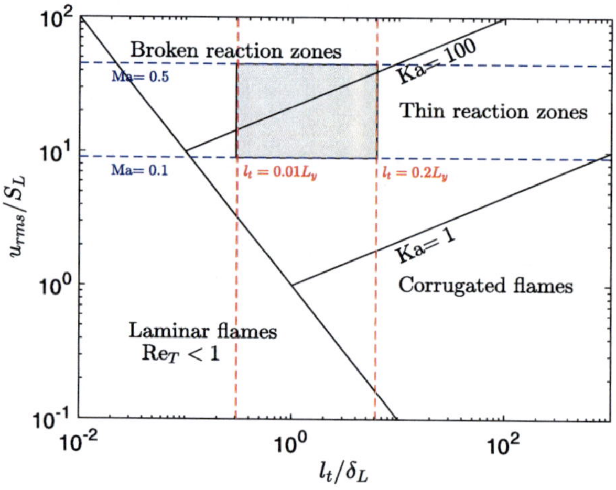

Fig. 1. Borghi plot to illustrate the area of investigation inside the premixed combustion diagram, fixed parameters for an example DNS computation: diffusive flame thickness δ_L and laminar flame speed s_L for a premixed H_2/air flame with equivalence ratio $\Phi = 1$ at $p = 1$bar, integral length scale l_t as part of the domain size $L_y = 2$ mm, turbulent Mach number is at least $\mathrm{Ma}_t = 0.1$

As mentioned, to investigate the turbulence-flame interaction at a specific point in the Borghi diagram, a turbulent field with a certain RMS-velocity and integral length scale is required. Therefore we need a turbulent forcing method capable of maintaining a turbulent field at these two parameters. Furthermore, the forcing method needs to work in the physical space because of the spatial formulation of our reactive DNS (direct numerical simulation) code. The utilization of locally adapted grids hinders the application of Fourier based methods.

For a Fourier transformation, we would have to interpolate the entire field to the finest grid level, which is prohibitive for high resolutions. And finally, the non-cubic domains cause difficulties in controlling the integral length scale, as discussed below.

A popular forcing method in physical space is proposed by Lundgren [11]. This method (Eq. 2) is easy to implement and depends only on local velocity values, which is a great advantage for adaptive meshes. Lundgren shows that his method is capable of computing statistically stationary turbulence with correct physical properties. Carroll and Blanquart [6] modified the linear forcing term to reduce fluctuations and therefore achieve a shorter transient time before getting statistically stationary turbulence. A further modification was proposed by Bassenne et al. [1] (Eq. 3). With this term, the statistically stationary state is reached even faster, and specific target values (turbulent kinetic energy or dissipation) can hold at nearly constant values. Similar to the original forcing term, it also works very well on adapted meshes. Global turbulence properties, used as inputs for this forcing term, can easily be computed on adapted grids with low computational costs. To summarize, the linear forcing can maintain specific turbulent kinetic energy (RMS-velocity), which is essential to research turbulence-flame interactions. In Sect. 3, we discuss the advantages and disadvantages of these different approaches for our specific problem.

The second important turbulent property is the integral length scale. Rosales and Meneveau [15] studied the differences between the linear forcing and the classical pseudospectral method. The pseudospectral forcing usually works with a band-limiting (low-wavenumber) approach, while the linear forcing term increases the turbulent kinetic energy at all wavenumbers. Rosales and Meneveau find that the integral length scale with Lundgren's method is approximately $l_t \approx 0.2L$ of the domain size L. Band-limited forcing yields $l_t \approx 0.4L$. Palmore and Desjardins [12] use a filtered velocity field to increase the reachable Taylor microscale Reynolds number. They use a low-pass filter on the velocity field and demonstrates that this method keeps the simplicity of implementing the linear forcing and allows the control of the Taylor microscale, which corresponds to controlling the integral length scale. In the work of Ketterl and Klein [9] another usage of a filtered velocity field can be found. The authors use the high-pass filtered velocity to control the integral length scale and turbulent kinetic energy fully. This approach offers the opportunity to study small-scale turbulence-flame interactions. Note that all of these works use cubic domains, which is not practical for researching flame fronts.

All of the above-discussed methods are developed for incompressible flows. Petersen and Livescu [13] show that the linear forcing method is capable of maintaining statistically stationary turbulence in a compressible flow. However, they also show that the original Lungren term is insufficient to control the ratio of dilatational to solenoidal kinetic energies and dissipation. Therefore a stationary state is not reachable. Petersen and Livescu propose splitting the forcing term, which needs a Fourier transformation of the flow field, making its application difficult on adapted grids. But Petersen and Livescu also report that at higher turbulent Mach number (in [13] greater than $\mathrm{Ma}_t = 0.1$), the ratio becomes more stable, but the long-time behavior is difficult to be predicted. We observe equivalent results at higher Mach numbers, and the dilatational to solenoidal ratio is very stable. Several other authors research compressible turbulence, e.g. [5,8,18]. Most authors do not use a linear forcing approach and study turbulence in cubic domains with a fixed mesh. Independently from the specific forcing, there exists statistically turbulence properties, for example, in terms of Mach number scaling, which can be used as validation.

In this report, we investigate turbulence forcing for a rectangular domain with a locally adapted mesh. For the designated investigation of compressible turbulent flames, the control of the RMS-velocity and the integral length scale (independently of the domain size) of compressible turbulence is necessary. For this, a spatial forcing with a quick convergence to a stationary state combined with a real-space filter is proposed.

The paper is organized as follows. In Sect. 2, we give a short overview of the most important equations and implementation details. Section 3, presents different test cases and discusses the advantages and disadvantages of the linear forcing method for turbulence-flame interactions studies. To this end, we focus on the requirements in terms of usage of a rectangular domain with an adapted grid. Finally, Sect. 4 gives a short outlook to a few computations in preparation for future turbulence-flame interactions. We take a closer look at important statistical properties.

2 Basis Equations and Numerical Details

2.1 Governing Equations

We study isotropic turbulence in a compressible inert gas, which is governed by the three-dimensional compressible Navier-Stokes equations in skew-symmetric form [14]. These equations are

$$\begin{aligned}\sqrt{\varrho}\partial_t\sqrt{\varrho} + \frac{1}{2}\partial_{x_i}\left(\varrho u_i\right) &= 0 \qquad (1)\\ \sqrt{\varrho}\partial_t\left(\sqrt{\varrho}u_i\right) + \frac{1}{2}\left(\partial_{x_j}\left(\varrho u_j u_i\right) + \varrho u_j \partial_{x_j} u_i\right) &= -\partial_{x_i} p + \partial_{x_j}\tau_{ij} + f_i\\ \frac{1}{\gamma-1}\partial_t p + \frac{\gamma}{\gamma-1}\partial_{x_i}\left(u_i p\right) &= \partial_{x_i}\left(\lambda\partial_{x_i} T\right) + u_i\partial_{x_i} p\\ &\quad - u_j\partial_{x_j}\tau_{ij} + \partial_{x_i} u_i\tau_{ij} + f_e,\end{aligned}$$

with density ϱ, velocity u_i, pressure p, temperature T, thermal conductivity λ and heat capacity ratio γ. The viscous tensor is $\tau_{ij} = -\frac{2}{3}\mu\partial x_k u_k\delta_{ij} + \mu\left(\partial_{x_j} u_i + \partial_{x_i} u_j\right)$ and we use the equation of state for an ideal gas. f_i and f_e are the turbulent forcing terms and to ensure total energy conservation $f_e = -f_i u_i$, analogous to [13]. The skew symmetric scheme is fully conservative and fluxes can be defined [14], by which it becomes analogous to a FV scheme. Its shock treatment was validated [4,14] .

2.2 Linear Forcing

The original Lundgren [11] forcing term is defined by a constant A and the velocity field u_i

$$f_i = Au_i. \tag{2}$$

The parameter A is an adjustable parameter. Analyzing the stationary turbulent kinetic energy equation, one can identify that A depends on the specific properties of the turbulent field, hence $A = (2\tau)^{-1}$ with eddy-turnover time $\tau = \frac{k}{\varepsilon}$. Due to the choice of A, a specific turbulent state is selected. If for the integral length scale $l_t = 0.2L$ [15] (domain size L) is assumed, the turbulent target values k and ε can be calculated with a fixed viscosity ν. After that, the Kolmogorov length scale is determined, and the necessary grid resolution could be calculated.

Bassenne et al. [1] modified the forcing term as:

$$f_i = \frac{\varepsilon(t) - G(k(t) - k_0)/t_0}{2k(t)} u_i. \tag{3}$$

k_0 is the target value for the turbulent kinetic energy and t_0 is the constant integral time. The parameters k and ε are the current time-dependent values for the turbulent field. All target values can compute similarly to the described method above. G is a dimensionless constant to ensure that the turbulent kinetic energy approaches the target value exponentially [1]. Note that larger values of G produce faster convergence and less fluctuating turbulence target values, but lead to stiffer equations and, therefore, eventually smaller time steps in order to avoid numerical errors.

Another modification of Lundgren's linear forcing approach is proposed by Ketterl and Klein [9]

$$f_i = A\tilde{u}_i. \tag{4}$$

Instead of using the entire velocity field and therefore inject kinetic energy at all velocity modes, Ketterl and Klein use a high-pass filtered field $\tilde{u}_i$. It is computed as the difference $\tilde{u}_i = u_i - \hat{u}_i$, with a typically LES convolution filter $\hat{u}_i$. We combine in our work the forcing term from Bassenne et al. with a filtered velocity analogous to Ketterl and Klein, where $\hat{u}_i$ is computed with a box filter

$$\hat{u}_i = \frac{1}{(2L_f + 1)^3} \sum_{j=-L_f,L_f} u_i(x+j, y+j, z+j), \tag{5}$$

with filter width L_f. This filter is easy to implement and most importantly, due to the filter width, capable of working at different grid refinement levels.

2.3 Compressible Turbulence

One important characteristic of compressible turbulence is revealed when the Navier-Stokes equations are split into a solenoidal and dilatational part with the Helmholtz decomposition. Erlebacher et al. [7] concludes that the solenoidal part corresponds to the incompressible part of the equations. With increasing turbulent Mach number, the dilatational part becomes larger, and therefore compressible effects occur in the flow field. Sakar et al. [16] propose a way to decompose the turbulent dissipation ε into solenoidal and dilatational parts without a full Helmholtz decomposition:

$$\varepsilon = \varepsilon_s + \varepsilon_d = 2\mu\omega_{ij}\omega_{ij} + \frac{4}{3}\mu\left(\partial_{x_k} u_k\right)^2, \tag{6}$$

with $\omega_{ij} = \frac{1}{2}(\partial_{x_i} u_j - \partial_{x_j} u_i)$. This computation of the dissipation ε is necessary to calculate the correct turbulent dissipation rate to adjust the forcing term. The incompressible definitions can be used for all other specific turbulent parameters.

2.4 Technical Details

We use the code WABBIT, which is introduced in [17]. On the RHS subroutine, we discretize Eq. 1 with 4th order schemes in space and time. The time step is adapted to meet the CFL criterium. We use an adaptive shock filter [3] to treat shocks at higher turbulent Mach numbers.

Forcing terms and statistical properties of the flow field are easy to implement because all of these parameters can be computed locally, which means on single blocks. Therefore only a few and simple MPI communications are needed, for example, to gather the statistical data after the local computation. The box filtered velocity is computed only every $n > 50$ time steps, which is sufficient to obtain statistical stationary turbulence. Large filter widths L_f are implemented due to a loop with smaller filter widths and regular ghost nodes synchronization.

3 Test Cases

3.1 Linear Forcing in Cubic Domains

The Lundgren forcing term (Eq. 2) is an attractive first choice, and due to its simple implementation, especially for using with adapted grids and reactive Navier-Stokes equations. With the specific parameter A, the turbulent properties are well determined. The integral length scale l_t correlates with the domain size L [15], and therefore l_t can only change due to a change in L, restricting its applicability. Too small domain sizes are not possible in our designated application, since the domain needs at least a size of a few flame widths. Larger domain sizes are easier to implement in principle, especially with adapted grids. However, these domains still lead to increasing computational costs.

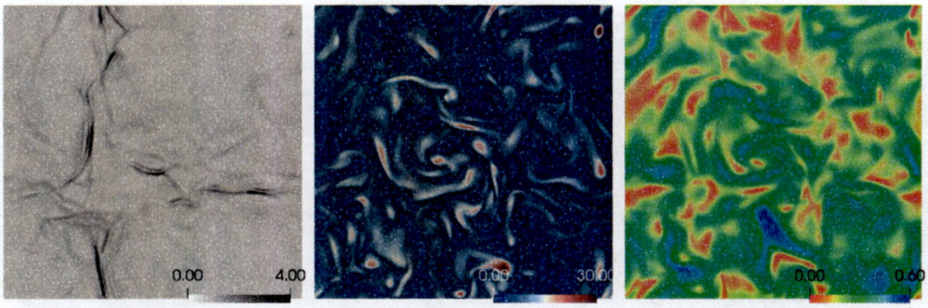

(a) (b) (c)

Fig. 2. Contour plots for a turbulent field with Lundgren forcing approach. (a) density gradient $\nabla\varrho$, (b) vorticity magnitude $|\omega|$, (c) local Mach Number Ma_{loc}, cubic domain $L = 2\pi$, resolution 256^3, fixed viscosity $\nu = 5.8\mathrm{e}{-3}$, $\mathrm{Re}_\lambda = 52$, $\mathrm{Ma}_t = 0.3$, $\varrho_0 = 1$, $p_0 = 16$

In Fig. 2 we show results for a classical Lundgren forcing case for a compressible regime. In the fluctuations of $\nabla\varrho$ shock-like structures dominate the flow field. The strongest shocks are not perfectly resolved, which can be seen from the oscillating (wave-like) structures in front of the shock. Note that this can be fixed by a stronger filtering. The vorticity ω (b) shows smoother structures than the density gradient. The size of these structures corresponds to the integral length scale of the flow. Typical structures like vortex kernels and filaments are well visible. In (c), the local Mach number is shown. The turbulent Mach number, which is calculated with the RMS-velocity, is for this case $\mathrm{Ma}_t = 0.3$. The local Mach number is due to velocity fluctuations much larger. For example from Fig. 11 one can observe, that locally a Mach number $\mathrm{Ma}_{loc} > 1$ is reached for a turbulent Mach number $\mathrm{Ma}_t > 0.3$.

As mentioned above, the linear forcing term can directly be used with adapted grids. A great benefit of this simple forcing term is the applicability to coarser grids, even for dynamically adapted grids, since the forcing parameter does not need to be changed for other grid levels. In Fig. 3 results for coarser grids are shown. It is worth noting that the target turbulent flow would not fit into the coarser grid, which means smaller scales simply not exist (well to observe in Fig. 3 (b–c)). But the statistical properties of the turbulent flow are well approximated, even though the filtering must be stronger for coarser grids. Note, that the modified term after Bassenne (Eq. 3) requires time-dependent values for k and ε. But these can be computed directly at a coarser mesh as well.

Figure 4 shows the time evolution of the statistical properties for $\tau = 60$ eddy-turnover times. The order of magnitude of the fluctuations corresponds to results in the literature. The very small fluctuation of the RMS-velocity u_{rms} for the Bassenne forcing term is clearly visible. The integral length scale l_t is

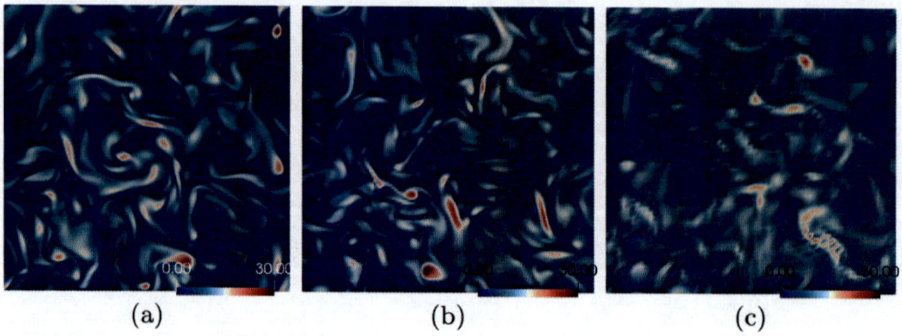

Fig. 3. Contour plots of vorticity magnitude $|\omega|$ for a turbulent field with Lundgren forcing approach. cubic domain $L = 2\pi$ with resolution (a) 256^3, (b) 128^3, (c) 64^3, fixed viscosity $\nu = 5.8\text{e}{-3}$, $\text{Re}_\lambda = 52$, $\text{Ma}_t = 0.3$, $\varrho_0 = 1$, $p_0 = 16$

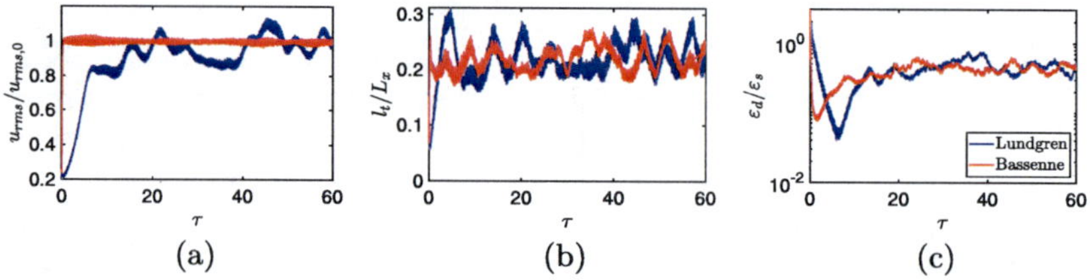

Fig. 4. Time evolution of important turbulent properties with Lundgren and Bassenne forcing approach. (a) RMS velocity u_{rms}, (b) integral length scale l_t, (c) ratio of dilatational to solenoidal dissipation $\frac{\varepsilon_d}{\varepsilon_s}$, cubic domain $L = 2\pi$ with resolution 256^3, fixed viscosity $\nu = 5.8\text{e}{-3}$, $\text{Re}_\lambda = 52$, $\text{Ma}_t = 0.3$, initial density $\varrho_0 = 1$, initial pressure $p_0 = 16$

approximately 0.2 of the domain size L. We observe for both methods (Lundgren, Bassenne) a constant ratio of solenoidal to dilatational dissipation if the turbulent Mach number is sufficiently large enough, as reported in [13]. Generally, in most cases, a constant ratio occurs with $\text{Ma}_t > 0.1$.

With increasing turbulent Mach number, also the fluctuations increase, especially with the original Lundgren forcing term. This can be observed in Fig. 5(a), which shows the accuracy of achieving the target value of the Taylor microscale Reynolds number Re_λ over the turbulent Mach number. For very large Mach numbers $\text{Ma}_t > 0.6$, these fluctuations further increase, and the flow field needs even more filtering. However this Mach number region is not attractive for our turbulence-flame interaction research, because for such large velocity fluctuations a spontaneous DDT could occur, and therefore we would need much more numerical/computational effort. (e.g. to resolve the detonation front) The energy spectra (Fig. 5(b)) show a sensible energy distribution over all modes and an excellent agreement between these two forcing methods.

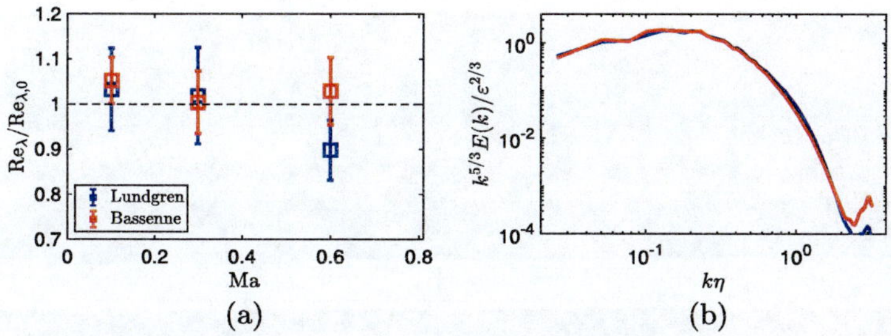

Fig. 5. Turbulent properties with Lundgren and Bassenne forcing approach. (a) Taylor microscale Reynolds number Re_λ, (b) normalized energy spectra, cubic domain $L = 2\pi$, resolution 256^3, fixed viscosity $\nu = 5.8\text{e}{-3}$, $\mathrm{Re}_\lambda = 52$, $\mathrm{Ma}_t = 0.3$, initial density $\varrho_0 = 1$, initial pressure $p_0 = 16$

3.2 Linear Forcing in Rectangular Domains

For the study of flame fronts, a rectangular non-cubic domain is very advantageous. Even a laminar flame needs a specific domain size to reach a stationary state. Turbulent flames typically need even more space to statistically converge since the flame speed is difficult to predict and can thus not fully be compensated with an inflow condition. Pure cubic domains thereby yield high computational costs. Unfortunately, as reported in [10], the linear forcing fails to reach for rectangular domains a statistically stationary state. The impact on the flow field could be very complicated to predict. In some cases, a few properties are in good agreement with the cubic reference case, and therefore the flow field would seem correct. Generally, the non-stationary behavior becomes more apparent with more elongated domains.

In Fig. 6, as an example, the velocity component U_y is shown at different times. In some areas of the velocity field, the fluctuations seem meaningful. But even in the first figure, the dominance of a mode in the elongated direction is observed. And this mode is becoming more dominant over time. The energy spectra in Fig. 7(c) show that the energy is wrongly distributed, especially at the lowest wavenumbers. Due to the energy conservation, the increasing energy in lower wavenumbers leads to decreasing energy at higher wavenumbers, which is very disadvantageous in studying turbulence-flame interactions because the small scales can enter the flame, respectively the reaction zone, and therefore significantly affect the flame speed.

The time evolution of the forcing parameter A is shown in Fig. 7. Note that first A is assumed constant for calculating the necessary parameters in the forcing term in the RHS-subroutine. However, looking at the (with time-dependent k and ε) computed value of A, the difference between the two domain shapes can be observed. For a cubic domain, the target value is reached and fluctuates similarly

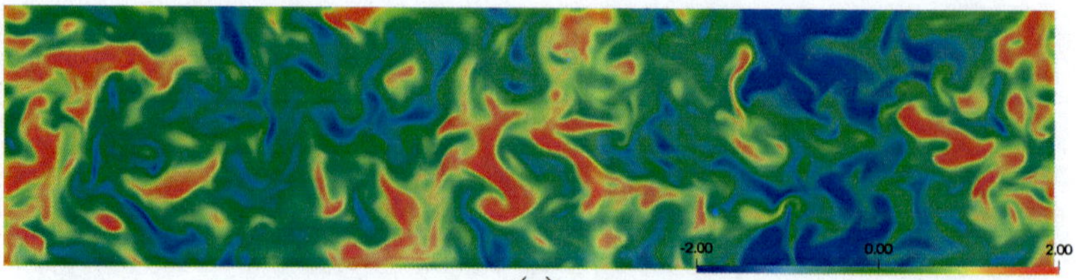

(a)

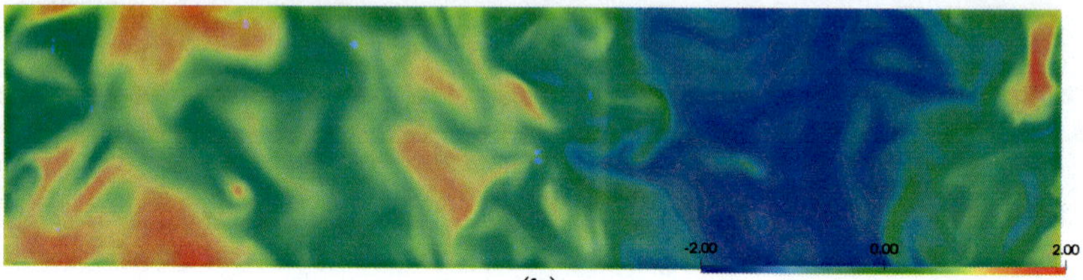

(b)

Fig. 6. Contour plots for velocity fluctuations u_y with Bassene forcing approach for a rectangular domain. (a) after $\tau = 10$ eddy-turnover times, (b) $\tau = 60$, rectangular domain 4:1. $L_y = 2\pi$ with resolution 1024×256^2, fixed viscosity $\nu = 5.8\text{e}{-3}$, $\text{Re}_\lambda = 52$, $\text{Ma}_t = 0.3$, initial density $\varrho_0 = 1$, initial pressure $p_0 = 16$

to all other turbulence properties. In a rectangular domain, on the other hand, the target value of A is missed. Additionally, even the ratio of solenoidal to dilatational dissipation decreases over time, which corresponds to an increasing solenoidal or decreasing dilatational part, which leads to less compressible effects in the flow field.

3.3 Linear Forcing with Box-Filtered Velocity in Rectangular Domains

One solution to reach a stationary and adequate turbulence, is the usage of a filtered velocity field for the forcing term. This approach is a usual method to control the integral length scale, especially when using Fourier space formulations. A sharp spectral filtered velocity, which removes the low wavenumber modes, works at fixed and equidistant grids. However, for a physical space solver, this method is not practical because of the high computational cost when transforming the entire velocity field into the Fourier space. Additionally, with an adapted mesh, the exact Fourier coefficient calculation is very expensive, maybe not affordable due to the computational limitations. It should be noted that an approximate computation of the Fourier transformation on the coarse grid did not solve the problem. From our experience, even a small part of any low wavenumber mode inside the forcing term leads to the above-described problem.

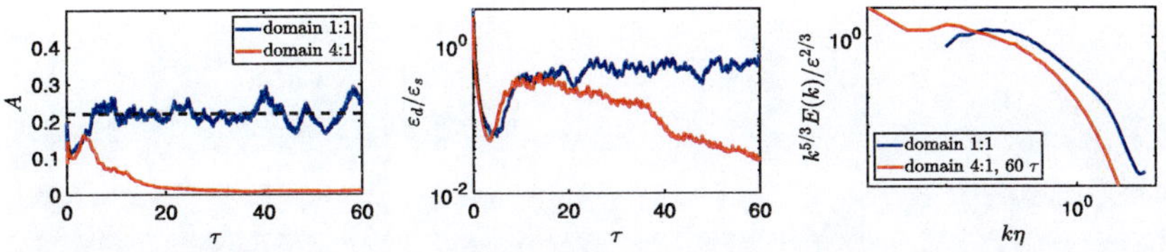

(a) (b) (c)

Fig. 7. Important turbulent properties with Bassenne forcing approach for a cubic/rectangular domain. (a) time evolution forcing parameter A, (b) time evolution ratio of dilatational to solenoidal dissipation $\frac{\varepsilon_d}{\varepsilon_s}$, (c) normalize energy spectra, rectangular domain 4:1. $L_y = 2\pi$ with resolution 1024×56^2, cubic domain accordingly, fixed viscosity $\nu = 5.8\text{e}{-3}$, $\text{Re}_\lambda = 52$, $\text{Ma}_t = 0.3$, initial density $\varrho_0 = 1$, initial pressure $p_0 = 16$

The integral length scale can be made smaller than 0.2 of the domain size L with a filtered velocity forcing term. This is an essential property concerning studying premixed turbulent combustion. Smaller integral length scales lead to higher Karlovitz numbers, which is, in general, a challenging field for research. Another significant advantage of this method (especially for the box-filtered method) is the fact that the filter works entirely locally. Therefore this forcing term can easily be used on adapted grids, without any Fourier space formulation. From a practical point of view, we can report that it is sufficient to filter the velocity field only every $n > 50$ time steps, which is essential when using large filter widths.

In Fig. 8 the results for different length scales in cubic domains with a non-adapted grid are shown. It is well observable that the turbulent structures and scales become smaller with decreasing integral length scales. This behaviour affects the vorticity kernels and the shock fronts, but generally, the vorticity magnitude and the shock strength increase with smaller length scales.

We find a linear correlation between the filter width L_f and the integral length scale l_t, which is shown in Fig. 9(b). In the energy spectra for the different filter widths, a decreasing integral length scale can be observed. If the turbulent Mach number is fixed for the different cases, then, due to a smaller integral length scale, one needs a finer grid (smaller Kolmogorov scale), and the Taylor microscale Reynolds number decreases. This is well visible in Fig. 9(c). From the time evolution of l_t (Fig. 9(a)) one can see that the statistical fluctuation is reduced for smaller length scales.

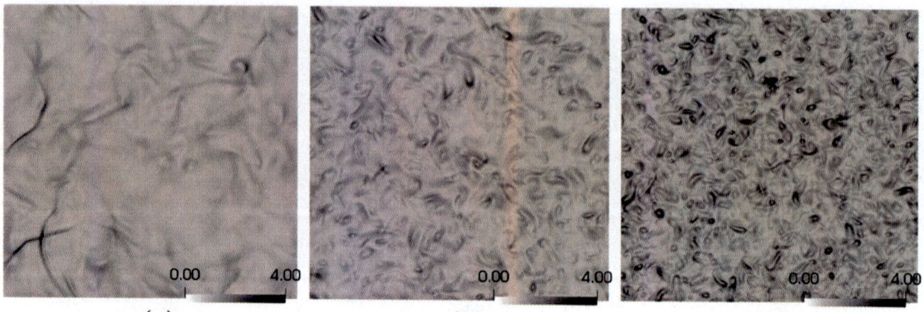

(a) (b) (c)

Fig. 8. Contour plots of density gradient $\nabla\varrho$ for a turbulent field with box filtered forcing approach. cubic domain $L = 2\pi$ (a) resolution 288^3, integral length scale $l_t = 0.8L_y$ at $\mathrm{Re}_\lambda = 41$, $L_f = 400$, (b) 416^3, $l_t = 0.2L_y$ at $\mathrm{Re}_\lambda = 21$, $L_f = 80$, (c) 512^3, $l_t = 0.1L_y$ at $\mathrm{Re}_\lambda = 15$, $L_f = 20$, fixed viscosity $\nu = 5.8\text{e}{-3}$, $\mathrm{Ma}_t = 0.3$, initial density $\varrho_0 = 1$, initial pressure $p_0 = 16$

4 Results

Finally, we present the results of our work in preparation to compute turbulence-flame interaction for the setup of the specific pulsed detonation combustor (PDC). Grid resolution and domain size are adapted to the parameters of premixed H_2/air flames. The integral length scale l_t and the turbulent Mach number Ma_t are chosen to reasonable values. It is planned to vary these parameters to study different regimes and characteristic points inside the Borghi diagram. In the center of the domain is the region of the finest mesh, which is approximately 2x of the thermal flame thickness δ_{th}. (see Fig. 10) The box-filtered forcing term is active there to maintain the turbulent field. The grid is coarsened outside this area, following the restrictions and rules of our code [17]. For this example, we use a mesh with a compression rate of ≈ 0.976, which saves 97% of the grid nodes compared to the corresponding equidistant mesh at the finest grid level. The area at the finest grid level could be quickly too small from a practical view,

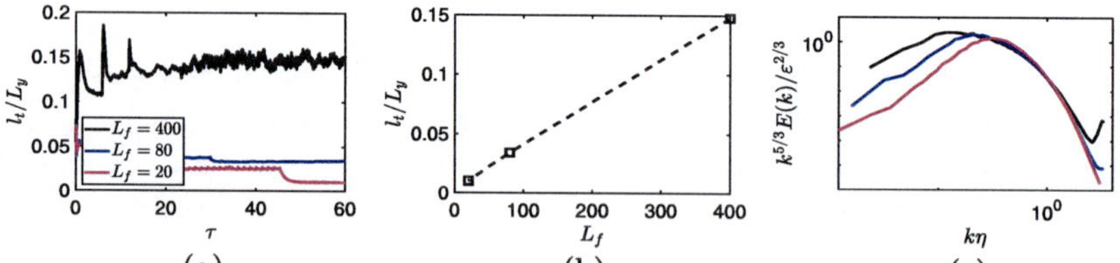

(a) (b) (c)

Fig. 9. Important turbulent properties with box filtered forcing approach for different filter widths L_f. cubic domain $L = 2\pi$ with resolution 256^3 (a) time evolution integral length scale l_t, (b) correlation l_t to L_f, (c) energy spectra, for $L_f = [400, 80, 20]$ with $\mathrm{Re}_\lambda = [41, 21, 15]$, fixed viscosity $\nu = 5.8\text{e}{-3}$, $\mathrm{Ma}_t = 0.3$, initial density $\varrho_0 = 1$, initial pressure $p_0 = 16$

especially for a wide distributed broken flame front. However, in this situation, we can adapt the grid, and due to the large velocity fluctuations, the turbulence needs only a short time to move to the newly adapted most refined grid level, which can be observed when turbulence enters the coarsened area of the grid. The fine turbulent scales, which can not exist on the coarse mesh, are suddenly removed, and overall the turbulent kinetic energy decays after that. We observe large-scale structures at the coarse grid level, which can have locally large velocity fluctuations. However, this can be handled due to specific customized filter steps, e.g., an adaptive filtering [3] to control shocks.

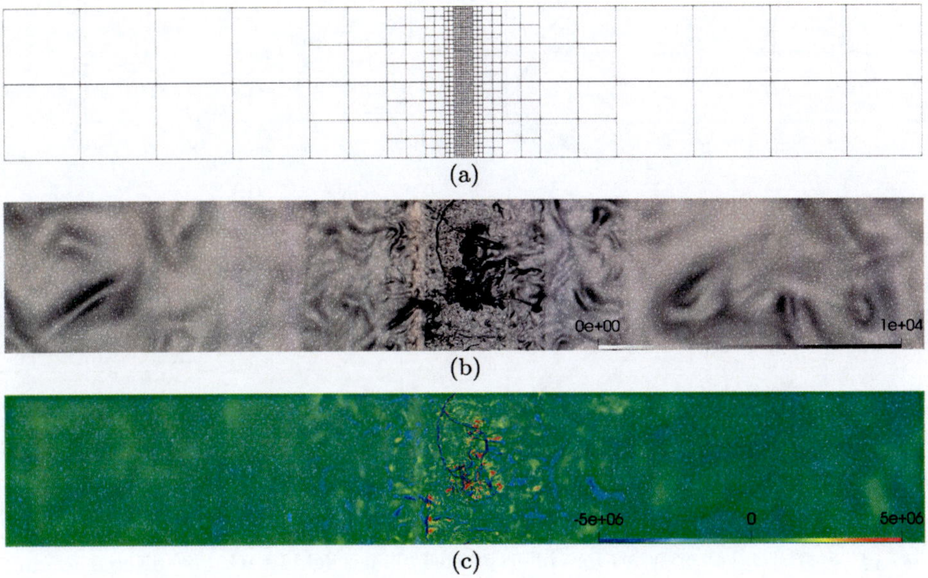

Fig. 10. Contour plots for a turbulent field with box filtered forcing approach in a rectangular domain (6:1) with an adapted grid. Domain $L_x = 12$ mm, $L_{y,z} = 2$ mm with block resolution 26^3, adaptation of 6 mesh levels (fixed mesh at level 6 corresponds to 9984×1664^2, compression rate ≈ 0.976), (a) grid, (b) density gradient $\nabla \varrho$, (c) velocity divergence Θ, fixed viscosity $\nu = 2.16\text{e}{-5}$, $\text{Ma}_t = 0.4$, initial density $\varrho_0 = 0.85$, initial pressure $p_0 = 1$ bar, $l_t = 0.1 L_y$, $\frac{\eta}{dx} = 1.03$

In the contour plots (Fig. 10), the different turbulent scales and structures can be seen. These flow characteristics are similar to the results from our test cases. Note that we choose the color bar scaling of the $\nabla \varrho$ to highlight the coarser parts of the grid. One can observe the decreasing magnitude of the density gradient from the finer mesh to the coarser level and also the disappearance of the small, turbulent scales. The velocity divergence contour plot is scaled to focus on the structures in the most refined mesh. The two typical flow field features, vorticity kernels and shock fronts can be observed in both plots. We do not show the time evolution of the turbulent properties, but these behave equivalent to the test cases. Generally, the box-filtered forcing method works very well for an adapted grid. A statistical stationary state is quickly reached, and the numerical effort is noticeably reduced due to the mesh adaptivity. It is advantageous not to start with the most refined grid level, rather than using a coarser mesh overall and adapt this step by step. For that procedure, the filter width L_f is easy to adapt to different mesh levels.

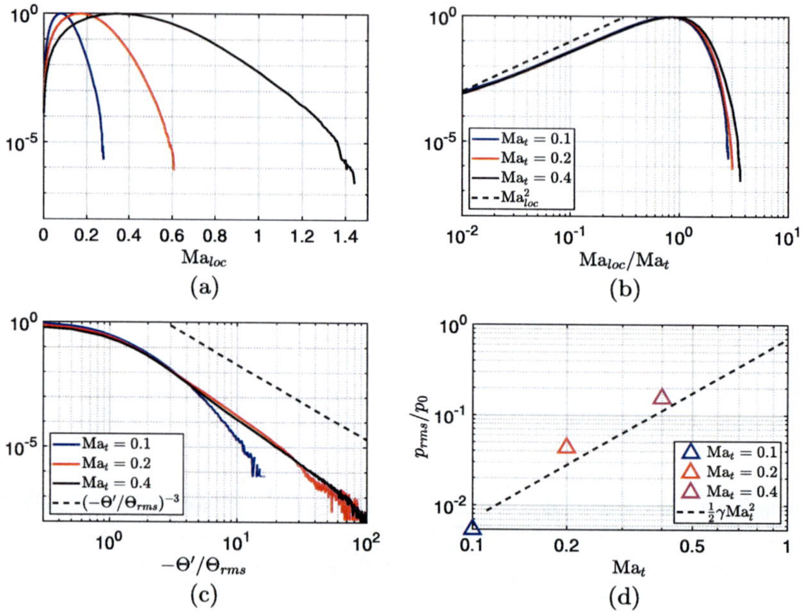

Fig. 11. Statistical properties for different turbulence fields with box filtered forcing approach in a rectangular domain, (a) PDF of local Mach number Ma_{loc}, (b) PDF of normalized local Mach number, (c) normalized velocity divergence Θ, (d) normalized RMS-pressure p_{rms} versus turbulent Mach number Ma_t

Different PDFs for computations of several turbulent Mach numbers are shown in Fig. 11. Excellent visible is the increasing local Mach number for increasing turbulent Mach number. Above $\mathrm{Ma}_t > 0.3$, the local Mach number becomes supersonic, according to the results from, e.g., [18]. The normalized Mach number curves collapse nearly to a single curve, and the values of $\mathrm{Ma}_{loc} < \mathrm{Ma}_t$ scales with Ma_t^2, which is also reported in the literature. Another statistically meaningful result can be observed for the normalized velocity divergence. The region of large divergence, which corresponds to the shock-like structures in the flow field, scales for $\mathrm{Ma} > 0.2$ with -3, similar to [18]. The pressure fluctuations are more significant for higher turbulent Mach numbers, and we observe (not shown), like in [8], an increase of the pressure fluctuations on the positive side for Mach numbers $\mathrm{Ma}_t > 0.2$. The RMS-pressure scales with Ma_t^2. Overall we see those important statistical properties correspond to results from the literature, and therefore we can conclude that our forcing method can reach and maintain a physically meaningful compressible turbulent flow field at an adapted non-cubic grid.

5 Conclusions

We discussed the applicability of the linear forcing method to adapted locally refined grids in rectangular domains. Our test cases show that the standard linear forcing term cannot create and maintain a physically meaningful turbulent field, even for an equidistant (uniform) mesh. We observed a highly increased turbulent kinetic energy for the low-wavenumber modes, which cause especially an unsteady evolution of the turbulent properties. Further, the integral length scale is not controllable, a key parameter for turbulent flame investigations. A solution to the described problem is the use of a filtered velocity field inside the forcing term. We use the box-filtering approach, which is simple to implement. For this, we observed complete control of the integral length scale on an adapted grid. Finally, we show a statistically good evolved turbulent field in preparation for our future work.

We plan to study turbulence-flame interactions in regimes at high Karlovitz numbers and compressible Mach numbers for our further work. The preliminary work of this report will be helpful in the creation of the necessary turbulent flow fields.

Acknowledgment. The authors gratefully acknowledge the support of the Deutsche Forschungsgemeinschaft (DFG) as part of collaborative research center CRC 1029 (project 200291049) "Substantial efficiency increase in gas turbines through direct use of coupled unsteady combustion and flow dynamics".

References

1. Bassenne, M., Urzay, J., Park, G.I., Moin, P.: Constant-energetics physical-space forcing methods for improved convergence to homogeneous-isotropic turbulence with application to particle-laden flows. Phys. Fluids **28**(3), 035114 (2016)
2. Bengoechea, S., Gray, J., Reiss, J., Moeck, J., Paschereit, O., Sesterhenn, J.: Detonation initiation in pipes with a single obstacle for mixtures of hydrogen and oxygen-enriched air. Combust. Flame **198**, 290–304 (2018)
3. Bogey, C., Cacqueray, N., Bailly, C.: A shock-capturing methodology based on adaptative spatial filtering for high-order non-linear computations. J. Comput. Phys. **228**, 1447–1465 (2009)
4. Brouwer, J., Reiss, J., Sesterhenn, J.: Conservative finite differences as an alternative to finite volume for compressible flows. In: Fuhrmann, J., Ohlberger, M., Rohde, C. (eds.) Finite Volumes for Complex Applications VII-Methods and Theoretical Aspects, pp. 169–176, Springer International Publishing, Cham (2014). https://doi.org/10.1007/978-3-319-05684-5_15
5. Campos, A., Morgan, B.: The effect of artificial bulk viscosity in simulations of forced compressible turbulence. J. Comput. Phys. **371**, 111–121 (2018)

6. Carroll, P., Blanquart, G.: A proposed modification to Lundgren's physical space velocity forcing method for isotropic turbulence. Phys. Fluids **25**, 5114 (2013)
7. Erlebacher, G., Hussaini, M., Sarkar, S., Kreiss, H.: The analysis and simulation of compressible turbulence. Theoret. Computat. Fluid Dyn. **2**, 2 (1990)
8. Jagannathan, S., Donzis, D.A.: Reynolds and mach number scaling in solenoidally-forced compressible turbulence using high-resolution direct numerical simulations. J. Fluid Mech. **789**, 669–707 (2016)
9. Ketterl, S., Klein, M.: A novel turbulent inflow data generation method and its application to the simulation of primary breakup. In: Salvetti, M.V., Armenio, V., Fröhlich, J., Geurts, B.J., Kuerten, H. (eds.) Direct and Large-Eddy Simulation XI. ES, vol. 25, pp. 229–235. Springer, Cham (2019). https://doi.org/10.1007/978-3-030-04915-7_31
10. Klein, M., Chakraborty, N., Ketterl, S.: A comparison of strategies for direct numerical simulation of turbulence chemistry interaction in generic planar turbulent premixed flames. Flow Turbul. Combust. **99**, 12 (2017)
11. Lundgren, T.: Linearly forced isotropic turbulence. Center Turbul. Res. Ann. Res. Briefs **2003**, 02 (2003)
12. Palmore, J.A., Desjardins, O.: Technique for forcing high reynolds number isotropic turbulence in physical space. Phys. Rev. Fluids **3**, 034605 (2018)
13. Petersen, M.R., Livescu, D.: Forcing for statistically stationary compressible isotropic turbulence. Phys. Fluids **22**(11), 116101 (2010)
14. Reiss, J., Sesterhenn, J.: A conservative, skew-symmetric finite difference scheme for the compressible Navier-Stokes equations. Comput. Fluids **101**, 08 (2013)
15. Rosales, C., Meneveau, C.: Linear forcing in numerical simulations of isotropic turbulence: physical space implementations and convergence properties. Phys. Fluids **17**(9), 095106 (2005)
16. Sarkar, S., Erlebacher, G., Hussaini, M.Y., Kreiss, H.O.: The analysis and modelling of dilatational terms in compressible turbulence. J. Fluid Mech. **227**, 473–493 (1991)
17. Sroka, M., Engels, T., Krah, P., Mutzel, S., Schneider, K., Reiss. J.: An open and parallel multiresolution framework using block-based adaptive grids. In: King, R. (ed.) Active Flow and Combustion Control 2018, pp. 305–319, Springer International Publishing, Cham (2019)
18. Wang, J., Gotoh, T., Watanabe, T.: Shocklet statistics in compressible isotropic turbulence. Phys. Rev. Fluids **2**, 023401 (2017)

Modal Decomposition of Flow Data via Gradient-Based Transport Optimization

Felix Black[1(✉)], Philipp Schulze[1], and Benjamin Unger[2]

[1] Institute of Mathematics, Technische Universität Berlin, Berlin, Germany
{black,pschulze}@math.tu-berlin.de
[2] Stuttgart Center for Simulation Science, Universität Stuttgart, Stuttgart, Germany
benjamin.unger@simtech.uni-stuttgart.de

Abstract. In the context of model reduction, we study an optimization problem related to the approximation of given data by a linear combination of transformed modes, called *transformed proper orthogonal decomposition* (tPOD). In the simplest case, the optimization problem reduces to a minimization problem well-studied in the context of proper orthogonal decomposition. Allowing transformed modes in the approximation renders this approach particularly useful to compress data with transported quantities, which are prevalent in many flow applications. We prove the existence of a solution to the infinite-dimensional optimization problem. Towards a numerical implementation, we compute the gradient of the cost functional and derive a suitable discretization in time and space. We demonstrate the theoretical findings with three numerical examples using a periodic shift operator as transformation.

Keywords: Nonlinear model order reduction · Transport-dominated phenomena · Transformed modes · Gradient-based optimization

1 Introduction

Projection-based model order reduction (MOR) typically relies on the fact that the solution manifold of a (parametrized) differential equation can be approximately embedded in a low-dimensional linear subspace. The best subspace of a given dimension, where best is understood as the minimal worst-case approximation error, is characterized by the Kolmogorov n-widths [14]. In practice, the minimizing subspace for the n-widths is difficult to compute. Instead, one relies on the proper orthogonal decomposition (POD) [11], which is typically combined with a greedy-search within the parameter domain, to get an approximate solution. In more detail, for given parameters $\mu_\sigma \in \mathbb{M}$ ($\sigma = 1, \ldots, \ell$), associated data samples $z(t, x; \mu_\sigma)$ with time variable $t \in \mathbb{T} := [0, T]$, space variable $x \in \Omega \subseteq \mathbb{R}^d$, and desired dimension $r \in \mathbb{N}$ of the low-dimensional subspace, POD determines orthonormal basis functions of a low-dimensional subspace solving the minimization problem

R. King and D. Peitsch (Eds.): AFCC 2021, NNFM 152, pp. 203–224, 2022.
https://doi.org/10.1007/978-3-030-90727-3_13

$$\begin{cases} \min \dfrac{1}{2} \displaystyle\sum_{\sigma=1}^{\ell} \int_0^T \left\| z(t,x;\mu_\sigma) - \sum_{i=1}^{r} \alpha_i(t;\mu_\sigma)\varphi_i(x) \right\|^2 \mathrm{d}t \\ \text{with } \alpha_i(t;\mu_\sigma) := \langle z(t,x;\mu_\sigma), \varphi_i(x) \rangle \quad \text{for } i = 1,\dots,r,\ \sigma = 1,\dots,\ell, \\ \text{s.t. } \langle \varphi_i, \varphi_j \rangle = \delta_{ij} \quad \text{for } i,j = 1,\dots,r. \end{cases} \tag{1}$$

If the n-widths, respectively the Hankel singular values for linear dynamical systems [33], decay fast, then one can expect to construct an effective reduced-order model (ROM) able to approximate the full dynamics with a small approximation error. Although one can show exponential decay for a large class of problems [18], it is well-known, see for instance [7,10], that the decay of the n-widths for flow problems is typically slow, thus conspiring against MOR. The main reason for slowly decaying n-widths is that flow problems often exhibit a strong space-time coupling, thus conspiring against the separation of space and time inherent to the definition of the Kolmogorov n-widths and POD approximation. Consequently, POD is often not able to produce accurate ROMs with a small number of modes.

To remedy this issue prevalent in transport-dominated phenomena, several strategies have been proposed in the literature. We refer to [7,12,15,22–24,27, 31,32,34] to name just a few. One promising approach, introduced in [16,26,30] and formalized in [3,6], is to replace the POD minimization problem with

$$\min \frac{1}{2} \sum_{\sigma=1}^{\ell} \int_0^T \left\| z(t,x;\mu_\sigma) - \sum_{i=1}^{r} \alpha_i(t;\mu_\sigma)\varphi_i(x - p_i(t;\mu_\sigma)) \right\|^2 \mathrm{d}t, \tag{2}$$

thus accounting explicitly for the transportation of quantities throughout the spatial domain. Let us emphasize that in the above formulation we may have $x - p_i(t;\mu_\sigma) \notin \Omega$. This may be resolved by a periodic domain, by an extrapolation approach [5], or by defining the modes on an extended domain $\widetilde{\Omega}$ [3, Sec. 7.2].

Consequently, the linear subspace in the Kolmogorov n-widths is replaced with a subspace able to adapt itself to the solution over time, hence rendering this a nonlinear approach, which we refer to as *transformed POD* (tPOD). Note that in contrast to the POD minimization problem (1), we do not require the modes to be orthogonal to each other. This is due to the fact that in the setting of (2), we would need orthogonality of $\varphi_i(x - p_i(t;\mu_\sigma))$ and $\varphi_j(x - p_j(t;\mu_\sigma))$ for all $i \neq j$, $t \in [0,T]$, and $\sigma = 1,\dots,\ell$, which is in general not a reasonable assumption, cf. [3, Ex. 4.4] for an illustrative example.

In the past years, there have been some attempts of solving discretized versions of (2) or related minimization problems. In [26], the authors propose a heuristic iterative method for computing a decomposition of a given snapshot matrix by an approximation ansatz as in (2). The numerical experiments indicate promising results, but it is not clear in which situations the proposed method actually determines an optimal solution. Another heuristic has been recently proposed in [5, sec. 5.2.1] and it is applied to snapshot data of a wildland fire simulation. This method is based on a decomposition of the snapshot matrix and involves a small number of singular value decompositions without requiring an iterative procedure. The numerical results presented in [5] demonstrate the effectiveness of this approach, but it is in general not optimal in the sense of the minimization problem (2). In contrast, the method introduced in [30] directly solves a fully discretized version of (2) by determining optimal modes φ and coefficients α, but assumes the paths p to be given or determined in a pre-processing step. A similar optimization problem has been proposed in [25] and aims at approximating the snapshot matrix by a sum of matrices representing different reference frames while achieving a fast singular value decay in each of the reference frames. The corresponding cost function is shown to be an upper bound for a fully discretized version of (2) and the cost functions coincide for the special case that only one reference frame is considered, i.e., if the same shift is applied to all modes in (2). Again, the paths are not considered as part of the optimization problem, but instead determined in a pre-processing step via peak or front tracking. On the contrary, the authors in [20] focus on determining optimal paths, whereas the determination of optimal ansatz functions and coefficients is not addressed. Moreover, to simplify the optimization problem, the paths are sought within a low-dimensional subspace consisting of predefined time-dependent library functions. As in the case of the other mentioned works, also the authors in [20] consider a fully discrete optimization problem.

We conclude that a gradient-based algorithm for the full optimization problem (2) is currently not available. Besides, a rigorous proof showing that (2) has a solution is missing in the literature. A notable exception is provided in [3, Thm. 4.6], albeit under the assumption that the path variables $p_i(t)$ are known a-priori. In this paper we aim to close this gap. Our main contributions are the following:

1. We show in Theorem 3 the existence of a minimizing solution for the optimization problem (6), which generalizes the minimization problem (2). Afterward, we reformulate the constrained minimization problem (6) as unconstrained problem (9) by adding appropriate penalty terms and conclude from Theorem 3 that also the unconstrained problem has a solution, cf. Corollary 1. In addition, Theorem 5 details that the solution of the unconstrained problem converges to the solution of the constraint problem.
2. We compute the gradient of the unconstrained problem in Theorem 7, which enables the use of gradient-based methods to solve (2). In this context, a remarkable finding is that the paths have to be sufficiently smooth (e.g. in $H^1(0, T)$), since otherwise some directional derivatives of the cost functional with respect to the paths may not exist, cf. Example 2.
3. We discuss the discretization of the gradient in space and time in Sect. 4 and explicitly compute the path-dependent inner products for the shift operator with periodic boundary conditions in Example 3. Finally, the effectiveness of gradient-based optimization is demonstrated for several examples in Sect. 5.

Notation. We denote the space of real $m \times n$ matrices by $\mathbb{R}^{m \times n}$ and the transpose of a matrix A is written as $A^\top$. Furthermore, for a vector with n entries all equal to one we use the symbol $\mathbf{1}_n$. Besides, for abbreviating diagonal and blockdiagonal matrices we use

$$\operatorname{diag}(a_1,\ldots,a_n) := \begin{bmatrix} a_1 & & \\ & \ddots & \\ & & a_n \end{bmatrix}, \qquad \operatorname{blkdiag}(A_1,\ldots,A_n) := \begin{bmatrix} A_1 & & \\ & \ddots & \\ & & A_n \end{bmatrix},$$

respectively, where $a_1,\ldots,a_n$ are scalars and $A_1,\ldots,A_n$ matrices of arbitrary size. For the Kronecker product of two matrices A and B we write $A \otimes B$. The space of square-integrable functions mapping from an interval (a,b) to a Banach space $\mathscr{X}$ is denoted by $L^2(a,b;\mathscr{X})$ and, similarly, the space of essentially bounded measurable functions by $L^\infty(a,b;\mathscr{X})$. Furthermore, we use $H^1(a,b;\mathscr{X})$ for the Sobolev subspace of functions in $L^2(a,b;\mathscr{X})$ possessing also a weak derivative in $L^2(a,b;\mathscr{X})$. The corresponding subspace consisting of $H^1(a,b;\mathscr{X})$ functions whose values at the boundaries a and b coincide is denoted by $H^1_{\mathrm{per}}(a,b;\mathscr{X})$. Besides, for the space of continuous functions from $[a,b]$ to $\mathscr{X}$ we use the symbol $C([a,b];\mathscr{X})$. For the special case $\mathscr{X} = \mathbb{R}$, we omit the last argument, i.e., we write, for instance, $L^2(a,b)$ instead of $L^2(a,b;\mathbb{R})$.

2 Preliminaries and Problem Formulation

To formalize the optimization problem (2), we introduce the following spaces and notation. Consider a real Hilbert space $(\mathscr{X}, \langle\cdot,\cdot\rangle_{\mathscr{X}})$ with induced norm $\|\cdot\|_{\mathscr{X}}$, and let $\mathscr{Y}$ denote a dense subspace of $\mathscr{X}$ that itself is a reflexive Banach space with norm $\|\cdot\|_{\mathscr{Y}}$. Our standing assumption is that we are minimizing the mean-squared distance to the data $z \in L^2(0,T;\mathscr{Y})$ in the Bochner space $L^2(0,T;\mathscr{X})$ with the additional requirement that the modes φ_i are elements of $\mathscr{Y}$.

To formalize the meaning of $\varphi_i(x - p_i(t))$ in (2), we follow the notation in [3] and introduce a family of linear and bounded operators $\mathcal{T}_i\colon \mathscr{P}_i \to \mathscr{B}(\mathscr{X})$ with real, finite-dimensional vector space $\mathscr{P}_i$, for which we postulate the following properties, taken from [3, Ass. 4.1].

Assumption 1. *For every $i = 1,\ldots,r$, every $\varphi_i \in \mathscr{Y}$, and every $p_i \in \mathscr{P}_i$, the operator $\mathcal{T}_i(p_i)$ is $\mathscr{Y}$-invariant, i.e., $\mathcal{T}_i(p_i)\mathscr{Y} \subseteq \mathscr{Y}$, and the mapping*

$$\mathscr{P}_i \to \mathscr{X}, \qquad p_i \mapsto \mathcal{T}_i(p_i)\varphi_i$$

is continuous.

A particular example for such a family of operators is given by the shift operator with periodic boundary conditions, see for instance [3, Ex. 5.2]. For further examples we refer to [2,15].

For the ease of presentation, we restrict ourselves to the case $\mathscr{P}_i = \mathbb{R}$, and emphasize that all results can be generalized to $\mathscr{P}_i = \mathbb{R}^{m_i}$ for some $m_i \in \mathbb{N}$. For

$$\mathscr{Z} := L^2(0,T;\mathbb{R}^r) \times H^1(0,T;\mathbb{R}^r) \times \mathscr{Y}^r \tag{3}$$

let us define the cost functional

$$J\colon \mathscr{X} \to \mathbb{R}, \qquad (\boldsymbol{\alpha}, \boldsymbol{p}, \boldsymbol{\varphi}) \mapsto \frac{1}{2}\left\| z - \sum_{i=1}^{r} \alpha_i \mathcal{T}_i\left(p_i\right)\varphi_i \right\|^2_{L^2(0,T;\mathscr{X})} \tag{4}$$

and for $C > 0$ the space

$$\mathcal{A}_C := \left\{(\boldsymbol{\alpha}, \boldsymbol{p}, \boldsymbol{\varphi}) \in \mathscr{X} \,\middle|\, \max\left\{\|\varphi_i\|_{\mathscr{Y}}, \|\alpha_i\|_{L^2(0,T)}, \|p_i\|_{H^1(0,T)}\right\} \le C\right\}, \tag{5}$$

where we use the notation $\boldsymbol{\alpha} = (\alpha_1, \ldots, \alpha_r)$ to denote the coefficients of $\boldsymbol{\alpha}$ and analogously for $\boldsymbol{p}$ and $\boldsymbol{\varphi}$. To ensure that the norm in (4) is defined, we invoke the following assumption, which is for instance satisfied if the family of operators $\mathcal{T}_i(\cdot)$ is uniformly bounded, cf. [3, Lem 4.2].

Assumption 2. *For every $i = 1, \ldots, r$, $\alpha_i \in L^2(0,T)$, $p_i \in H^1(0,T)$, and every $\varphi_i \in \mathscr{Y}$, we assume*

$$\alpha_i \mathcal{T}_i(p_i)\varphi_i \in L^2(0,T;\mathscr{X}).$$

With these preparations, the constrained minimization problem that we are interested in takes the form

$$\min_{(\boldsymbol{\alpha}, \boldsymbol{p}, \boldsymbol{\varphi})} J(\boldsymbol{\alpha}, \boldsymbol{p}, \boldsymbol{\varphi}), \quad \text{s. t. } (\boldsymbol{\alpha}, \boldsymbol{p}, \boldsymbol{\varphi}) \in \mathcal{A}_C. \tag{6}$$

Note that we have set $\ell = 1$ in (2) to simplify the notation. We emphasize that it is straightforward to generalize all results to $\ell > 1$.

3 Main Results

As first main result, we discuss the existence of a solution for the optimization problem (6), thus generalizing [3, Thm. 4.6] to include the path variables.

Theorem 3. *Assume that the reflexive Banach space $\mathscr{Y}$ is compactly embedded into $\mathscr{X}$, and let $z \in L^2(0,T;\mathscr{Y})$. Furthermore, let the family of transformation operators satisfy Assumptions 1 and 2. Then the constraint minimization problem (6) has a solution for every $C > 0$.*

Proof. The proof follows along the lines of the proof of [3, Thm. 4.6], with slight modifications to account for the optimization with respect to the path variables. Let $C > 0$. We first observe that the optimization problem possesses a finite infimum $J^\star \ge 0$. This follows directly from $J \ge 0$ and $(0,0,0) \in \mathcal{A}_C$. We may thus choose a sequence $(\boldsymbol{\alpha}^k, \boldsymbol{p}^k, \boldsymbol{\varphi}^k)_{k\in\mathbb{N}} \in \mathcal{A}_C$ satisfying

$$\lim_{k\to\infty} J(\boldsymbol{\alpha}^k, \boldsymbol{p}^k, \boldsymbol{\varphi}^k) = J^\star.$$

Additionally, we have

$$\begin{aligned}\|(\boldsymbol{\alpha}^k, \boldsymbol{p}^k, \boldsymbol{\varphi}^k)\|^2_{L^2(0,T;\mathbb{R}^r)\times H^1(0,T;\mathbb{R}^r)\times\mathscr{Y}^r} \\ = \|\boldsymbol{\alpha}^k\|^2_{L^2(0,T;\mathbb{R}^r)} + \|\boldsymbol{p}^k\|^2_{H^1(0,T;\mathbb{R}^r)} + \|\boldsymbol{\varphi}^k\|^2_{\mathscr{Y}^r} \le 3rC^2\end{aligned}$$

for all $k \in \mathbb{N}$, such that the Eberlein-Šmuljan theorem [35, Thm. 21.D] ensures the existence of a weakly convergent subsequence $(\boldsymbol{\alpha}^{k_n}, \boldsymbol{p}^{k_n}, \boldsymbol{\varphi}^{k_n})_{n\in\mathbb{N}} \subseteq \mathcal{A}_C$ with weak limit $(\boldsymbol{\alpha}^\star, \boldsymbol{p}^\star, \boldsymbol{\varphi}^\star) \in \mathcal{A}_C$, cf. [35, Prop. 21.23 (c)]. Due to the compact embeddings $\mathscr{Y} \hookrightarrow \mathscr{X}$ and $H^1(0,T;\mathbb{R}^r) \hookrightarrow L^2(0,T;\mathbb{R}^r)$, we conclude that $(\boldsymbol{\varphi}^{k_n})_{n\in\mathbb{N}}$ and $(\boldsymbol{p}^{k_n})_{n\in\mathbb{N}}$ converge strongly in $\mathscr{X}$ and $L^2(0,T;\mathbb{R}^r)$ to $\boldsymbol{\varphi}^\star$ and $\boldsymbol{p}^\star$, respectively, cf. [35, Prop. 21.35]. Using [29, Thm. 3.12], we conclude the existence of a subsequence, for which we use the same indexing, such that $(\boldsymbol{p}^{k_n})_{n\in\mathbb{N}}$ converges pointwise to $\boldsymbol{p}^\star$ for almost all $t \in (0,T)$.

For the next part of the proof, we introduce the mapping

$$\beta\colon \mathscr{Z} \to L^2(0,T;\mathscr{X}), \qquad (\boldsymbol{\alpha}, \boldsymbol{p}, \boldsymbol{\varphi}) \mapsto \sum_{i=1}^{r} \alpha_i \mathcal{T}_i(p_i)\varphi_i \tag{7}$$

with $\mathscr{Z}$ as defined in (3) and notice

$$J(\boldsymbol{\alpha}, \boldsymbol{p}, \boldsymbol{\varphi}) = \frac{1}{2}\|z - \beta(\boldsymbol{\alpha}, \boldsymbol{p}, \boldsymbol{\varphi})\|^2_{L^2(0,T;\mathscr{X})}.$$

If $\beta(\boldsymbol{\alpha}^{k_n}, \boldsymbol{p}^{k_n}, \boldsymbol{\varphi}^{k_n})$ converges weakly to $\beta(\boldsymbol{\alpha}^\star, \boldsymbol{p}^\star, \boldsymbol{\varphi}^\star)$, then the weak sequential lower semicontinuity of the norm, see for instance [35, Prop. 21.23 (c)], implies that $(\boldsymbol{\alpha}^\star, \boldsymbol{p}^\star, \boldsymbol{\varphi}^\star)$ is a minimizer of J. It thus remains to show that $\beta(\boldsymbol{\alpha}^{k_n}, \boldsymbol{p}^{k_n}, \boldsymbol{\varphi}^{k_n})$ converges weakly to $\beta(\boldsymbol{\alpha}^\star, \boldsymbol{p}^\star, \boldsymbol{\varphi}^\star)$.

To this end, we observe that

$$\begin{aligned}
&\|\mathcal{T}_i(p_i^{k_n}(t))\varphi_i^{k_n} - \mathcal{T}_i(p_i^\star(t))\varphi^\star\|_{\mathscr{X}} \\
&\le \|\mathcal{T}_i(p_i^{k_n}(t))\varphi_i^{k_n} - \mathcal{T}_i(p_i^\star(t))\varphi_i^{k_n}\|_{\mathscr{X}} + \|\mathcal{T}_i(p_i^\star(t))\varphi_i^{k_n} - \mathcal{T}_i(p_i^\star(t))\varphi_i^\star\|_{\mathscr{X}}
\end{aligned}$$

together with Assumption 1 and the strong convergence of $(\varphi_i^{k_n})_{n\in\mathbb{N}}$ in $\mathscr{X}$ implies

$$\|\mathcal{T}_i(p_i^{k_n}(t))\varphi_i^{k_n} - \mathcal{T}_i(p_i^\star(t))\varphi^\star\|_{\mathscr{X}} \to 0 \qquad \text{for } n \to \infty$$

for $i = 1, \ldots, r$ and almost all $t \in (0,T)$. Let $f \in L^2(0,T;\mathscr{X})$. Then clearly

$$\langle f(t), \mathcal{T}_i(p_i^{k_n}(t))\varphi_i^{k_n}\rangle_{\mathscr{X}} \to \langle f(t), \mathcal{T}_i(p_i^\star(t))\varphi_i^\star\rangle_{\mathscr{X}} \qquad \text{for } n \to \infty$$

for $i = 1, \ldots, r$ and almost all $t \in (0,T)$ such that [35, Prop 21.23 (j)] implies

$$\sum_{i=1}^{r} \left\langle \alpha_i^{k_n}, \langle f, \mathcal{T}_i(p_i^{k_n})\varphi_i^{k_n}\rangle_{\mathscr{X}} \right\rangle_{L^2(0,T)} \to \sum_{i=1}^{r} \langle \alpha_i^\star, \langle f, \mathcal{T}_i(p_i^\star)\varphi_i^\star\rangle_{\mathscr{X}}\rangle_{L^2(0,T)}$$

for $n \to \infty$ and thus

$$\beta(\boldsymbol{\alpha}^{k_n}, \boldsymbol{p}^{k_n}, \boldsymbol{\varphi}^{k_n}) \rightharpoonup \beta(\boldsymbol{\alpha}^\star, \boldsymbol{p}^\star, \boldsymbol{\varphi}^\star) \qquad \text{for } n \to \infty,$$

which completes the proof. □

For numerical methods, it may be easier to work with unconstrained optimization problems. To this end, we use a penalty method, see for instance [17, Cha. 13.1], i.e., we add the constraint equation with a penalty parameter to the cost functional. In more detail, we assume for $C > 0$ a penalty functional

$$\Lambda_C \colon \mathscr{X} \to \mathbb{R} \tag{8}$$

with the following properties available.

Assumption 4. *The penalty functional (8) is continuous, weakly sequentially lower semicontinuous, non-negative and has the following properties:*

- *We have $\Lambda_C(\boldsymbol{\alpha}, \boldsymbol{p}, \boldsymbol{\varphi}) = 0$ if, and only if, $(\boldsymbol{\alpha}, \boldsymbol{p}, \boldsymbol{\varphi}) \in \mathcal{A}_C$.*
- *For any sequence $(\boldsymbol{\alpha}^k, \boldsymbol{p}^k, \boldsymbol{\varphi}^k)$ with*

$$\max\{\|\boldsymbol{\alpha}^k\|_{L^2(0,T;\mathbb{R}^r)}, \|\boldsymbol{p}^k\|_{H^1(0,T;\mathbb{R}^r)}, \|\boldsymbol{\varphi}^k\|_{\mathscr{Y}^r}\} \to \infty \qquad \text{for } k \to \infty,$$

we have $\Lambda_C(\boldsymbol{\alpha}^k, \boldsymbol{p}^k, \boldsymbol{\varphi}^k) \to \infty$ for $k \to \infty$.

Example 1. The penalty functional

$$\Lambda_C(\boldsymbol{\alpha}, \boldsymbol{p}, \boldsymbol{\varphi}) := \sum_{i=1}^{r} \max\{0, \max\{\|\alpha_i\|_{L^2(0,T)}, \|p_i\|_{H^1(0,T)}, \|\varphi_i\|_{\mathscr{Y}}\} - C\}$$

satisfies Assumption 4.

The penalized cost functional is then given as

$$\widetilde{J}_C(\boldsymbol{\alpha}, \boldsymbol{p}, \boldsymbol{\varphi}, \lambda) := J(\boldsymbol{\alpha}, \boldsymbol{p}, \boldsymbol{\varphi}) + \lambda \Lambda_C(\boldsymbol{\alpha}, \boldsymbol{p}, \boldsymbol{\varphi})$$

with penalty coefficient $\lambda > 0$. The associated (unconstrained) minimization problem is thus given by

$$\min_{(\boldsymbol{\alpha}, \boldsymbol{p}, \boldsymbol{\varphi}) \in \mathscr{X}} \widetilde{J}_C(\boldsymbol{\alpha}, \boldsymbol{p}, \boldsymbol{\varphi}, \lambda) \tag{9}$$

with $\mathscr{X}$ as defined in (3) and given $\lambda > 0$.

Corollary 1. *Let the assumptions of Theorem 3 and Assumption 4 be satisfied. Then for any $C > 0$ and any $\lambda > 0$ the optimization problem (9) has a solution.*

Proof. Similarly as in the proof of Theorem 3, we conclude the existence of a finite infimum, such that we can choose a minimizing sequence $(\boldsymbol{\alpha}^k, \boldsymbol{p}^k, \boldsymbol{\varphi}^k) \in \mathscr{X}$. Due to Assumption 4, we deduce that $(\boldsymbol{\alpha}^k, \boldsymbol{p}^k, \boldsymbol{\varphi}^k)_{k \in \mathbb{N}}$ is bounded in $\mathscr{X}$, i.e., there exists some $\widetilde{C} > 0$ such that $(\boldsymbol{\alpha}^k, \boldsymbol{p}^k, \boldsymbol{\varphi}^k) \in \mathcal{A}_{\widetilde{C}}$ for all $k \in \mathbb{N}$. The remaining proof thus follows along the lines of the proof of Theorem 3. □

Theorem 5. *Let $(\lambda^k)_{k \in \mathbb{N}} \subseteq \mathbb{R}$ denote a non-decreasing sequence of positive numbers with $\lim_{k \to \infty} \lambda^k = \infty$, and let $C > 0$. For $k \in \mathbb{N}$, let $(\boldsymbol{\alpha}^k, \boldsymbol{p}^k, \boldsymbol{\varphi}^k) \in \mathscr{X}$ denote a solution of (9) with penalty parameter λ^k. If the assumptions of Corollary 1 are satisfied, then any limit point of $(\boldsymbol{\alpha}^k, \boldsymbol{p}^k, \boldsymbol{\varphi}^k)_{k \in \mathbb{N}}$ is a solution of (6).*

Proof. The proof follows along the lines of the proof of the main theorem in [17, Cha. 13.1]. Let $(\boldsymbol{\alpha}^\star, \boldsymbol{p}^\star, \boldsymbol{\varphi}^\star) \in \mathcal{A}_C$ denote a minimizer of (6) with minimum $J^\star$. Then for every $k \in \mathbb{N}$ we have

$$\begin{aligned}\widetilde{J}_C(\boldsymbol{\alpha}^k, \boldsymbol{p}^k, \boldsymbol{\varphi}^k, \lambda^k) &\le \widetilde{J}_C(\boldsymbol{\alpha}^{k+1}, \boldsymbol{p}^{k+1}, \boldsymbol{\varphi}^{k+1}, \lambda^k)\\ &\le \widetilde{J}_C(\boldsymbol{\alpha}^{k+1}, \boldsymbol{p}^{k+1}, \boldsymbol{\varphi}^{k+1}, \lambda^{k+1})\end{aligned}$$

and

$$J(\boldsymbol{\alpha}^k, \boldsymbol{p}^k, \boldsymbol{\varphi}^k) \le \widetilde{J}_C(\boldsymbol{\alpha}^k, \boldsymbol{p}^k, \boldsymbol{\varphi}^k, \lambda^k) \le \widetilde{J}_C(\boldsymbol{\alpha}^\star, \boldsymbol{p}^\star, \boldsymbol{\varphi}^\star, \lambda^k) = J(\boldsymbol{\alpha}^\star, \boldsymbol{p}^\star, \boldsymbol{\varphi}^\star) = J^\star.$$

Thus $(\widetilde{J}_C(\boldsymbol{\alpha}^k, \boldsymbol{p}^k, \boldsymbol{\varphi}^k, \lambda^k))_{k\in\mathbb{N}}$ is a monotone sequence bounded above by $J^\star$. We thus set

$$\widetilde{J}_C^\star := \lim_{k\to\infty} \widetilde{J}_C(\boldsymbol{\alpha}^k, \boldsymbol{p}^k, \boldsymbol{\varphi}^k, \lambda^k) \le J^\star. \tag{10}$$

Let $(\boldsymbol{\alpha}^{k_n}, \boldsymbol{p}^{k_n}, \boldsymbol{\varphi}^{k_n})$ denote a convergent subsequence with limit $(\boldsymbol{\alpha}^\dagger, \boldsymbol{p}^\dagger, \boldsymbol{\varphi}^\dagger)$ and set

$$J^\dagger := \lim_{n\to\infty} J(\boldsymbol{\alpha}^{k_n}, \boldsymbol{p}^{k_n}, \boldsymbol{\varphi}^{k_n}) = J(\boldsymbol{\alpha}^\dagger, \boldsymbol{p}^\dagger, \boldsymbol{\varphi}^\dagger), \tag{11}$$

using the continuity of J. Subtracting (10) from (11) yields

$$\lim_{n\to\infty} \lambda^{k_n} \Lambda_C(\boldsymbol{\alpha}^{k_n}, \boldsymbol{p}^{k_n}, \boldsymbol{\varphi}^{k_n}) = \widetilde{J}_C^\star - J^\dagger.$$

Assumption 4 and $\lambda^{k_n} \to \infty$ for $n \to \infty$ together with the continuity of Λ_C thus implies

$$\Lambda_C(\boldsymbol{\alpha}^\dagger, \boldsymbol{p}^\dagger, \boldsymbol{\varphi}^\dagger) = \lim_{n\to\infty} \Lambda_C(\boldsymbol{\alpha}^{k_n}, \boldsymbol{p}^{k_n}, \boldsymbol{\varphi}^{k_n}) = 0,$$

showing $(\boldsymbol{\alpha}^\dagger, \boldsymbol{p}^\dagger, \boldsymbol{\varphi}^\dagger) \in \mathcal{A}_C$. We conclude

$$J^\dagger = \lim_{n\to\infty} J(\boldsymbol{\alpha}^{k_n}, \boldsymbol{p}^{k_n}, \boldsymbol{\varphi}^{k_n}) \le J^\star,$$

which completes the proof. □

Although (9) is an unconstrained optimization problem, we still have to choose a suitable constant $C > 0$ for the admissible set. Let us emphasize that the proofs of Theorem 3 and Corollary 1 heavily depend on the fact that we have bounded sequences, which is the main reason for the constant $C > 0$ in the admissible set (5). However, we observed faster convergence in our numerical experiments when considering the unconstrained minimization problem without penalization. For this reason and the sake of a concise presentation, we consider in the following only the unconstrained optimization problem (9) with penalty parameter $\lambda = 0$. Nevertheless, we emphasize that adding the derivatives of the penalty terms to the gradient formulas is straightforward as long as the partial Fréchet derivatives of Λ_C are available.

To solve the optimization problem (9) with penalty parameter $\lambda = 0$ numerically, we employ a gradient-based algorithm and thus have to compute the gradient of the objective function (4). It is easy to see that the directional derivatives of J with respect to the coefficient function $\boldsymbol{\alpha} \in L^2(0,T;\mathbb{R}^r)$ and the modes $\boldsymbol{\varphi} \in \mathscr{Y}^r$ in directions $\boldsymbol{d} \in L^2(0,T;\mathbb{R}^r)$ and $\boldsymbol{h} \in \mathscr{Y}^r$, respectively, are given by

$$\partial_{\boldsymbol{\alpha},\boldsymbol{d}} J(\boldsymbol{\alpha},\boldsymbol{p},\boldsymbol{\varphi}) = \sum_{i=1}^{r} \left\langle \sum_{j=1}^{r} \alpha_j \mathcal{T}_j(p_j)\varphi_j - z, d_i \mathcal{T}_i(p_i)\varphi_i \right\rangle_{L^2(0,T;\mathscr{X})}, \tag{12a}$$

$$\partial_{\boldsymbol{\varphi},\boldsymbol{h}} J(\boldsymbol{\alpha},\boldsymbol{p},\boldsymbol{\varphi}) = \sum_{i=1}^{r} \left\langle \sum_{j=1}^{r} \alpha_j \mathcal{T}_j(p_j)\varphi_j - z, \alpha_i \mathcal{T}_i(p_i)h_i \right\rangle_{L^2(0,T;\mathscr{X})}. \tag{12b}$$

The situation is slightly different for the partial derivative with respect to the path variable. First of all, we have to ensure that the transformed modes are differentiable (with respect to the path variable), i.e., we have to evoke the following assumption.

Assumption 6. *For every $\varphi_i \in \mathscr{Y}$ and every $i = 1,\ldots,r$, the mapping*

$$\mathbb{R} \to \mathscr{X}, \qquad p_i \mapsto \mathcal{T}_i(p_i)\varphi_i,$$

is continuously differentiable with derivatives in $\mathscr{X}$. For $p_i \in \mathbb{R}$ we denote the derivative by $\frac{\partial}{\partial p_i}\mathcal{T}_i(p_i)\varphi_i \in \mathscr{X}$ and assume $\alpha_i \frac{\partial}{\partial p_i}\mathcal{T}_i(p_i)\varphi_i \in L^2(0,T;\mathscr{X})$ for all $\alpha_i \in L^2(0,T)$ and all $p_i \in H^1(0,T)$.

In this case, the directional derivative in direction $\boldsymbol{g} \in H^1(0,T;\mathbb{R}^r)$ is given as

$$\partial_{\boldsymbol{p},\boldsymbol{g}} J(\boldsymbol{\alpha},\boldsymbol{p},\boldsymbol{\varphi}) = \sum_{i=1}^{r} \left\langle \sum_{j=1}^{r} \alpha_j \mathcal{T}_j(p_j)\varphi_j - z, \alpha_i \left[\frac{\partial}{\partial p_i}\mathcal{T}_i(p_i)\varphi_i\right] g_i \right\rangle_{L^2(0,T;\mathscr{X})}. \tag{12c}$$

Note that the Sobolev embedding theorems, see for instance [35, Thm. 21.A.(d)], imply $g_i \in C([0,T]) \subseteq L^\infty(0,T)$, such that (12c) is defined.

Theorem 7. *Let the transformation operators satisfy Assumptions 1, 2, and 6. Let $(\boldsymbol{\alpha},\boldsymbol{p},\boldsymbol{\varphi}) \in \mathscr{Z}$ and assume*

$$\mathcal{T}_i(p_i)\varphi_i \in L^\infty(0,T;\mathscr{X}), \tag{13a}$$

$$\alpha_i \|\mathcal{T}_i(p_i)\| \in L^2(0,T) \tag{13b}$$

for $i = 1,\ldots,r$, then the partial Fréchet derivatives of the cost functional J (defined in (4)) with respect to the coefficients, paths, and modes at $(\boldsymbol{\alpha},\boldsymbol{p},\boldsymbol{\varphi}) \in \mathscr{Z}$ are given by

$$\partial_{\boldsymbol{\alpha}} J(\boldsymbol{\alpha},\boldsymbol{p},\boldsymbol{\varphi})(\boldsymbol{d}) := \partial_{\boldsymbol{\alpha},\boldsymbol{d}} J(\boldsymbol{\alpha},\boldsymbol{p},\boldsymbol{\varphi}), \qquad \forall \boldsymbol{d} \in L^2(0,T;\mathbb{R}^r), \tag{14a}$$

$$\partial_{\boldsymbol{p}} J(\boldsymbol{\alpha},\boldsymbol{p},\boldsymbol{\varphi})(\boldsymbol{g}) := \partial_{\boldsymbol{p},\boldsymbol{g}} J(\boldsymbol{\alpha},\boldsymbol{p},\boldsymbol{\varphi}), \qquad \forall \boldsymbol{g} \in H^1(0,T;\mathbb{R}^r), \tag{14b}$$

$$\partial_{\boldsymbol{\varphi}} J(\boldsymbol{\alpha},\boldsymbol{p},\boldsymbol{\varphi})(\boldsymbol{h}) := \partial_{\boldsymbol{\varphi},\boldsymbol{h}} J(\boldsymbol{\alpha},\boldsymbol{p},\boldsymbol{\varphi}), \qquad \forall \boldsymbol{h} \in \mathscr{Y}^r, \tag{14c}$$

with directional derivatives as defined in (12).

Proof. It suffices to show that J is partially Fréchet differentiable with respect to the coefficients, paths, and modes. Let $(\boldsymbol{\alpha}, \boldsymbol{p}, \boldsymbol{\varphi}), (\boldsymbol{d}, \boldsymbol{g}, \boldsymbol{h}) \in \mathscr{X}$. Using (13a) we obtain

$$J(\boldsymbol{\alpha}+\boldsymbol{d}, \boldsymbol{p}, \boldsymbol{\varphi}) - J(\boldsymbol{\alpha}, \boldsymbol{p}, \boldsymbol{\varphi}) - \partial_{\boldsymbol{\alpha},\boldsymbol{d}} J(\boldsymbol{\alpha}, \boldsymbol{p}, \boldsymbol{\varphi}) = \frac{1}{2} \left\| \sum_{i=1}^{r} d_i \mathcal{T}_i(p_i)\varphi_i \right\|^2_{L^2(0,T;\mathscr{X})}$$
$$\leq \frac{1}{2} \max_{i=1,\dots,r} \|\mathcal{T}_i(p_i)\varphi_i\|^2_{L^\infty(0,T;\mathscr{X})} \left(\sum_{i=1}^{r} \|d_i\|_{L^2(0,T)} \right)^2$$
$$\leq \frac{r^2}{2} \max_{i=1,\dots,r} \|\mathcal{T}_i(p_i)\varphi_i\|^2_{L^\infty(0,T;\mathscr{X})} \|\boldsymbol{d}\|^2_{L^2(0,T;\mathbb{R}^r)}$$

and thus

$$\lim_{\|\boldsymbol{d}\|_{L^2(0,T;\mathbb{R}^r)} \to 0} \frac{|J(\boldsymbol{\alpha}+\boldsymbol{d}, \boldsymbol{p}, \boldsymbol{\varphi}) - J(\boldsymbol{\alpha}, \boldsymbol{p}, \boldsymbol{\varphi}) - \partial_{\boldsymbol{\alpha},\boldsymbol{d}} J(\boldsymbol{\alpha}, \boldsymbol{p}, \boldsymbol{\varphi})|}{\|\boldsymbol{d}\|_{L^2(0,T;\mathbb{R}^r)}} = 0.$$

We conclude that J is Fréchet differentiable with respect to the coefficients with Fréchet derivative as in (14a). For the partial derivative with respect to the modes we obtain

$$J(\boldsymbol{\alpha}, \boldsymbol{p}, \boldsymbol{\varphi}+\boldsymbol{h}) - J(\boldsymbol{\alpha}, \boldsymbol{p}, \boldsymbol{\varphi}) - \partial_{\boldsymbol{\varphi},\boldsymbol{h}} J(\boldsymbol{\alpha}, \boldsymbol{p}, \boldsymbol{\varphi}) = \frac{1}{2} \left\| \sum_{i=1}^{r} \alpha_i \mathcal{T}_i(p_i) h_i \right\|^2_{L^2(0,T;\mathscr{X})}$$
$$\leq \frac{1}{2} \int_0^T \left(\sum_{i=1}^{r} |\alpha_i(t)| \|\mathcal{T}_i(p_i(t))\| \|h_i\|_{\mathscr{X}} \right)^2 \mathrm{d}t$$
$$\leq \frac{\|\boldsymbol{h}\|^2_{\mathscr{X}^r}}{2} \int_0^T \left(\sum_{i=1}^{r} |\alpha_i(t)| \|\mathcal{T}_i(p_i(t))\| \right)^2 \mathrm{d}t.$$

Using (13b), we observe that the integral is finite. Similarly as before, we thus conclude that J is Fréchet differentiable with respect to the modes with Fréchet derivative as in (14c). We conclude our proof for the partial derivative with respect to the path variable. Note that the Sobolev embedding theorem [1, Thm 4.12, Part I, Case A] implies that the Sobolev space $H^1(0,T)$ is continuously embedded into the space $L^\infty(0,T)$, i.e., there exists a constant $\gamma > 0$ independent of g_i, such that $\|g_i\|_{L^\infty(0,T)} \leq \gamma \|g_i\|_{H^1(0,T)}$. We define

$$f_i(p_i, \varphi_i, g_i) := \mathcal{T}_i(p_i + g_i)\varphi_i - \mathcal{T}_i(p_i)\varphi_i - \left[\frac{\partial}{\partial p_i} \mathcal{T}_i(p_i) \right] g_i.$$

For $g_i \equiv 0$ we have $f_i(p_i, \varphi_i, g_i) = 0$ for almost all $t \in (0,T)$. For $\|g_i\|_{H^1(0,T)} \neq 0$, let us define $\widehat{\mathbb{T}}_i := \{t \in (0,T) \mid g_i(t) \neq 0\}$. Then

$$\frac{\langle z, \alpha_i f_i(p_i, \varphi_i, g_i) \rangle_{L^2(0,T;\mathscr{X})}}{\|g_i\|_{H^1(0,T)}} \leq \gamma \int_{\widehat{\mathbb{T}}_i} \alpha_i(t) \left\langle z(t), \frac{f_i(p_i(t), \varphi_i, g_i(t))}{|g_i(t)|} \right\rangle_{\mathscr{X}} \mathrm{d}t.$$

From Assumption 6 we conclude

$$\lim_{\|g_i\|_{H^1(0,T)}\to 0} \frac{\langle z, \alpha_i f_i(p_i, \varphi_i, g_i)\rangle_{L^2(0,T;\mathscr{X})}}{\|g_i\|_{H^1(0,T)}} = 0,$$

and thus

$$\lim_{\|\boldsymbol{g}\|_{H^1(0,T;\mathbb{R}^r)}\to 0} \frac{\sum_{i=1}^r \langle z, \alpha_i f_i(p_i, \varphi_i, g_i)\rangle_{L^2(0,T;\mathscr{X})}}{\|\boldsymbol{g}\|_{H^1(0,T;\mathbb{R}^r)}} = 0.$$

Furthermore, using β as defined in (7), we obtain

$$\begin{aligned}
&\tfrac{1}{2}\|\beta(\boldsymbol{\alpha}, \boldsymbol{p}+\boldsymbol{g}, \boldsymbol{\varphi})\|^2_{L^2(0,T;\mathscr{X})} - \tfrac{1}{2}\|\beta(\boldsymbol{\alpha}, \boldsymbol{p}, \boldsymbol{\varphi})\|^2_{L^2(0,T;\mathscr{X})} \\
&\qquad - \sum_{j=1}^r \left\langle \beta(\boldsymbol{\alpha}, \boldsymbol{p}, \boldsymbol{\varphi}), \alpha_j \left[\tfrac{\partial}{\partial p_j}\mathcal{T}_j(p_j)\varphi_j\right] g_j \right\rangle_{L^2(0,T;\mathscr{X})} \\
&\quad = \tfrac{1}{2}\|\beta(\boldsymbol{\alpha}, \boldsymbol{p}+\boldsymbol{g}, \boldsymbol{\varphi}) - \beta(\boldsymbol{\alpha}, \boldsymbol{p}, \boldsymbol{\varphi}) + \beta(\boldsymbol{\alpha}, \boldsymbol{p}, \boldsymbol{\varphi})\|^2_{L^2(0,T;\mathscr{X})} \\
&\qquad - \tfrac{1}{2}\|\beta(\boldsymbol{\alpha}, \boldsymbol{p}, \boldsymbol{\varphi})\|^2_{L^2(0,T;\mathscr{X})} \\
&\qquad - \sum_{j=1}^r \left\langle \beta(\boldsymbol{\alpha}, \boldsymbol{p}, \boldsymbol{\varphi}), \alpha_j \left[\tfrac{\partial}{\partial p_j}\mathcal{T}_j(p_j)\varphi_j\right] g_j \right\rangle_{L^2(0,T;\mathscr{X})} \\
&\quad = \tfrac{1}{2}\|\beta(\boldsymbol{\alpha}, \boldsymbol{p}+\boldsymbol{g}, \boldsymbol{\varphi}) - \beta(\boldsymbol{\alpha}, \boldsymbol{p}, \boldsymbol{\varphi})\|^2_{L^2(0,T;\mathscr{X})} \\
&\qquad + \sum_{j=1}^r \langle \beta(\boldsymbol{\alpha}, \boldsymbol{p}, \boldsymbol{\varphi}), \alpha_j f_j(p_j, \varphi_j, g_j)\rangle_{L^2(0,T;\mathscr{X})}.
\end{aligned}$$

Similarly as before, we obtain

$$\lim_{\|\boldsymbol{g}\|_{H^1(0,T;\mathbb{R}^r)}\to 0} \frac{1}{2}\frac{\|\beta(\boldsymbol{\alpha}, \boldsymbol{p}+\boldsymbol{g}, \boldsymbol{\varphi}) - \beta(\boldsymbol{\alpha}, \boldsymbol{p}, \boldsymbol{\varphi})\|^2_{L^2(0,T;\mathscr{X})}}{\|\boldsymbol{g}\|_{H^1(0,T;\mathbb{R}^r)}} = 0,$$

$$\lim_{\|\boldsymbol{g}\|_{H^1(0,T;\mathbb{R}^r)}\to 0} \frac{\langle \beta(\boldsymbol{\alpha}, \boldsymbol{p}, \boldsymbol{\varphi}), \sum_{j=1}^r \alpha_j f_j(p_j, \varphi_j, g_j)\rangle_{L^2(0,T;\mathscr{X})}}{\|\boldsymbol{g}\|_{H^1(0,T;\mathbb{R}^r)}} = 0.$$

Combining the previous results, we thus infer

$$\lim_{\|\boldsymbol{g}\|_{H^1(0,T;\mathbb{R}^r)}\to 0} \frac{|J(\boldsymbol{\alpha}, \boldsymbol{p}+\boldsymbol{g}, \boldsymbol{\varphi}) - J(\boldsymbol{\alpha}, \boldsymbol{p}, \boldsymbol{\varphi}) - \partial_{\boldsymbol{p},\boldsymbol{g}} J(\boldsymbol{\alpha}, \boldsymbol{p}, \boldsymbol{\varphi})|}{\|\boldsymbol{g}\|_{H^1(0,T;\mathbb{R}^r)}} = 0,$$

which concludes the proof. □

Remark 1. If the family of transformation operators is uniformly bounded, i.e., there exists some $\overline{C} > 0$ such that

$$\|\mathcal{T}_i(p_i)\| \leq \overline{C} \qquad \text{for all } p_i \in \mathbb{R},$$

then it is easy to see that condition (13) is satisfied. Note that in this case Assumption 2 is also satisfied, cf. [3, Lem 4.2]. An example for such a family of operators is (again) the periodic shift operator.

Let us emphasize that it is essential for the directional derivative $\partial_{\boldsymbol{p},\boldsymbol{g}} J$ to have the path variable and associated directions in $H^1(0,T;\mathbb{R}^r)$. The following example details that if we take a direction in $L^2(0,T;\mathbb{R}^r)$, then the directional derivative may not be finite.

Example 2. Consider the shift operator $\mathcal{T}(p)\varphi = \varphi(\cdot - p)$ with periodic embedding into the spaces $\mathscr{X} := L^2(0,2\pi)$ and $\mathscr{Y} := H^1_{\text{per}}(0,2\pi)$, cf. [3, Ex. 4.3 and 5.12]. It is well-known, that the shift operator is a semi-group with generator $-\frac{\partial}{\partial x}$, see for instance [9, Sec. II.2.10]. Let $z(t,x) = t^{-1/3}\cos(x)$, $r = 1$, $p_1 \equiv 0$, and $\varphi_1(x) = \sin(x)$. Then for any $\alpha_1, g_1 \in L^2(0,T)$, we obtain

$$\begin{aligned}\partial_{\boldsymbol{p},\boldsymbol{g}} J(\alpha_1, p_1, \varphi_1) &= -\left\langle \alpha_1\varphi_1 - z, \alpha_1 \tfrac{\partial}{\partial x}\varphi_1 g_1 \right\rangle_{L^2(0,T;\mathscr{X})} \\ &= \langle z, \alpha_1 \tfrac{\partial}{\partial x}\varphi_1 g_1\rangle_{L^2(0,T;\mathscr{X})} = \|\cos(\cdot)\|^2_{\mathscr{X}} \int_0^T t^{-1/3}\alpha_1(t) g_1(t)\,\mathrm{d}t.\end{aligned}$$

We notice that for $\alpha_1(t) = g_1(t) = t^{-1/3}$ we have $\alpha_1, g_1 \in L^2(0,T)$ but the product $t^{-1/3}\alpha_1 g_1$ is not in $L^1(0,T)$. We conclude $\partial_{\boldsymbol{p},\boldsymbol{g}} J(\boldsymbol{\alpha},\boldsymbol{p},\boldsymbol{\varphi}) \notin \mathbb{R}$.

Remark 2. To ensure $\boldsymbol{p} \in H^1(0,T;\mathbb{R}^r)$ during a (numerical) optimization, we may choose a suitable low-dimensional subspace with continuously differentiable basis functions, such as the space of polynomials with given maximal degree. The associated gradient is easily computed from Theorem 7 via the chain rule. Besides the reduced computational cost, such an approach yields an interpretable representation for the wave speeds. We refer to [20] for a similar idea in a fully discretized setting.

4 Discretization

Towards a numerical implementation, we derive discretized versions of the partial derivatives from Theorem 7. To shorten notation, we introduce for $(\boldsymbol{\alpha},\boldsymbol{p},\boldsymbol{\varphi}) \in \mathscr{Z}$ and $\boldsymbol{h} \in \mathscr{Y}^r$ the quantities

$$v_i(\boldsymbol{\alpha},\boldsymbol{p},\boldsymbol{\varphi}) := \left\langle \sum_{j=1}^r \alpha_j \mathcal{T}_j(p_j)\varphi_j - z, \mathcal{T}_i(p_i)\varphi_i \right\rangle_{\mathscr{X}}, \tag{15a}$$

$$\xi_i(\boldsymbol{\alpha},\boldsymbol{p},\boldsymbol{\varphi}) := \left\langle \sum_{j=1}^r \alpha_j \mathcal{T}_j(p_j)\varphi_j - z, \alpha_i \left[\tfrac{\partial}{\partial p_i}\mathcal{T}_i(p_i)\varphi_i\right] \right\rangle_{\mathscr{X}}, \tag{15b}$$

$$\mu_i(\boldsymbol{\alpha},\boldsymbol{p},\boldsymbol{\varphi},\boldsymbol{h}) := \left\langle \sum_{j=1}^r \alpha_j \mathcal{T}_j(p_j)\varphi_j - z, \alpha_i \mathcal{T}_i(p_i)h_i \right\rangle_{\mathscr{X}}, \tag{15c}$$

for $i = 1,\ldots,r$.

We start our exposition with the discretization with respect to time. To this end, consider a time grid $0 = t_0 < t_1 < \ldots < t_m = T$ and associated quadrature rule defined by weights $\omega_\ell \geq 0$ for $\ell = 0, \ldots, m$. The approximation of the directional derivative of J with respect to $\boldsymbol{\alpha}$ is thus given by

$$\begin{aligned}
\partial_{\boldsymbol{\alpha},\boldsymbol{d}} J(\boldsymbol{\alpha},\boldsymbol{p},\boldsymbol{\varphi}) &= \sum_{i=1}^{r} \int_0^T v_i(\boldsymbol{\alpha}(t),\boldsymbol{p}(t),\boldsymbol{\varphi}) d_i(t)\,\mathrm{d}t \\
&\approx \sum_{i=1}^{r} \sum_{k=0}^{m} w_k v_i(\boldsymbol{\alpha}(t_k),\boldsymbol{p}(t_k),\boldsymbol{\varphi}) d_i(t_k) \\
&= \left(\boldsymbol{v}^m(\boldsymbol{\alpha},\boldsymbol{p},\boldsymbol{\varphi})\right)^\top (I_r \otimes W)\, \boldsymbol{d}^m,
\end{aligned}$$

where $\otimes$ is the Kronecker product and

$$\begin{aligned}
W &:= \operatorname{diag}(w_0,\ldots,w_m) \in \mathbb{R}^{(m+1)\times(m+1)}, \\
\boldsymbol{v}_i^m(\boldsymbol{\alpha},\boldsymbol{p},\boldsymbol{\varphi}) &:= \left[v_i(\boldsymbol{\alpha}(t_0),\boldsymbol{p}(t_0),\boldsymbol{\varphi}) \cdots v_i(\boldsymbol{\alpha}(t_m),\boldsymbol{p}(t_m),\boldsymbol{\varphi})\right]^\top \in \mathbb{R}^{m+1}, \\
\boldsymbol{v}^m(\boldsymbol{\alpha},\boldsymbol{p},\boldsymbol{\varphi}) &:= \left[\boldsymbol{v}_1^m(\boldsymbol{\alpha},\boldsymbol{p},\boldsymbol{\varphi})^\top \cdots \boldsymbol{v}_r^m(\boldsymbol{\alpha},\boldsymbol{p},\boldsymbol{\varphi})^\top\right]^\top \in \mathbb{R}^{r(m+1)}, \\
\boldsymbol{d}_i^m &:= \left[d_i(t_0) \cdots d_i(t_m)\right]^\top \in \mathbb{R}^{m+1}, \\
\boldsymbol{d}^m &:= \left[(\boldsymbol{d}_1^m)^\top \cdots (\boldsymbol{d}_r^m)^\top\right]^\top \in \mathbb{R}^{r(m+1)}.
\end{aligned}$$

The time-discrete approximation of the partial derivative is thus given by

$$\partial_{\boldsymbol{\alpha}} J^m(\boldsymbol{\alpha},\boldsymbol{p},\boldsymbol{\varphi}) := \left(\boldsymbol{v}^m(\boldsymbol{\alpha},\boldsymbol{p},\boldsymbol{\varphi})\right)^\top (I_r \otimes W) \in \mathbb{R}^{1\times r(m+1)}.$$

Analogously, the time-discrete approximation of the partial derivative of J with respect to the path variables, i.e., $\partial_{\boldsymbol{p}} J(\boldsymbol{\alpha},\boldsymbol{p},\boldsymbol{\varphi})$, is given by

$$\partial_{\boldsymbol{p}} J^m(\boldsymbol{\alpha},\boldsymbol{p},\boldsymbol{\varphi}) := \left(\boldsymbol{\xi}^m(\boldsymbol{\alpha},\boldsymbol{p},\boldsymbol{\varphi})\right)^\top (I_r \otimes W) \in \mathbb{R}^{1\times r(m+1)},$$

with $\boldsymbol{\xi}^m(\boldsymbol{\alpha},\boldsymbol{p},\boldsymbol{\varphi})$ defined analogously as $\boldsymbol{v}^m(\boldsymbol{\alpha},\boldsymbol{p},\boldsymbol{\varphi})$. In the same fashion, we obtain the time-discrete approximation for the directional derivative with respect to the mode variables as

$$\partial_{\boldsymbol{\varphi},\boldsymbol{h}} J^m(\boldsymbol{\alpha},\boldsymbol{p},\boldsymbol{\varphi}) := \left(\boldsymbol{\mu}^m(\boldsymbol{\alpha},\boldsymbol{p},\boldsymbol{\varphi},\boldsymbol{h})\right)^\top (I_r \otimes W)\, \mathbf{1}_{r(m+1)} \in \mathbb{R},$$

where we denote by $\mathbf{1}_{r(m+1)} \in \mathbb{R}^{r(m+1)}$ the vector with all entries equal to 1, and $\boldsymbol{\mu}^m$ defined analogously as $\boldsymbol{v}^m$.

For the spatial discretization, let $\mathscr{Y}_n$ denote an n-dimensional subspace of $\mathscr{Y}$ with basis functions $\psi_1,\ldots,\psi_n \in \mathscr{Y}$. Let us define for $i,j = 1,\ldots,r$ and $p_i, p_j \in \mathbb{R}$ the matrices $M_{i,j}, N_{i,j}, F_i, G_i \in \mathbb{R}^{n\times n}$ via

$$\begin{aligned}
\left[M_{i,j}(p_i,p_j)\right]_{k,\ell} &:= \langle \mathcal{T}_j(p_j)\psi_k, \mathcal{T}_i(p_i)\psi_\ell \rangle_{\mathscr{X}}, && \text{(16a)} \\
\left[N_{i,j}(p_i,p_j)\right]_{k,\ell} &:= \langle \mathcal{T}_j(p_j)\psi_k, \tfrac{\partial}{\partial p_i}\mathcal{T}_i(p_i)\psi_\ell \rangle_{\mathscr{X}}, && \text{(16b)} \\
\left[F_i(p_i)\right]_{k,\ell} &:= \langle \psi_k, \mathcal{T}_i(p_i)\psi_\ell \rangle_{\mathscr{X}}, && \text{(16c)} \\
\left[G_i(p_i)\right]_{k,\ell} &:= \langle \psi_k, \tfrac{\partial}{\partial p_i}\mathcal{T}_i(p_i)\psi_\ell \rangle_{\mathscr{X}}, && \text{(16d)}
\end{aligned}$$

for $k, \ell = 1, \ldots, r$. For the data $z \in L^2(0,T;\mathscr{Y})$, the modes $\varphi_i \in \mathscr{Y}$, and directions $h_i \in \mathscr{Y}$, we consider the approximations

$$z(t) \approx \sum_{\ell=1}^{n} \widehat{z}_\ell(t)\psi_\ell, \qquad \varphi_i \approx \sum_{\ell=1}^{n} \widehat{\varphi}_{i,\ell}\psi_\ell, \qquad h_i \approx \sum_{\ell=1}^{n} \widehat{h}_{i,\ell}\psi_\ell,$$

with

$$\widehat{\boldsymbol{z}}(t) := [\widehat{z}_1(t) \ \cdots \ \widehat{z}_n(t)]^\top \in \mathbb{R}^n, \qquad \widehat{\boldsymbol{\varphi}}_i := [\widehat{\varphi}_{i,1} \ \cdots \ \widehat{\varphi}_{i,n}]^\top \in \mathbb{R}^n,$$
$$\widehat{\boldsymbol{h}}_i := [\widehat{h}_{i,1} \ \cdots \ \widehat{h}_{i,n}]^\top \in \mathbb{R}^n.$$

With these preparations, we obtain the spatial approximation of the inner products in (15) as

$$\widehat{v}_i(\boldsymbol{\alpha},\boldsymbol{p},\boldsymbol{\varphi}) := \left(\sum_{j=1}^{r} \alpha_j \widehat{\boldsymbol{\varphi}}_j^\top M_{i,j}(p_i,p_j) - \widehat{\boldsymbol{z}}^\top F_i(p_i)\right)\widehat{\boldsymbol{\varphi}}_i \in \mathbb{R},$$

$$\widehat{\xi}_i(\boldsymbol{\alpha},\boldsymbol{p},\boldsymbol{\varphi}) := \alpha_i\left(\sum_{j=1}^{r} \widehat{\boldsymbol{\varphi}}_j^\top N_{i,j}(p_i,p_j) - \widehat{\boldsymbol{z}}^\top G_i(p_i)\right)\widehat{\boldsymbol{\varphi}}_i \in \mathbb{R},$$

$$\widehat{\mu}_i(\boldsymbol{\alpha},\boldsymbol{p},\boldsymbol{\varphi}) := \alpha_i\left(\sum_{j=1}^{r} \alpha_j \widehat{\boldsymbol{\varphi}}_j^\top M_{i,j}(p_i,p_j) - \widehat{\boldsymbol{z}}^\top F_i(p_i)\right) \in \mathbb{R}^{1\times n}.$$

We thus obtain the space- and time-discretized partial derivatives as

$$\partial_{\boldsymbol{\alpha}} \widehat{J}^m(\boldsymbol{\alpha},\boldsymbol{p},\boldsymbol{\varphi}) := (\widehat{\boldsymbol{v}}^m(\boldsymbol{\alpha},\boldsymbol{p},\boldsymbol{\varphi}))^\top (I_r \otimes W) \in \mathbb{R}^{1\times(m+1)r},$$
$$\partial_{\boldsymbol{p}} \widehat{J}^m(\boldsymbol{\alpha},\boldsymbol{p},\boldsymbol{\varphi}) := (\widehat{\boldsymbol{\xi}}^m(\boldsymbol{\alpha},\boldsymbol{p},\boldsymbol{\varphi}))^\top (I_r \otimes W) \in \mathbb{R}^{1\times(m+1)r},$$
$$\partial_{\boldsymbol{\varphi}} \widehat{J}^m(\boldsymbol{\alpha},\boldsymbol{p},\boldsymbol{\varphi}) := \mathbf{1}_{r(m+1)}^\top (I_r \otimes W)\,\mathcal{M} \in \mathbb{R}^{1\times nr},$$

with $\mathcal{M} := \operatorname{blkdiag}(\widehat{\mu}_1^m(\boldsymbol{\alpha},\boldsymbol{p},\boldsymbol{\varphi}),\ldots,\widehat{\mu}_r^m(\boldsymbol{\alpha},\boldsymbol{p},\boldsymbol{\varphi}))$.

We conclude this section with a specific computation of the quantities depending on the inner products for the periodic shift operator and P_1 finite elements.

Example 3. Let us assume we have a one-dimensional domain $\Omega = (0,1)$ and a corresponding equidistant grid of step size $h := \frac{1}{n}$. We discretize $\mathscr{Y} = H^1_{\text{per}}(\Omega)$ via periodic P_1 finite element functions. For $\mathscr{X} = L^2(0,1)$ and shift operator with periodic embedding, we observe

$$M_{i,j}(p_i,p_j) = F_i(p_i - p_j) \qquad \text{and} \qquad N_{i,j}(p_i,p_j) = G_i(p_i - p_j).$$

For $p_i = qh + \tilde{p}_i$ with $q \in \mathbb{Z}$ and $\tilde{p}_i \in [0,h)$ we obtain

$$\langle \psi_k, \mathcal{T}_i(p_i)\psi_\ell\rangle = \begin{cases} \frac{1}{h^2}\left(\frac{2}{3}(h-\tilde{p}_i)^3 + \tilde{p}_i(h-\tilde{p}_i)^2 + \tilde{p}_i h(h-\tilde{p}_i) + \frac{1}{6}\tilde{p}_i^3\right), & \text{if } \ell = k-q,\\ \frac{1}{6h^2}(h-\tilde{p}_i)^3, & \text{if } \ell = k-q+1,\\ \frac{1}{h^2}\left(\frac{1}{6}(h-\tilde{p}_i)^3 - \frac{1}{3}\tilde{p}_i^3 + h^2\tilde{p}_i\right), & \text{if } \ell = k-q-1,\\ \frac{1}{6h^2}\tilde{p}_i^3, & \text{if } \ell = k-q-2,\\ 0, & \text{otherwise.}\end{cases}$$

For further details, including the computation of $\langle \psi_k, \frac{\partial}{\partial p_i} \mathcal{T}_i(p_i)\psi_\ell \rangle_{\mathscr{X}}$, we refer to the preprint version of this manuscript [4].

5 Numerical Examples

For our numerical examples, we use an equidistant time grid $t_i := i\tau$ with step size $\tau > 0$. The weights for the time integration are chosen based on the trapezoidal rule. For the discretization in space, we also use an equidistant grid and follow Example 3 for approximating the inner products occurring in the cost functional and the gradient. In particular, we use the periodic shift operator for all numerical examples. The optimization itself is carried out with the MATLAB® package GRANSO with default settings, see [8], which is based on a quasi-Newton solver.

For notational convenience, we assumed so far that there is exactly one mode per transformation operator. In practice, it is often more reasonable to cluster the modes into different reference frames, see, for instance, [3, sec. 7.1]. Thus, we use the clustered approximation ansatz

$$z \approx \sum_{\rho=1}^{q} \sum_{i=1}^{r_\rho} \alpha_{\rho,i} \mathcal{T}_\rho (p_\rho) \varphi_{\rho,i} \tag{18}$$

for the following numerical experiments and emphasize that this only requires a minor and straightforward modification of the gradient. We denote the approximation based on our optimization results with tPOD. Furthermore, we use dashed lines in the plots to display the (optimized) path variables. Finally, note that even though the data for the numerical examples stems from partial differential equations, our main concern in this work is the approximation and compression of any given data. We thus forego a thorough treatment of the actual values in the following pseudocolor plots.

To ensure reproducibility of the conducted experiments, the code for the numerical examples is publicly available under https://doi.org/10.5281/zenodo.5471404.

5.1 Viscous Burgers' Equation

We consider the one-dimensional viscous Burgers' equation

$$\tfrac{\partial}{\partial t} z(t,x) = \tfrac{1}{\mathrm{Re}} \tfrac{\partial^2}{\partial x^2} z(t,x) - z(t,x) \tfrac{\partial}{\partial x} z(t,x), \qquad (t,x) \in (0,2) \times (0,1), \tag{19}$$

and, following [19], use the analytical solution

$$z(t,x) = \frac{x}{t+1} \left(1 + \sqrt{\frac{t+1}{\exp\left(\frac{\mathrm{Re}}{8}\right)}} \exp\left(\mathrm{Re} \frac{x^2}{4t+4}\right) \right)^{-1} \tag{20}$$

with Reynolds number $\mathrm{Re} = 1000$ for our experiment. Let us emphasize that (20) is not periodic in x. Nevertheless, the shock front (as depicted in Fig. 1a) stays

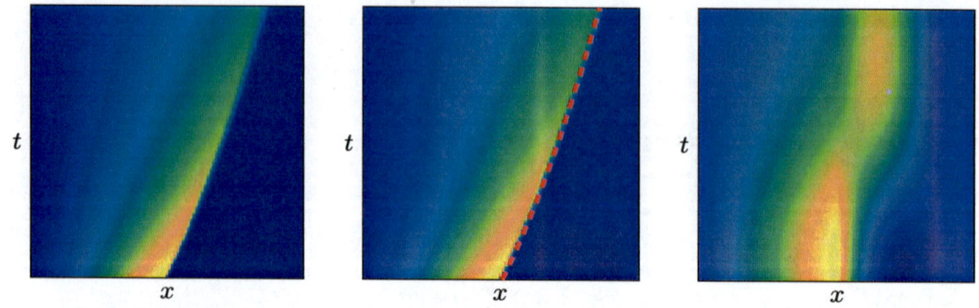

(a) snapshot data (b) tPOD $r = 2$ (c) POD $r = 2$

Fig. 1. Burgers' equation – comparison of original data (a), tPOD approximation (b), and POD approximation (c). The red dashed line in (b) denotes the optimized path variable.

within the computational domain and the solution is approximately zero at the boundaries, allowing us to treat this problem as quasi-periodic. We test our algorithm with data obtained from the analytical solution, evaluated on a grid with 100 equidistant intervals in space and time, respectively. We use a single frame, i.e., $q = 1$ in (18), and $r_1 = r = 2$ modes, and supply the first snapshot and the zero vector as starting values for the modes. The corresponding coefficients are initialized as a constant function with value 1. For the path, we start with a straight line given by $p(t) = \frac{37}{200}t$. The results are depicted in Fig. 1, detailing that already with $r = 2$, an accurate approximation with a relative L^2 error of less than 3% can be achieved, while the POD approximation is not able to reproduce the shock front. Besides, we compare the relative L^2 errors of the approximations for different mode numbers in Table 1, detailing the superior approximation capabilities of tPOD for this test case.

Table 1. Burgers' equation – comparison of relative L^2 errors

r	tPOD	POD
1	1.224×10^{-1}	4.510×10^{-1}
2	2.910×10^{-2}	2.863×10^{-1}
3	1.328×10^{-2}	2.111×10^{-1}
4	8.453×10^{-3}	1.662×10^{-1}
5	6.821×10^{-3}	1.354×10^{-1}

5.2 Nonlinear Schrödinger Equation

In this section, we consider the nonlinear Schrödinger equation

$$\begin{aligned} \mathrm{i}\tfrac{\partial}{\partial t} z(t,x) &= -\tfrac{1}{2}\tfrac{\partial^2}{\partial x^2} z(t,x) + \kappa |z(t,x)|^2 z(t,x), \\ z(t,0) &= 2\mathrm{sech}(x+7)\exp(2\mathrm{i}x) + 2\mathrm{sech}(x-7)\exp(-2\mathrm{i}x), \end{aligned}$$

as presented in [20, Example 4]. We compute a solution using the code from [20] on a uniform grid of 501×1024 points on the domain $\mathbb{T} \times \Omega = [0, 2\pi] \times [-15, 15]$. The absolute value of the numerical solution is presented in Fig. 2a.

We initialize our algorithm by assuming an approximation with two frames, each with two modes. As starting value for the modes, we use

$$\varphi_{\rho,1}(x) = 2\text{sech}(x - (-1)^{\rho}7)\exp(-(-1)^{\rho}2\text{i}x) \qquad \text{and} \qquad \varphi_{\rho,2} \equiv 0$$

for $\rho = 1, 2$. The coefficients are initialized as constants, with value 1 at each time point. For the initial paths, we use $p_1(t) = 2t$ and $p_2(t) = -2t$. The corresponding approximation and the absolute error are presented in Figs. 2b and c. We observe that the error results mainly from the complicated wave dynamics in the middle of the spatial and time domain, whereas the transported wave profiles are captured accurately. Let us emphasize that the error is very localized such that it can be captured with only a few additional POD modes. We notice that the optimizer does not keep the linear path over the whole time domain. Instead, as

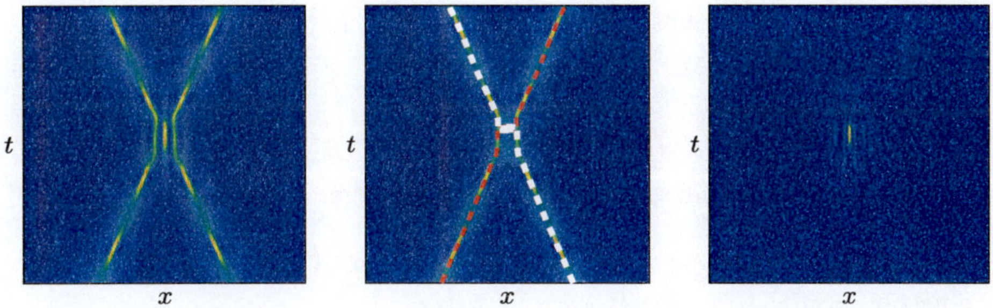

(a) snapshot data (b) tPOD $r = 4$ (c) absolute error

Fig. 2. Nonlinear Schrödinger equation – original data (a), tPOD approximation with $r = 4$ total modes (b), and absolute error (c) for initial paths $p_1(t) = 2t$ and $p_2(t) = -2t$. The red and white dashed lines in (b) denote the optimization results for the first and second path variable, respectively.

depicted in Fig. 2b, in the middle of the computational domain, the paths jump between the wavefronts. Inspecting the snapshot matrix in the co-moving frame along the path $p_1(t) = 2t$ in Fig. 3 provides a possible explanation: the vertical wavefront features an offset after the two waves have crossed. The optimizer needs to account for this offset, which explains the jump. Let us emphasize that with a different initialization for the path variables, the optimizer finds another local minimum with a similar approximation quality. Using piecewise linear paths as depicted in Fig. 4a, we observe that the resulting optimized path smoothes out the edges of the initial path in the middle of the domain (cf. Fig. 4c) and

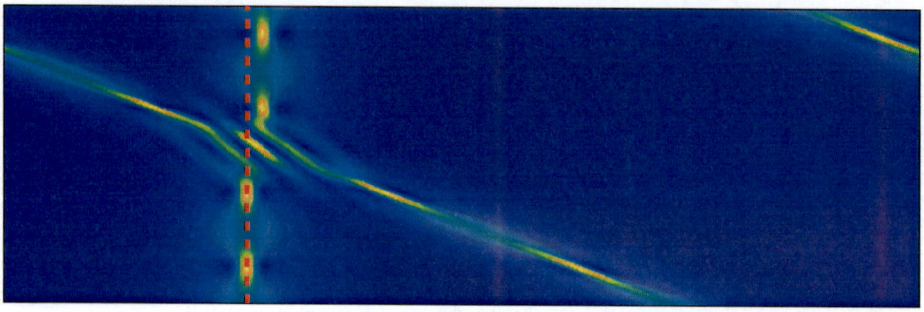

Fig. 3. Nonlinear Schrödinger equation – transformation of Fig. 2a to the co-moving frame along the path $p_1(t) = 2t$, i.e., applying the shift operator along the path $-p_1(t)$ to the original snapshot data in Fig. 2a, such that the wave front originally traveling to the right becomes stationary. For illustration purposes, the image is stretched to clearly present the offset in the vertical wave profile after the two waves have crossed.

does not feature any jumps. It is smooth and tracks the wavefronts as if the waves reflect off each other.

5.3 FitzHugh–Nagumo Wave Train

We follow [13] and consider the FitzHugh–Nagumo model given by

$$\begin{aligned} \tfrac{\partial}{\partial t}u_1(t,x) &= \nu\tfrac{\partial^2}{\partial x^2}u_1(t,x) - u_2(t,x) + u_1(t,x)(1-u_1(t,x))(u_1(t,x)-a), \\ \tfrac{\partial}{\partial t}u_2(t,x) &= \epsilon(bu_1(t,x) - u_2(t,x)), \end{aligned} \tag{21}$$

with spatial domain $(0, 500)$ and time interval $(0, 1000)$. The partial differential equation (21) is closed by periodic boundary conditions and the initial condition

$$u_1(0,x) = \tfrac{1}{2}\left(1+\sin\left(\tfrac{\pi}{50}x\right)\right), \quad u_2(0,x) = \tfrac{1}{2}\left(1+\cos\left(\tfrac{\pi}{50}x\right)\right).$$

For the parameter values, we choose $\nu = 1$, $a = -0.1$, $\epsilon = 0.05$, and $b = 0.3$. The spatial discretization of (21) is performed via a central sixth-order finite-difference scheme with mesh width $h = 0.5$ and for the time integration we use MATLAB®'s `ode45` function based on a time grid with step size 1. The corresponding numerical solution for the variable u_1 is depicted in Fig. 5a.

For the optimization we consider only the data of the variable u_1 and use an approximation with one reference frame to account for the traveling wave train. Furthermore, we reduce the computational complexity by considering the optimization problem only in terms of the path, whereas the coefficients and modes are computed in each iteration via a truncated singular value decomposition of the snapshot matrix shifted into the co-moving reference frame. Here we exploit that the periodic shift operator is isometric, such that we can solve the optimization problem via classical POD with transformed data, cf. [3, Thm. 4.8]. As starting value for the path, we choose a linear function in t with slope

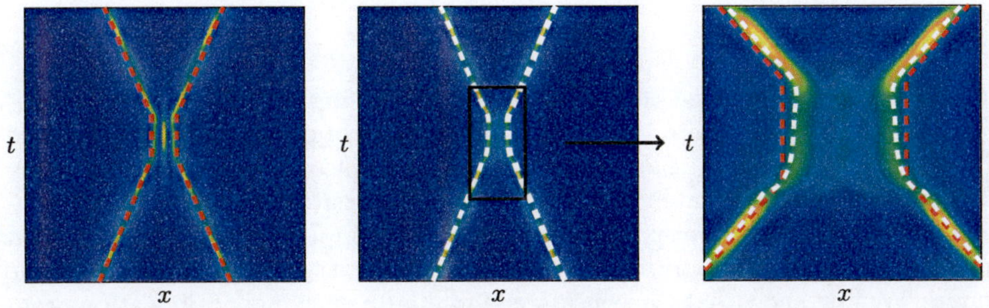

(a) initial paths (b) tPOD $r = 4$ (c) zoom: initial vs. optimized path

Fig. 4. Nonlinear Schrödinger equation – comparison of original data (a) with piecewise linear initialization of the path variables (red dashed lines) and tPOD approximation (b) with $r = 4$ total modes and optimized path variables (white dashed lines). Figure 4c shows a zoom into Fig. 4b and illustrates that the optimized path variables (white dashed lines) differ from the initial path variables (red dashed lines).

1.04, which we determined by inspecting the first and the last snapshot of the original data. The corresponding approximation obtained from the optimization procedure is depicted in Fig. 5b. As reference approximation, we consider a POD

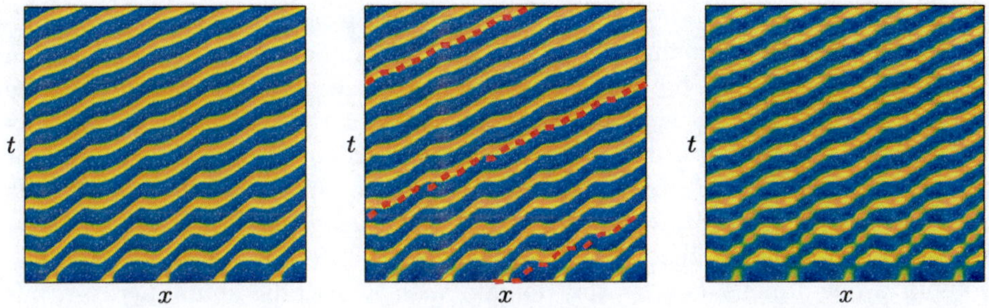

(a) snapshot data (b) tPOD $r = 4$ (c) POD $r = 4$

Fig. 5. FitzHugh–Nagumo model – comparison of original data for u_1 (a), tPOD approximation with $r = 4$ modes (b), and POD approximation with $r = 4$ modes (c). The red dashed line in (b) denotes the optimized path variable.

approximation with the same number of modes in Fig. 5c. The corresponding total relative errors are 15% for the approximation based on shifted modes and 31% for the POD approximation. We note that in contrast to the Burgers test case considered in Sect. 5.1, the traveling wave train can be better approximated by POD due to the lack of a traveling shock wave. Correspondingly, the difference between tPOD and POD approximation for the considered FitzHugh–Nagumo test case is less striking than the one observed for the example in Sect. 5.1.

6 Summary

In this paper, we analyze the problem of determining an optimal approximation of given snapshot data by a linear combination of dynamically transformed modes. This data compression can, for instance, be used for model order reduction of transport-dominated systems [3]. As optimization parameters, we consider the modes, the corresponding coefficients or amplitudes, and the so-called path variables, which parameterize the coordinate transforms applied to the modes. We first show that the considered infinite-dimensional optimization problem possesses a minimizing solution if the admissible set is constrained such that the optimization parameters are norm bounded. Afterward, we derive a corresponding unconstrained optimization problem by adding an appropriate penalization term and show that the unconstrained problem also has a solution. Furthermore, we demonstrate that if the penalization coefficient tends to infinity, each limit point of the corresponding sequence of minimizers is a solution to the original constrained optimization problem. To derive a gradient-based optimization procedure, we compute the partial Fréchet derivatives of the unconstrained cost functional and discuss their space and time discretization. Finally, we apply the optimization procedure to some numerical test cases and observe that the optimized decompositions are significantly more accurate than corresponding approximations obtained by the classical POD with the same number of modes.

After full discretization, the optimization problem still features a large number of optimization parameters scaling with the number of grid points in space and time. Thus, an interesting future research direction is to investigate approaches for reducing the computational complexity of the optimization procedure, for instance, by using multigrid optimization techniques [21], or by making use of low-dimensional parametrizations of the optimization parameters, cf. Remark 2. Such a parametrization seems to be especially promising for reducing the complexity in the path variables since our numerical experiments revealed that the optimization procedure is sensitive with respect to the paths. Furthermore, let us emphasize that, although we have only discussed applications in a one-dimensional spatial domain with a periodic shift operator, our framework is not restricted to this case. Thus, another promising direction for the future is to explore the applicability to problems with higher-dimensional spatial domains using different transformation operators, see e.g. [15,28,31] for some contributions in this direction.

Acknowledgments. The work of F. Black is supported by the Deutsche Forschungsgemeinschaft (DFG, German Research Foundation) Collaborative Research Center (CRC) 1029 *Substantial efficiency increase in gas turbines through direct use of coupled unsteady combustion and flow dynamics*, project number 200291049. P. Schulze acknowledges funding by the DFG CRC Transregio 154 *Mathematical Modelling, Simulation and Optimization Using the Example of Gas Networks*, project number 239904186. B. Unger acknowledges funding from the DFG under Germany's Excellence Strategy – EXC 2075 – 390740016 and is thankful for support by the Stuttgart Center for Simulation Science (SimTech).

References

1. Adams, R.A., Fournier, J.J.F.: Sobolev Spaces, 2nd edn. Elsevier, Oxford (2003)
2. Beyn, W.J., Thümmler, V.: Freezing solutions of equivariant evolution equations. SIAM J. Appl. Dyn. Syst. **3**(2), 85–116 (2004)
3. Black, F., Schulze, P., Unger, B.: Projection-based model reduction with dynamically transformed modes. ESAIM Math. Model. Numer. Anal. **54**(6), 2011–2043 (2020)
4. Black, F., Schulze, P., Unger, B.: Decomposition of flow data via gradient-based transport optimization. ArXiv preprint 2107.03481, ArXiv (2021)
5. Black, F., Schulze, P., Unger, B.: Efficient wildland fire simulation via nonlinear model order reduction. Fluids **6**(8), 280 (2021)
6. Black, F., Schulze, P., Unger, B.: Model order reduction with dynamically transformed modes for the wave equation. PAMM **20**(1), e202000321 (2021)
7. Cagniart, N., Maday, Y., Stamm, B.: Model order reduction for problems with large convection effects. In: Chetverushkin, B.N., Fitzgibbon, W., Kuznetsov, Y.A., Neittaanmäki, P., Periaux, J., Pironneau, O. (eds.) Contributions to Partial Differential Equations and Applications, pp. 131–150. Springer, Cham (2019). https://doi.org/10.1007/978-3-319-78325-3_10
8. Curtis, F.E., Mitchell, T., Overton, M.L.: A BFGS-SQP method for nonsmooth, nonconvex, constrained optimization and its evaluation using relative minimization profiles. Optim. Methods Softw. **32**(1), 148–181 (2017)
9. Engel, K.J., Nagel, R.: One-Parameter Semigroups for Linear Evolution Equations. Springer, New York (2000). https://doi.org/10.1007/b97696
10. Greif, C., Urban, K.: Decay of the Kolmogorov n-width for wave problems. Appl. Math. Lett. **96**, 216–222 (2019)
11. Gubisch, M., Volkwein, S.: Chapter 1: Proper orthogonal decomposition for linear-quadratic optimal control. In: Benner, P., Cohen, A., Ohlberger, M., Willcox, K. (eds.) Model Reduction and Approximation, pp. 3–63. SIAM, Philadelphia (2017)
12. Iollo, A., Lombardi, D.: Advection modes by optimal mass transfer. Phys. Rev. E **89**(2), 022923 (2014)
13. Koch, J.: Data-driven modeling of nonlinear traveling waves. Chaos **31**(4), 043128 (2021)
14. Kolmogoroff, A.: Über die beste Annäherung von Funktionen einer gegebenen Funktionenklasse. Ann. Math. **37**(1), 107–110 (1936)
15. Krah, P., Sroka, M., Reiss, J.: Model order reduction of combustion processes with complex front dynamics. In: Vermolen, F.J., Vuik, C. (eds.) Numerical Mathematics and Advanced Applications ENUMATH 2019, pp. 803–811. Springer, Cham (2021). https://doi.org/10.1007/978-3-030-55874-1_79
16. Lee, K., Carlberg, K.T.: Model reduction of dynamical systems on nonlinear manifolds using deep convolutional autoencoders. J. Comput. Phys. **404**, 108973 (2020)
17. Luenberger, D.G., Ye, Y.: Linear and Nonlinear Programming. ISORMS, vol. 116. Springer, New York (2008). https://doi.org/10.1007/978-0-387-74503-9_15
18. Maday, Y., Patera, A.T., Turinici, G.: A priori convergence theory for reduced-basis approximations of single-parameter elliptic partial differential equations. J. Sci. Comput. **17**(1), 437–446 (2002)
19. Mauli, R., Lusch, B., Balaprakash, P.: Reduced-order modeling of advection-dominated systems with recurrent neural networks and convolutional autoencoders. Phys. Fluids **33**, 037106 (2021)

20. Mendible, A., Brunton, S.L., Aravkin, A.Y., Lowrie, W., Kutz, J.N.: Dimensionality reduction and reduced-order modeling for traveling wave physics. Theoret. Comput. Fluid Dyn. **34**(4), 385–400 (2020). https://doi.org/10.1007/s00162-020-00529-9
21. Nash, S.G.: A multigrid approach to discretized optimization problems. Optim. Methods Softw. **14**(1–2), 99–116 (2000)
22. Nonino, M., Ballarin, F., Rozza, G., Maday, Y.: Overcoming slowly decaying Kolmogorov n-width by transport maps: application to model order reduction of fluid dynamics and fluid structure interaction problems. ArXiv preprint 1911.06598 (2019)
23. Ohlberger, M., Rave, S.: Nonlinear reduced basis approximation of parameterized evolution equations via the method of freezing. C. R. Acad. Sci. Paris **351**(23–24), 901–906 (2013)
24. Peherstorfer, B.: Model reduction for transport-dominated problems via online adaptive bases and adaptive sampling. SIAM J. Sci. Comput. **42**(5), A2803–A2836 (2020)
25. Reiss, J.: Optimization-based modal decomposition for systems with multiple transports. SIAM J. Sci. Comput. **43**(3), A2079–A2101 (2021)
26. Reiss, J., Schulze, P., Sesterhenn, J., Mehrmann, V.: The shifted proper orthogonal decomposition: a mode decomposition for multiple transport phenomena. SIAM J. Sci. Comput. **40**(3), A1322–A1344 (2018)
27. Rim, D., Moe, S., LeVeque, R.J.: Transport reversal for model reduction of hyperbolic partial differential equations. SIAM/ASA J. Uncertain. Quantif. **6**(1), 118–150 (2018)
28. Rim, D., Peherstorfer, B., Mandli, K.T.: Manifold approximations via transported subspaces: model reduction for transport-dominated problems. ArXiv preprint 1912.13024v2 (2020)
29. Rudin, W.: Real and Complex Analysis. ISORMS, Springer, Singapore (2018). https://doi.org/10.1007/978-981-13-2886-2_4
30. Schulze, P., Reiss, J., Mehrmann, V.: Model reduction for a pulsed detonation combuster via shifted proper orthogonal decomposition. In: King, R. (ed.) Active Flow and Combustion Control 2018, pp. 271–286. Springer, Cham (2019). https://doi.org/10.1007/978-3-319-98177-2_17
31. Taddei, T.: A registration method for model order reduction: data compression and geometry reduction. SIAM J. Sci. Comput. **42**(2), A997–A1027 (2020)
32. Taddei, T., Perotto, S., Quarteroni, A.: Reduced basis techniques for nonlinear conservation laws. ESAIM Math. Model. Numer. Anal. **49**(3), 787–814 (2015)
33. Unger, B., Gugercin, S.: Kolmogorov n-widths for linear dynamical systems. Adv. Comput. Math. **45**(5–6), 2273–2286 (2019)
34. Welper, G.: Transformed snapshot interpolation with high resolution transforms. SIAM J. Sci. Comput. **42**(4), A2037–A2061 (2020)
35. Zeidler, E.: Nonlinear Functional Analysis and its Applications IIA: Linear Monotone Operators. Springer, New York (1990). https://doi.org/10.1007/978-1-4612-0985-0

Towards Data-Driven Model Reduction of the Navier-Stokes Equations Using the Loewner Framework

Alejandro N. Diaz and Matthias Heinkenschloss(✉)

Department of Computational and Applied Mathematics, MS-134, Rice University, 6100 Main Street, Houston, TX 77005-1892, USA
{and5,heinken}@rice.edu

Abstract. The Loewner framework is extended to compute reduced order models (ROMs) for systems governed by the incompressible Navier-Stokes (NS) equations. For quadratic ordinary differential equations (ODEs) it constructs a ROM directly from measurements of transfer function components derived from an expansion of the system's input-to-output map. Given measurements, no explicit access to the system is required to construct the ROM.

To extend the Loewner framework, the NS equations are transformed into ODEs by projecting onto the subspace defined by the incompressibility condition. This projection is used theoretically, but avoided computationally. This paper presents the overall approach. Currently, transfer function measurements are obtained via computational simulations; obtaining them from experiments is an open issue. Numerical results show the potential of the Loewner framework, but also reveal possible lack of stability of the ROM. A possible approach, which currently requires access to the NS system, to deal with these instabilities is outlined.

Keywords: Model reduction · Loewner framework · Navier-Stokes equations · Data-driven

1 Introduction

This paper extends the data-driven Loewner framework to construct a reduced order model (ROM) for systems governed by the semi-discretized incompressible Navier-Stokes (NS) equations. These computationally inexpensive ROMs are useful in applications that require many queries of the system, such as optimal design, optimal control, or uncertainty quantification, which would be prohibitively expensive with the original, high dimensional, computationally expensive full order model (FOM). The Loewner framework constructs a ROM that is given by a Petrov-Galerkin projection of the FOM. However, unlike traditional Petrov-Galerkin projection ROMs, the Loewner ROM is computed directly from measurements of transfer function components associated with the FOM. In

R. King and D. Peitsch (Eds.): AFCC 2021, NNFM 152, pp. 225–239, 2022.
https://doi.org/10.1007/978-3-030-90727-3_14

particular, the Loewner framework is non-intrusive. It only requires these transfer function component measurements, but it does not need explicit access to the FOM or explicit application of the Petrov-Galerkin projections. The transfer function components of the Loewner ROM typically well approximate the corresponding transfer function components of the FOM near the frequencies at which measurements were taken to construct the Loewner ROM. Thus, the overall quality of the Loewner ROM depends on the measurements on which it is built. This is the first application of the Loewner framework to the semi-discretized incompressible NS equations, which in the terminology of ROMs is a quadratic semi-explicit differential algebraic equation (DAE) system.

The development of the Loewner framework started with the paper [16]. The tutorial paper [6] and the book [2] provide the state-of-the-art of the Loewner framework. The papers [5,12], extend the Loewner framework to so-called bilinear and quadratic ordinary differential equation (ODE) systems, and [3] discusses the application of the Loewner framework to systems governed by Burgers' equation. The Loewner framework is extended in [4] to a class of linear semi-explicit DAE systems, which includes the Oseen equations. This paper builds on [3,4].

To extend the existing Loewner framework to the semi-discretized NS equations we first project these equations onto the subspace defined by the discrete incompressibility constraints to express the semi-discretized NS equations as a quadratic ODE system. This projection has already been used, e.g., in [1,4,7,9,14]. Then the quadratic ODE system is expanded into a system of infinitely many linear equations. Such expansions are discussed, e.g., in the book [17] and they have been applied to develop ROMs for bilinear or quadratic bilinear systems in, e.g., [1,3,5,8–10,12,13]. The papers [3,5,8,10,12] consider ODE systems, not DAEs. The papers [1,9] also consider projection based ROM for the NS equations, and the ROMs are also designed to approximate transfer function components of the FOM. However, the transfer function components used in [1,9] are different from those used here, and the ROM approaches in [1,9] require explicit access to components of the NS equations and are intrusive.

This paper describes the extension of the Loewner framework to a class of semi-discretized incompressible NS systems, and numerically explores potential stability issues of the resulting Loewner ROM and possible modifications of this ROM to avoid them.

2 Loewner Framework for the Navier-Stokes Equations

We state the semi-discretized NS system, and its projection onto the subspace defined by the discrete incompressibility conditions. This projection transforms the Navier-Stokes system into a quadratic ODE system. We then review the Loewner framework for this projected ODE system.

2.1 Navier-Stokes System

Let $\mathbf{v} : (0, T) \rightarrow \mathbb{R}^{n_v}$ and $\mathbf{p} : (0, T) \rightarrow \mathbb{R}^{n_p}$ be the semi-discretized velocities and pressures, respectively. Furthermore, let $\mathbf{E}_{11} \in \mathbb{R}^{n_v \times n_v}$ be symmetric positive definite, $\mathbf{A}_{11} \in \mathbb{R}^{n_v \times n_v}$, let $\mathbf{A}_{12}^T \in \mathbb{R}^{n_p \times n_v}$ be of rank $n_p < n_v$, $\mathbf{b} \in \mathbb{R}^{n_v}$, $\mathbf{Q} \in \mathbb{R}^{n_v \times n_v^2}$, and $\mathbf{c} \in \mathbb{R}^{n_v}$. We consider the following system with input $\mathbf{g} : (0,T) \to \mathbb{R}$ and output $\mathbf{y} : (0,T) \to \mathbb{R}$,

$$\mathbf{E}_{11}\frac{d}{dt}\mathbf{v}(t) = \mathbf{A}_{11}\mathbf{v}(t) + \mathbf{Q}(\mathbf{v}(t) \otimes \mathbf{v}(t)) + \mathbf{A}_{12}\mathbf{p}(t) + \mathbf{b}\,\mathbf{g}(t), \quad t \in (0,T), \quad \text{(1a)}$$

$$\mathbf{0} = \mathbf{A}_{12}^T\mathbf{v}(t), \qquad t \in (0,T), \quad \text{(1b)}$$

with homogeneous initial condition $\mathbf{v}(0) = \mathbf{0}$, and with the output equation

$$\mathbf{y}(t) = \mathbf{c}^T\mathbf{v}(t), \qquad t \in (0,T). \quad \text{(1c)}$$

Semi-discretization of the Navier-Stokes equations leads to (1), but the Loewner framework presented in this paper can, of course, be applied to any system governed by DAEs (1). Because of page limitations we only consider single-input-single-output (SISO) systems (1). We will elaborate elsewhere on how to generalize the approach to multiple-input-multiple-output (MIMO) systems by interpolating transfer function components along tangential directions.

2.2 Transformation into Quadratic ODE System

Next, we project (1) onto the subspace defined by the discrete incompressibility conditions (1b) to write (1) as an ODE system. As in [14], define the projection

$$\boldsymbol{\Pi} = \mathbf{I} - \mathbf{A}_{12}(\mathbf{A}_{12}^T\mathbf{E}_{11}^{-1}\mathbf{A}_{12})^{-1}\mathbf{A}_{12}^T\mathbf{E}_{11}^{-1}.$$

It can be verified that $\boldsymbol{\Pi}^2 = \boldsymbol{\Pi}, \boldsymbol{\Pi}\mathbf{E}_{11} = \mathbf{E}_{11}\boldsymbol{\Pi}^T$, $\text{null}(\boldsymbol{\Pi}) = \text{range}(\mathbf{A}_{12})$, and $\text{range}(\boldsymbol{\Pi}) = \text{null}(\mathbf{A}_{12}^T\mathbf{E}_{11}^{-1})$, i.e. $\boldsymbol{\Pi}$ is an $\mathbf{E}_{11}$-orthogonal projection. The properties of $\boldsymbol{\Pi}$ imply that

$$\mathbf{A}_{12}^T\mathbf{v}(t) = \mathbf{0} \quad \text{if and only if} \quad \boldsymbol{\Pi}^T\mathbf{v}(t) = \mathbf{v}(t). \quad \text{(2)}$$

Next we express $\mathbf{p}$ in terms of $\mathbf{v}$ and project onto the constraint (1b). Specifically, we premultiply (1a) by $\mathbf{A}_{12}^T\mathbf{E}_{11}^{-1}$, then use (1b) and solve the resulting equation for $\mathbf{p}$ to obtain

$$\mathbf{p}(t) = -\,(\mathbf{A}_{12}^T\mathbf{E}_{11}^{-1}\mathbf{A}_{12})^{-1}\mathbf{A}_{12}^T\mathbf{E}_{11}^{-1}\Big(\mathbf{A}_{11}\mathbf{v}(t) + \mathbf{Q}(\mathbf{v}(t) \otimes \mathbf{v}(t)) + \mathbf{b}\mathbf{g}(t)\Big). \quad \text{(3)}$$

Now insert (3) into (1), apply (2), and use $\boldsymbol{\Pi}\mathbf{A}_{12}(\mathbf{A}_{12}^T\mathbf{E}_{11}^{-1}\mathbf{A}_{12})^{-1} = \mathbf{0}$ to write (1) as

$$\boldsymbol{\Pi}\mathbf{E}_{11}\boldsymbol{\Pi}^T\frac{d}{dt}\mathbf{v}(t) = \boldsymbol{\Pi}\mathbf{A}_{11}\boldsymbol{\Pi}^T\mathbf{v}(t) + \boldsymbol{\Pi}\mathbf{Q}(\boldsymbol{\Pi}^T\mathbf{v}(t) \otimes \boldsymbol{\Pi}^T\mathbf{v}(t)) + \boldsymbol{\Pi}\mathbf{b}\mathbf{g}(t), \quad \text{(4a)}$$

$$\mathbf{y}(t) = \mathbf{c}^T\boldsymbol{\Pi}^T\mathbf{v}(t), \quad \text{(4b)}$$

with initial condition $\boldsymbol{\Pi}^T\mathbf{v}(0) = \mathbf{0}$. This is a dynamical system in the $n_v - n_p$ dimensional subspace $\text{null}(\boldsymbol{\Pi})$ and (4a,b) has to be solved for $\boldsymbol{\Pi}^T\mathbf{v} = \mathbf{v}$.

As in [14], this is made explicit by decomposing $\boldsymbol{\Pi} = \boldsymbol{\Theta}_l \boldsymbol{\Theta}_r^T$ with $\boldsymbol{\Theta}_l, \boldsymbol{\Theta}_r \in \mathbb{R}^{n_v \times (n_v - n_p)}$ satisfying $\boldsymbol{\Theta}_l^T \boldsymbol{\Theta}_r = \mathbf{I}$. Substituting this decomposition into (4) and using the Kronecker product property

$$\mathbf{HX} \otimes \mathbf{KZ} = (\mathbf{H} \otimes \mathbf{K})(\mathbf{X} \otimes \mathbf{Z}) \tag{5}$$

shows that $\widetilde{\mathbf{v}} = \boldsymbol{\Theta}_l^T \mathbf{v} \in \mathbb{R}^{n_v - n_p}$ satisfies

$$\widetilde{\mathbf{E}} \frac{d}{dt} \widetilde{\mathbf{v}}(t) = \widetilde{\mathbf{A}} \widetilde{\mathbf{v}}(t) + \widetilde{\mathbf{Q}}(\widetilde{\mathbf{v}}(t) \otimes \widetilde{\mathbf{v}}(t)) + \widetilde{\mathbf{b}}\mathbf{g}(t), \qquad t \in (0, T), \tag{6a}$$

$$\mathbf{y}(t) = \widetilde{\mathbf{c}}^T \widetilde{\mathbf{v}}(t), \qquad t \in (0, T), \tag{6b}$$

with initial conditions $\widetilde{\mathbf{v}}(0) = \mathbf{0}$, where

$$\widetilde{\mathbf{E}} := \boldsymbol{\Theta}_r^T \mathbf{E}_{11} \boldsymbol{\Theta}_r \in \mathbb{R}^{(n_v - n_p) \times (n_v - n_p)}, \qquad \widetilde{\mathbf{A}} := \boldsymbol{\Theta}_r^T \mathbf{A}_{11} \boldsymbol{\Theta}_r \in \mathbb{R}^{(n_v - n_p) \times (n_v - n_p)},$$
$$\widetilde{\mathbf{b}} := \boldsymbol{\Theta}_r^T \mathbf{b} \in \mathbb{R}^{n_v - n_p}, \qquad \widetilde{\mathbf{c}} := \boldsymbol{\Theta}_r^T \mathbf{c} \in \mathbb{R}^{n_v - n_p},$$
$$\widetilde{\mathbf{Q}} := \boldsymbol{\Theta}_r^T \mathbf{Q}(\boldsymbol{\Theta}_r \otimes \boldsymbol{\Theta}_r) \in \mathbb{R}^{(n_v - n_p) \times (n_v - n_p)^2}.$$

Since $\widetilde{\mathbf{E}}$ is invertible, (6) is an ODE. The DAE system (1) and the ODE system (6) are equivalent. Specifically, the transfer function components of (1) are identical to the transfer function components of (6). The matrices $\boldsymbol{\Theta}_l, \boldsymbol{\Theta}_r \in \mathbb{R}^{n_v \times (n_v - n_p)}$ are expensive to compute, and (6) is a dense large-scale system. Therefore, (6) is used only theoretically. Ultimately, computations are performed using quantities that arise in the original system (1).

2.3 Expansion of Quadratic ODE System into a Linear System

Now we expand the quadratic ODEs (6) into a system of infinitely many linear ODEs. The approach used is the so-called variational approach in [17, Ch. 3].

Let $\widetilde{\mathbf{v}}(\mathbf{g}; \cdot)$ denote the solution of (6a,b) with input $\mathbf{g}$. The variational approach computes an expansion of the solution in terms of its derivatives with respect to $\mathbf{g}$. The solution of (6a,b) with input $\alpha\mathbf{g}$, $\alpha \in \mathbb{R}$, is given by

$$\widetilde{\mathbf{v}}(\alpha\mathbf{g}; \cdot) = \widetilde{\mathbf{v}}(\mathbf{0} + \alpha\mathbf{g}; \cdot) = \widetilde{\mathbf{v}}(\mathbf{0}; \cdot) + \sum_{l=1}^{\infty} \alpha^l \widetilde{\mathbf{v}}_l = \sum_{l=1}^{\infty} \alpha^l \widetilde{\mathbf{v}}_l, \tag{7}$$

where $\widetilde{\mathbf{v}}_l$ is the l-th derivative of the solution map $\mathbf{g} \mapsto \widetilde{\mathbf{v}}(\mathbf{g}; \cdot)$ at $\mathbf{0}$ evaluated in the direction $\mathbf{g}$, and we have used that the solution with zero input is zero, $\widetilde{\mathbf{v}}(\mathbf{0}; \cdot) = \mathbf{0}$. The proof that the series in the right hand side of (7) converges and is equal to $\widetilde{\mathbf{v}}(\alpha\mathbf{g}; \cdot)$ is given, e.g., in [17, Appendix 3.1].

Consider (6) with $\mathbf{g}$ replaced by $\alpha\mathbf{g}$ and insert the expansion (7) to arrive at

$$\sum_{l=1}^{\infty} \alpha^l \widetilde{\mathbf{E}} \frac{d}{dt} \widetilde{\mathbf{v}}_l(t) = \sum_{l=1}^{\infty} \alpha^l \widetilde{\mathbf{A}} \widetilde{\mathbf{v}}_l(t) + \alpha \widetilde{\mathbf{b}}\mathbf{g}(t) + \sum_{m,l=1}^{\infty} \alpha^{l+m} \widetilde{\mathbf{Q}}(\widetilde{\mathbf{v}}_l(t) \otimes \widetilde{\mathbf{v}}_m(t)), \tag{8a}$$

$$\mathbf{y}(t) = \sum_{l=1}^{\infty} \alpha^l \widetilde{\mathbf{c}}^T \widetilde{\mathbf{v}}_l(t), \tag{8b}$$

and $\sum_{l=1}^{\infty} \alpha^l \widetilde{\mathbf{v}}_l(0) = \mathbf{0}$. Equating powers of α gives the infinite system of ODEs

$$\widetilde{\mathbf{E}}\frac{d}{dt}\widetilde{\mathbf{v}}_1(t) = \widetilde{\mathbf{A}}\widetilde{\mathbf{v}}_1(t) + \widetilde{\mathbf{b}}\mathbf{g}(t), \qquad \widetilde{\mathbf{v}}_1(0) = \mathbf{0}, \tag{9a}$$

$$\widetilde{\mathbf{E}}\frac{d}{dt}\widetilde{\mathbf{v}}_l(t) = \widetilde{\mathbf{A}}\widetilde{\mathbf{v}}_l(t) + \sum_{j=1}^{l-1} \widetilde{\mathbf{Q}}(\widetilde{\mathbf{v}}_j(t) \otimes \widetilde{\mathbf{v}}_{l-j}(t)), \qquad \widetilde{\mathbf{v}}_l(0) = \mathbf{0}, \qquad l \geq 2, \tag{9b}$$

with output equations

$$\mathbf{y}_1(t) = \widetilde{\mathbf{c}}^T\widetilde{\mathbf{v}}_1(t), \qquad \mathbf{y}_l(t) = \widetilde{\mathbf{c}}^T\widetilde{\mathbf{v}}_l(t), \qquad l \geq 2. \tag{9c}$$

Recall that the solution of the ODE

$$\widetilde{\mathbf{E}}\frac{d}{dt}\widetilde{\mathbf{v}}(t) = \widetilde{\mathbf{A}}\widetilde{\mathbf{v}}(t) + \mathbf{f}(t), \qquad t \in (0, T), \qquad \widetilde{\mathbf{v}}(0) = \mathbf{0}$$

is given by $\widetilde{\mathbf{v}}(t) = \int_0^t e^{\widetilde{\mathbf{E}}^{-1}\widetilde{\mathbf{A}}(t-\tau)}\widetilde{\mathbf{E}}^{-1}\mathbf{f}(\tau)d\tau = \int_0^t e^{\widetilde{\mathbf{E}}^{-1}\widetilde{\mathbf{A}}\tau}\widetilde{\mathbf{E}}^{-1}\mathbf{f}(t-\tau)d\tau$ $= \int_0^t \widetilde{\mathbf{E}}^{-1}e^{\widetilde{\mathbf{A}}\widetilde{\mathbf{E}}^{-1}\tau}\mathbf{f}(t-\tau)d\tau$. Applying this to (9a,b) yields

$$\widetilde{\mathbf{v}}_1(t) = \int_0^t \widetilde{\mathbf{E}}^{-1}e^{\widetilde{\mathbf{A}}\widetilde{\mathbf{E}}^{-1}t_1}\widetilde{\mathbf{b}}\mathbf{g}(t-t_1)dt_1, \tag{10a}$$

$$\widetilde{\mathbf{v}}_l(t) = \int_0^t \widetilde{\mathbf{E}}^{-1}e^{\widetilde{\mathbf{A}}\widetilde{\mathbf{E}}^{-1}t_l}\sum_{j=1}^{l-1}\widetilde{\mathbf{Q}}\big(\widetilde{\mathbf{v}}_j(t-t_l) \otimes \widetilde{\mathbf{v}}_{l-j}(t-t_l)\big)dt_l, \qquad l \geq 2. \tag{10b}$$

Next we use (9c) and (10) to write $\mathbf{y}$ in terms of $\mathbf{g}$. We focus on the first two terms in the series, as the resulting expressions become increasingly complex and tedious to compute. Using (10) and the Kronecker product property (5) gives

$$\begin{aligned}
\widetilde{\mathbf{v}}_2(t) &= \int_0^t \widetilde{\mathbf{E}}^{-1}e^{\widetilde{\mathbf{A}}\widetilde{\mathbf{E}}^{-1}t_2}\widetilde{\mathbf{Q}}(\widetilde{\mathbf{v}}_1(t-t_2) \otimes \widetilde{\mathbf{v}}_1(t-t_2))dt_2 \\
&= \int_0^t \int_0^{t-t_2} \int_0^{t-t_2} \widetilde{\mathbf{E}}^{-1}e^{\widetilde{\mathbf{A}}\widetilde{\mathbf{E}}^{-1}t_2}\widetilde{\mathbf{Q}}\Big(\widetilde{\mathbf{E}}^{-1}e^{\widetilde{\mathbf{A}}\widetilde{\mathbf{E}}^{-1}t_1}\widetilde{\mathbf{b}} \otimes \widetilde{\mathbf{E}}^{-1}e^{\widetilde{\mathbf{A}}\widetilde{\mathbf{E}}^{-1}t_3}\widetilde{\mathbf{b}}\Big) \\
&\qquad \times \Big(\mathbf{g}(t-t_1-t_2) \otimes \mathbf{g}(t-t_2-t_3)\Big)dt_1dt_3dt_2.
\end{aligned} \tag{11}$$

Substituting (10a) and (11) into the output equation (8c) yields

$$\mathbf{y}_1(t) = \int_0^t \mathbf{h}_1(t_1)\,\mathbf{g}(t-t_1)dt_1,$$

$$\mathbf{y}_2(t) = \int_0^t \int_0^{t-t_3} \int_0^{t-t_3} \mathbf{h}_2(t_1, t_2, t_3)\big(\mathbf{g}(t-t_1-t_2) \otimes \mathbf{g}(t-t_2-t_3)\big)dt_1dt_2dt_3,$$

etc., where

$$\mathbf{h}_1(t_1) = \widetilde{\mathbf{c}}^T\widetilde{\mathbf{E}}^{-1}e^{\widetilde{\mathbf{A}}\widetilde{\mathbf{E}}^{-1}t_1}\widetilde{\mathbf{b}}, \tag{12a}$$

$$\mathbf{h}_2(t_1, t_2, t_3) = \widetilde{\mathbf{c}}^T\widetilde{\mathbf{E}}^{-1}e^{\widetilde{\mathbf{A}}\widetilde{\mathbf{E}}^{-1}t_3}\widetilde{\mathbf{Q}}(\widetilde{\mathbf{E}}^{-1}e^{\widetilde{\mathbf{A}}\widetilde{\mathbf{E}}^{-1}t_1}\widetilde{\mathbf{b}} \otimes \widetilde{\mathbf{E}}^{-1}e^{\widetilde{\mathbf{A}}\widetilde{\mathbf{E}}^{-1}t_2}\widetilde{\mathbf{b}}), \tag{12b}$$

etc. Lastly we take the (multidimensional) Laplace transforms of the kernel function components (12) to compute the transfer function components

$$\mathbf{H}_1(s_1) = \widetilde{\mathbf{c}}^T \widetilde{\boldsymbol{\Phi}}(s_1)\widetilde{\mathbf{b}}, \quad \mathbf{H}_2(s_1, s_2, s_3) = \widetilde{\mathbf{c}}^T \widetilde{\boldsymbol{\Phi}}(s_1)\widetilde{\mathbf{Q}} \left(\widetilde{\boldsymbol{\Phi}}(s_2)\widetilde{\mathbf{b}} \otimes \widetilde{\boldsymbol{\Phi}}(s_3)\widetilde{\mathbf{b}} \right), \quad (13)$$

where $\widetilde{\boldsymbol{\Phi}}(s) := (s\widetilde{\mathbf{E}} - \widetilde{\mathbf{A}})^{-1}$.

2.4 Transfer Function Interpolation for Quadratic ODE Systems

The Loewner framework is related to interpolation based ROM approaches that construct a ROM whose transfer function components interpolate the FOM transfer function components at desired points. See, e.g., [1,2,8,10,12,13]. A review of these interpolation based ROM approaches is useful to understand the Loewner framework. Different transfer function components are interpolated in [1,8,10,12,13]. We follow [12].

We truncate the expansion (8) after $l = 2$ terms, and we seek a ROM so that the FOM transfer function components (13) that correspond to this truncation are interpolated at selected points by the corresponding ROM transfer function components. These interpolation conditions are derived for linear systems, e.g., in [2, Sect. 3.3], and for quadratic bilinear systems in [12]. They require a particular grouping of the frequencies at which the transfer functions are interpolated. Some benefits of this grouping are only reaped in the MIMO case (not considered in this paper) in which the transfer functions (13) are matrix valued, and interpolation conditions hold for the matrices multiplied from the left or from the right by so-called tangential directions, but not for the entire matrix valued transfer functions. To be consistent with the literature, e.g., [2,12], we apply these groupings even though we only consider the SISO case.

We assume that the total number of frequencies is a multiple of four, $4\bar{k}$, and we split the interpolation points into two disjoint sets $\mathcal{S}_\mu \subset \mathbb{C}$ and $\mathcal{S}_\lambda \subset \mathbb{C}$, each containing $k = 2\bar{k}$ points. This split could be avoided in the SISO case, but is needed in the MIMO case where the frequencies in $\mathcal{S}_\mu$ and $\mathcal{S}_\lambda$ are associated with the left and the right tangential directions respectively. Within $\mathcal{S}_\mu$ and $\mathcal{S}_\lambda$ we arrange the interpolation points as

$$\mathcal{S}_\mu = \cup_{1 \le j \le \bar{k}} \{\mu_1^{(j)}, \mu_2^{(j)}\}, \quad \mathcal{S}_\lambda = \cup_{1 \le j \le \bar{k}} \{\lambda_1^{(j)}, \lambda_2^{(j)}\}. \quad (14)$$

For each j, we will obtain interpolation conditions at combinations of $\mu_1^{(j)}, \mu_2^{(j)}$, $\lambda_1^{(j)}, \lambda_2^{(j)}$. See (22) below.

The ROM is constructed using a Petrov-Galerkin projection with projection matrices given by the so-called generalized controllability and generalized observability matrices. The generalized controllability matrix is

$$\widetilde{\mathcal{R}} = \left[\widetilde{\mathcal{R}}^{(1)}, \ \widetilde{\mathcal{R}}^{(2)}, \ \ldots, \ \widetilde{\mathcal{R}}^{(\bar{k})} \right] \in \mathbb{C}^{(n_v - n_p) \times k}, \quad (15)$$

where

$$\widetilde{\mathcal{R}}^{(j)} = \left[\widetilde{\boldsymbol{\Phi}}(\lambda_1^{(j)})\, \widetilde{\mathbf{b}}, \ \widetilde{\boldsymbol{\Phi}}(\lambda_2^{(j)}) \widetilde{\mathbf{Q}} (\widetilde{\boldsymbol{\Phi}}(\lambda_1^{(j)}) \widetilde{\mathbf{b}} \otimes \widetilde{\boldsymbol{\Phi}}(\lambda_1^{(j)}) \widetilde{\mathbf{b}}) \right] \in \mathbb{C}^{(n_v - n_p) \times 2}, \quad (16)$$

$j = 1, \ldots, \bar{k}$. The generalized observability matrix is

$$\widetilde{\mathcal{O}} = \left[(\widetilde{\mathcal{O}}^{(1)})^T, (\widetilde{\mathcal{O}}^{(2)})^T, \ldots, (\widetilde{\mathcal{O}}^{(\bar{k})})^T \right]^T \in \mathbb{C}^{k \times (n_v - n_p)}, \tag{17}$$

where, for $j = 1, \ldots, \bar{k}$,

$$\widetilde{\mathcal{O}}^{(j)} = \begin{bmatrix} \widetilde{\mathbf{c}}^T \widetilde{\boldsymbol{\Phi}}(\mu_1^{(j)}) \\ \widetilde{\mathbf{c}}^T \widetilde{\boldsymbol{\Phi}}(\mu_1^{(j)}) \widetilde{\mathbf{Q}} \left(\widetilde{\boldsymbol{\Phi}}(\lambda_1^{(j)}) \widetilde{\mathbf{b}} \otimes \widetilde{\boldsymbol{\Phi}}(\mu_2^{(j)}) \right) \end{bmatrix} \in \mathbb{C}^{2 \times (n_v - n_p)}. \tag{18}$$

Projecting (6) from the left and the right by the generalized observability matrix (17) and the generalized controllability matrix (15), respectively, gives a ROM whose transfer function components have the desired interpolation properties. Consider the ROM

$$\widehat{\widetilde{\mathbf{E}}} \frac{d}{dt} \widehat{\widetilde{\mathbf{v}}}(t) = \widehat{\widetilde{\mathbf{A}}} \widehat{\widetilde{\mathbf{v}}}(t) + \widehat{\widetilde{\mathbf{Q}}} (\widehat{\widetilde{\mathbf{v}}}(t) \otimes \widehat{\widetilde{\mathbf{v}}}(t)) + \widehat{\widetilde{\mathbf{b}}} \mathbf{g}(t), \tag{19a}$$

$$\widehat{\widetilde{\mathbf{y}}}(t) = \widehat{\widetilde{\mathbf{c}}}^T \widehat{\widetilde{\mathbf{v}}}(t) \tag{19b}$$

of state dimension k, where

$$\widehat{\widetilde{\mathbf{E}}} = \widetilde{\mathcal{O}} \widetilde{\mathbf{E}} \widetilde{\mathcal{R}}, \quad \widehat{\widetilde{\mathbf{A}}} = \widetilde{\mathcal{O}} \widetilde{\mathbf{E}} \widetilde{\mathcal{R}}, \quad \widehat{\widetilde{\mathbf{Q}}} = \widetilde{\mathcal{O}} \widetilde{\mathbf{Q}} (\widetilde{\mathcal{R}} \otimes \widetilde{\mathcal{R}}), \quad \widehat{\widetilde{\mathbf{b}}} = \widetilde{\mathcal{O}} \widetilde{\mathbf{b}}, \quad \widehat{\widetilde{\mathbf{c}}} = \widetilde{\mathcal{R}}^T \widetilde{\mathbf{c}}. \tag{20}$$

Analogous to (13), the first two transfer function components of (19) are

$$\widehat{\mathbf{H}}_1(s_1) = \widehat{\widetilde{\mathbf{c}}}^T \widehat{\widetilde{\boldsymbol{\Phi}}}(s_1) \widehat{\widetilde{\mathbf{b}}}, \qquad \widehat{\mathbf{H}}_2(s_1, s_2, s_3) = \widehat{\widetilde{\mathbf{c}}}^T \widehat{\widetilde{\boldsymbol{\Phi}}}(s_1) \widehat{\widetilde{\mathbf{Q}}} \left(\widehat{\widetilde{\boldsymbol{\Phi}}}(s_2) \widehat{\widetilde{\mathbf{b}}} \otimes \widehat{\widetilde{\boldsymbol{\Phi}}}(s_3) \widehat{\widetilde{\mathbf{b}}} \right), \tag{21}$$

where $\widehat{\widetilde{\boldsymbol{\Phi}}}(s) = (s\widehat{\widetilde{\mathbf{E}}} - \widehat{\widetilde{\mathbf{A}}})^{-1}$. If $s\widehat{\widetilde{\mathbf{E}}} - \widehat{\widetilde{\mathbf{A}}}$ is invertible for all $s \in \mathcal{S}_\mu \cup \mathcal{S}_\lambda$, then the interpolation conditions

$$\mathbf{H}_1(\lambda_1^{(j)}) = \widehat{\mathbf{H}}_1(\lambda_1^{(j)}), \qquad \mathbf{H}_1(\mu_1^{(j)}) = \widehat{\mathbf{H}}_1(\mu_1^{(j)}), \tag{22a}$$

$$\mathbf{H}_2(\lambda_2^{(j)}, \lambda_1^{(j)}, \lambda_1^{(j)}) = \widehat{\mathbf{H}}_2(\lambda_2^{(j)}, \lambda_1^{(j)}, \lambda_1^{(j)}), \tag{22b}$$

$$\mathbf{H}_2(\mu_1^{(j)}, \lambda_1^{(j)}, \mu_2^{(j)}) = \widehat{\mathbf{H}}_2(\mu_1^{(j)}, \lambda_1^{(j)}, \mu_2^{(j)}), \tag{22c}$$

for all $j = 1, \ldots, \bar{k}$, and additional interpolation conditions are satisfied. This fact follows from [12, Lemma 3.1].

2.5 The Loewner Framework for Quadratic ODE Systems

As written, the ROM (19) requires the explicit projection by $\widetilde{\mathcal{O}}$ and $\widetilde{\mathcal{R}}$. The crucial observation that underlies the Loewner framework is that the matrices in (20) can be computed directly from measurements of the transfer function components, but without explicit projection.

The Loewner matrix $\mathbb{L}$ and the shifted Loewner matrix $\mathbb{L}_s$ are defined as

$$\mathbb{L} = -\widetilde{\mathcal{O}} \widetilde{\mathbf{E}} \widetilde{\mathcal{R}} \in \mathbb{C}^{k \times k}, \quad \mathbb{L}_s = -\widetilde{\mathcal{O}} \widetilde{\mathbf{A}} \widetilde{\mathcal{R}} \in \mathbb{C}^{k \times k} \tag{23}$$

and can be expressed directly in terms of measurements. Unfortunately, the expression of all terms in $\mathbb{L}$ and $\mathbb{L}_s$ becomes involved very quickly. Therefore we only consider one case to illustrate the idea and refer to [12] for a detailed discussion. Recall (15)–(18). The Loewner matrix $\mathbb{L}$ in (23) contains entries

$$\begin{aligned}
&\widetilde{\mathbf{c}}^T(\mu_1^{(j)}\widetilde{\mathbf{E}} - \widetilde{\mathbf{A}})^{-1}\widetilde{\mathbf{E}}(\lambda_1^{(l)}\widetilde{\mathbf{E}} - \widetilde{\mathbf{A}})^{-1}\widetilde{\mathbf{b}} \\
&= \frac{1}{\mu_1^{(j)} - \lambda_1^{(l)}}\Big(\widetilde{\mathbf{c}}^T(\mu_1^{(j)}\widetilde{\mathbf{E}} - \widetilde{\mathbf{A}})^{-1}(\mu_1^{(j)} - \lambda_1^{(l)})\widetilde{\mathbf{E}}(\lambda_1^{(l)}\widetilde{\mathbf{E}} - \widetilde{\mathbf{A}})^{-1}\widetilde{\mathbf{b}}\Big) \\
&= \frac{1}{\mu_1^{(j)} - \lambda_1^{(l)}}\Big(\widetilde{\mathbf{c}}^T(\mu_1^{(j)}\widetilde{\mathbf{E}} - \widetilde{\mathbf{A}})^{-1}\Big((\mu_1^{(j)}\widetilde{\mathbf{E}} - \widetilde{\mathbf{A}}) - (\lambda_1^{(l)}\widetilde{\mathbf{E}} - \widetilde{\mathbf{A}})\Big)(\lambda_1^{(l)}\widetilde{\mathbf{E}} - \widetilde{\mathbf{A}})^{-1}\widetilde{\mathbf{b}}\Big) \\
&= \frac{1}{\mu_1^{(j)} - \lambda_1^{(l)}}\Big(\mathbf{H}_1(\lambda_1^{(l)}) - \mathbf{H}_1(\mu_1^{(j)})\Big),
\end{aligned}$$

which are written in terms of measurements of the transfer function component $\mathbf{H}_1$. As mentioned before, the representation of other entries of $\mathbb{L}$ and entries of $\mathbb{L}_s$ follows the same ideas, but is more involved. See [12] for details.

So far we assumed that the 'right' amount of data is available so that $\widetilde{\mathcal{O}}$ and $\widetilde{\mathcal{R}}$ have full row and column rank, respectively, and the ROM (19) is well-posed. Another advantage of the Loewner framework is that larger amounts of data can be used. Since the ROMs can be directly computed from data, using all the measurements available intuitively leads to a better ROM. Specifically, the Loewner and shifted Loewner matrices $\mathbb{L}, \mathbb{L}_s \in \mathbb{C}^{k\times k}$ are computed from data and the SVD is used to extract the relevant data. Consider the (short) SVDs

$$[\mathbb{L} \;\; \mathbb{L}_s] = \mathbf{Y}_1\Sigma_1\mathbf{X}_1^*, \qquad \begin{bmatrix}\mathbb{L}\\ \mathbb{L}_s\end{bmatrix} = \mathbf{Y}_2\Sigma_2\mathbf{X}_2^*, \tag{24}$$

where $\Sigma_1 \in \mathbb{R}^{k\times 2k}$, $\Sigma_2 \in \mathbb{R}^{2k\times k}$, $\mathbf{Y}_1, \mathbf{X}_2 \in \mathbb{C}^{k\times k}$. The matrices $\mathbf{Y}, \mathbf{X} \in \mathbb{C}^{k\times r}$ are obtained by selecting the first r columns of the matrices $\mathbf{Y}_1$ and $\mathbf{X}_2$, we define

$$\widetilde{\mathbf{V}} = \widetilde{\mathcal{R}}\mathbf{X} \in \mathbb{C}^{(n_v-n_p)\times r}, \quad \widetilde{\mathbf{W}} = \widetilde{\mathcal{O}}^*\mathbf{Y} \in \mathbb{C}^{(n_v-n_p)\times r}, \tag{25}$$

The reduced order model that matches the desired interpolation/approximation property of the transfer function components is

$$\widehat{\widetilde{\mathbf{E}}} = -\mathbf{Y}^*\mathbb{L}\mathbf{X} = \widetilde{\mathbf{W}}^*\widetilde{\mathbf{E}}\widetilde{\mathbf{V}}, \qquad \widehat{\widetilde{\mathbf{A}}} = -\mathbf{Y}^*\mathbb{L}_s\mathbf{X}^* = \widetilde{\mathbf{W}}^*\widetilde{\mathbf{A}}\widetilde{\mathbf{V}}, \tag{26a}$$

$$\widehat{\widetilde{\mathbf{Q}}} = \widetilde{\mathbf{W}}^*\widetilde{\mathbf{Q}}(\widetilde{\mathbf{V}}\otimes\widetilde{\mathbf{V}}), \qquad \widehat{\widetilde{\mathbf{b}}} = \widetilde{\mathbf{W}}^*\,\widetilde{\mathbf{b}}, \qquad \widehat{\widetilde{\mathbf{c}}^T} = \widetilde{\mathbf{c}}^T\,\widetilde{\mathbf{V}}. \tag{26b}$$

We have written (26) in terms of projections with $\widetilde{\mathbf{V}}, \widetilde{\mathbf{W}}$, but these quantities can be computed from measurements and $\mathbf{Y}, \mathbf{X}$. See [12].

2.6 Computational Details

The projection matrices $\widetilde{\mathbf{V}}, \widetilde{\mathbf{W}}$ and the ROM (26) are complex, but we can obtain real projection matrices $\widetilde{\mathbf{V}}, \widetilde{\mathbf{W}}$ and corresponding ROMs if the sets of interpolation points $\mathcal{S}_\mu \subset \mathbb{C}$ and $\mathcal{S}_\lambda \subset \mathbb{C}$ contain also the conjugate complex data. See [2, Appendix A.1] or [6, p. 360]. This is what we do in our implementation.

As we have mentioned earlier, the projection of the Navier-Stokes system (1) to obtain the ODE system (6) is a theoretical tool, but this projection is not carried out explicitly. The approaches in [3,4,14] can be extended to our case, and this extension is applied in the following computations. Because of space limitations, we will expand on these computational details elsewhere.

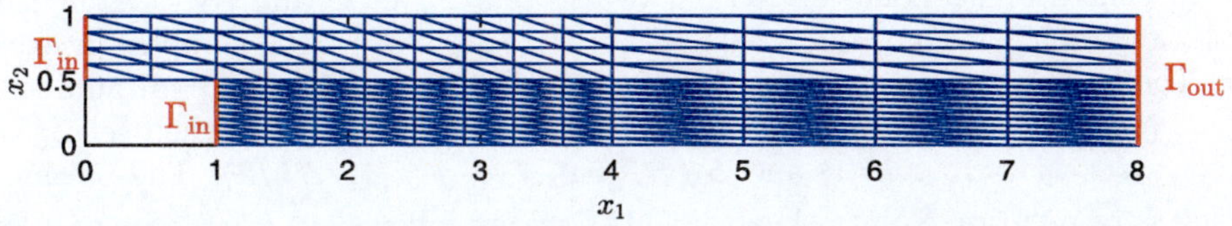

Fig. 1. Channel with backward facing step and coarse grid

3 Numerical Example

We extend an example modeled after [4,14], where model reduction of the Oseen equation is considered. Let $\Omega \subset \mathbb{R}^2$ be the backward facing step geometry shown in Fig. 1. The boundary is decomposed into segments $\Gamma_{\text{out}}, \Gamma_d, \Gamma_{\text{in}}$, where $\Gamma_{\text{out}} = \{8\} \times (0,1)$ is the outflow boundary, inputs are applied on $\Gamma_{\text{in}} = \{0\} \times (1/2,1) \cup \{1\} \times (0,1/2)$, and the velocities are set to zero on $\Gamma_D = \partial\Omega \setminus (\Gamma_{\text{in}} \cup \Gamma_{\text{out}})$. Consider the Navier-Stokes equations

$$\partial_t v(x,t) - \nu \Delta v(x,t) + v(x,t) \cdot \nabla v(x,t) + \nabla p(x,t) = 0, \text{ in } \Omega \times (0,T), \quad (27a)$$

$$\nabla \cdot v(x,t) = 0, \text{ in } \Omega \times (0,T), \quad (27b)$$

$$(\nabla v(x,t) - p(x,t)I)n(x) + \frac{1}{\delta} v(x,t) = \frac{1}{\delta} g_{\text{in}}(x,t), \text{ on } \Gamma_{\text{in}} \times (0,T), \quad (27c)$$

$$v(x,t) = 0, \text{ on } \Gamma_D \times (0,T), \quad (27d)$$

$$(\nabla v(x,t) - p(x,t)I)n(x) = 0, \text{ on } \Gamma_{\text{out}} \times (0,T), \quad (27e)$$

where $\nu > 0$ is the viscosity and $\delta > 0$. The Robin boundary condition (27c) can be viewed as penalized version of a Dirichlet boundary condition as $\delta \to 0$.

We assume that the boundary input in (27c) is parameterized as

$$\mathbf{g}_{\text{in}}(x,t) = \mathbf{g}_1(t) \begin{pmatrix} \sin(2\pi(x_2 - 1/2)) \\ 0 \end{pmatrix} \text{ on } \{0\} \times (1/2,1) \quad (28)$$

and $\mathbf{g}_{\text{in}}(x,t) = 0$ on $\{1\} \times (0,1/2)$.

Our output is the integral of the curl of the velocity, often used in the context of control of vorticity,

$$y(t) = \int_{\Omega_{\text{obs}}} -\partial_{x_2} v_1(x,t) + \partial_{x_1} v_2(x,t) dx$$

over the subdomain $\Omega_{\text{obs}} = (1,3) \times (0,1/2)$ behind the backward facing step.

We use a $P_1 - P_2$ Taylor-Hood discretization (see, e.g., [11]) to arrive at the semi-discrete equations (1). The grid used in our computations is obtained from the coarse grid shown in Fig. 1 by uniform refinement. The used grid leads to $n_v = 11,489$, $n_p = 2,929$ degrees of freedom for the velocities and pressures, respectively. We use a viscosity of $\nu = 1/50$.

To generate the Loewner ROMs we use sets $\mathcal{S}_\mu \subset \mathbb{C}$ and $\mathcal{S}_\lambda \subset \mathbb{C}$ of left and right interpolation points with either $k = 50$ or $k = 100$ samples in each set. The frequencies are computed as follows: We first generate k real numbers $\omega_1, \dots, \omega_k \in \mathbb{R}$, logarithmically spaced between 10 and 10^3, and then set $\mathcal{S}_\mu = \{\pm\omega_{2j-1}i \; : \; j = 1, \dots k/2\}$ and $\mathcal{S}_\lambda = \{\pm\omega_{2j}i \; : \; j = 1, \dots k/2\}$. The pairings of left interpolation points shown in (14) are as follows: $\mu_1^{(1)}, \mu_2^{(1)} = \omega_1 i, \omega_3 i$, $\mu_1^{(2)}, \mu_2^{(2)} = -\omega_1 i, -\omega_3 i$, $\mu_1^{(3)}, \mu_2^{(3)} = \omega_5 i, \omega_7 i$, $\mu_1^{(4)}, \mu_2^{(4)} = -\omega_5 i, -\omega_7 i$, etc. The right interpolation points are paired analogously.

The transfer function measurements are obtained via computational simulations; obtaining them from experiments is an open issue. Actually, in our computations we generate the generalized controllability and observability matrices. However, we do not generate them from (15), (17) but instead extend the approaches in [14] to compute them in terms of the original system (1).

Figure 2 shows the normalized singular values $\sigma_j^{(1)}/\sigma_1^{(1)}$ and $\sigma_j^{(2)}/\sigma_1^{(2)}$ of the matrices in (24) generated with $k = 50$ and with $k = 100$ frequency samples. For Loewner matrices generated with fixed number k of frequency samples the normalized singular values $\sigma_j^{(1)}/\sigma_1^{(1)}$ and $\sigma_j^{(2)}/\sigma_1^{(2)}$ are very similar, which has also been observed in other applications. Also, the decay of the normalized singular values for the matrices (24) generated with $k = 50$ and with $k = 100$ frequency samples is similar.

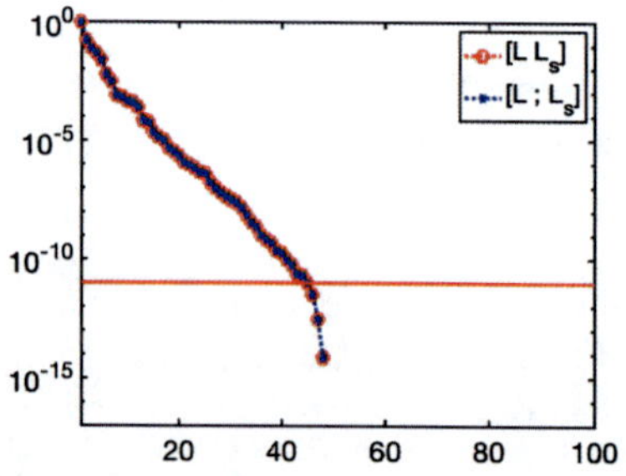

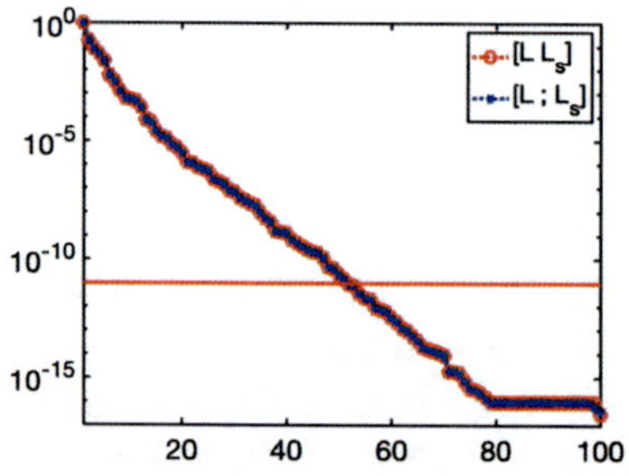

Fig. 2. Normalized singular values of the matrices (24) generated with $k = 50$ (left plot) and $k = 100$ (right plot) frequency samples each in $\mathcal{S}_\mu$ and in $\mathcal{S}_\lambda$. Solid line indicates tol $= 10^{-11}$ used to determine Loewner ROMs.

The size of the Loewner ROM is computed as the smallest r such that

$$\sigma_{r+1}^{(1)}/\sigma_{r+1}^{(1)} \leq \text{tol}.$$

We use tol $= 10^{-11}$ which gives Loewner ROM sizes $r = 45$ when $k = 50$ samples are used and $r = 51$ when $k = 100$ samples are used to generate the data.

To compare the FOM (1) and with the ROM (26) we perform time domain simulations over $[0, T] = [0, 2\pi]$ with input $\mathbf{g}_1(t) = \sin(t)$. The FOM and the ROM are solved numerically using the backward Euler method in time with time step size $T/100$.

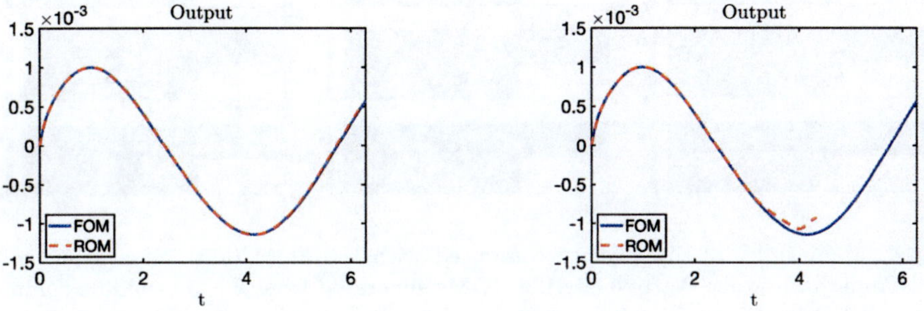

Fig. 3. System outputs computed with the standard Loewner Petrov-Galerkin ROMs generated from $k = 50$ (left plot) and $k = 100$ (right plot) frequency samples.

Figure 3 compares the FOM output against the outputs of the Loewner ROMs generated with $k = 50$ and with $k = 100$ frequency samples, respectively. The output of the Loewner ROM generated with $k = 50$ frequency samples is in excellent agreement with the FOM output (left plot). However, interestingly, when more data, i.e., $k = 100$ frequency samples are used to generate the Loewner ROM, then the ROM and FOM outputs begin to differ around $t = 3.5$. In fact, at time $t_{70} \approx 4.34$ Newton's method used to solve the equation in the backward Euler time stepping for the ROM fails to converge, and the ROM simulation is terminated (right plot in Fig. 3). It is illustrative to look at the velocities generated by the FOM and the Loewner ROMs. The magnitude of the velocities generated by the FOM and the Loewner ROMs are shown in Fig. 4. These plots indicate that the Loewner ROM generated with $k = 100$ frequency samples becomes unstable (right column in Fig. 4). There are also noticeable differences between the velocities generated by the FOM and by the Loewner ROM generated with $k = 50$ frequency samples (middle column in Fig. 4). However, since the Loewner ROM aims to approximate the input-to-output map $\mathbf{g} \mapsto \mathbf{y}$ of the system, only the state information (here the velocities) needed in the input-to-output map is well approximated, but the entire system states may not be well approximated.

The stability properties of the Loewner ROM, and the source of the instability of the Loewner ROM in this case are still under investigation. However, instability in the Loewner ROM when applied to Burgers' equation was also observed in [3]. The standard Loewner approach is based on a Petrov-Galerkin projection with projection matrices $\widetilde{\mathbf{V}} \neq \widetilde{\mathbf{W}}$. In [3] it was demonstrated that the stability

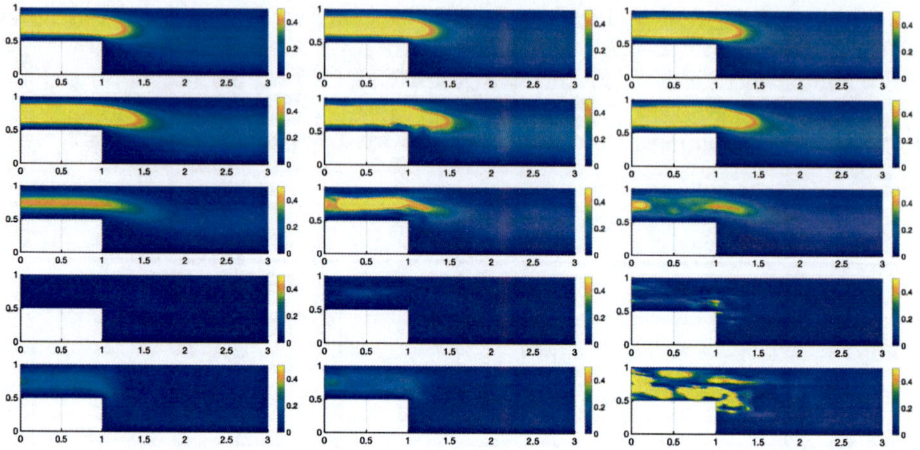

Fig. 4. Magnitudes of the velocity computed with the FOM (left column) and with the standard Loewner Petrov-Galerkin ROMs generated from $k = 50$ (middle column) and $k = 100$ (right column) frequency samples. Magnitudes of the velocities at times $t = 1.19$, $t = 1.82$, $t = 2.45$, $t = 3.08$, $t = 3.71$ are shown top to bottom.

properties can be maintained if the projection matrices $\widetilde{\mathbf{V}}, \widetilde{\mathbf{W}} \in \mathbb{R}^{(n_v-n_p)\times r}$ are merged into one larger matrix $[\widetilde{\mathbf{V}}, \widetilde{\mathbf{W}}] \in \mathbb{R}^{(n_v-n_p)\times 2r}$ (more precisely, an orthonormal basis of the columns of $[\widetilde{\mathbf{V}}, \widetilde{\mathbf{W}}]$ is computed to ensure that the resulting matrix is full rank), and this matrix is used to construct a Galerkin ROM. This resulting ROM is referred to as a Loewner Galerkin ROM. Actually, in the context of the Navier-Stokes equations the use of Galerkin ROMs may allow the extension of stability estimates for semidiscrete finite element approximations (see, e.g., [15, Sect. 9.2]) to the Galerkin ROM.

We merge the matrices $\widetilde{\mathbf{V}} \neq \widetilde{\mathbf{W}}$ computed using the standard Loewner approach with $k = 100$ frequency samples into a projection matrix $[\widetilde{\mathbf{V}}, \widetilde{\mathbf{W}}] \in \mathbb{R}^{(n_v-n_p)\times 2r}$ and construct a Loewner Galerkin ROM of size $2r = 102$. Figure 5 shows that the output of the Loewner Galerkin ROM is in excellent agreement with the output of the FOM.

The magnitude of the velocities generated by the FOM and the Loewner ROMs are shown in Fig. 4. In this case the velocities of the Loewner Galerkin ROM approximate the FOM velocities well at all simulation times.

While these results indicate that the Loewner Galerkin ROM performs well when the standard Loewner Petrov-Galerkin ROM suffers from instabilities, the computation of the Loewner Galerkin ROM is intrusive and requires projection of the system with $[\widetilde{\mathbf{V}}, \widetilde{\mathbf{W}}]$. Analysis of the source of instability in the standard Loewner Petrov-Galerkin ROM and possible remedies, which preserve the non-intrusive data-driven nature of this approach, are part of ongoing research.

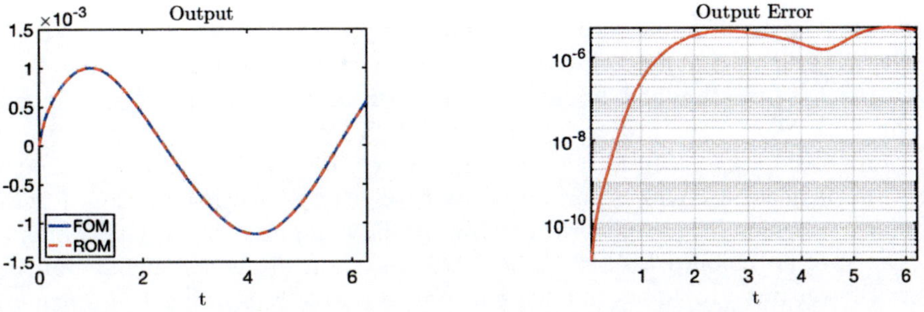

Fig. 5. System output computed with the Loewner Galerkin ROM generated from $k = 100$ frequency samples is in excellent agreement with the FOM output.

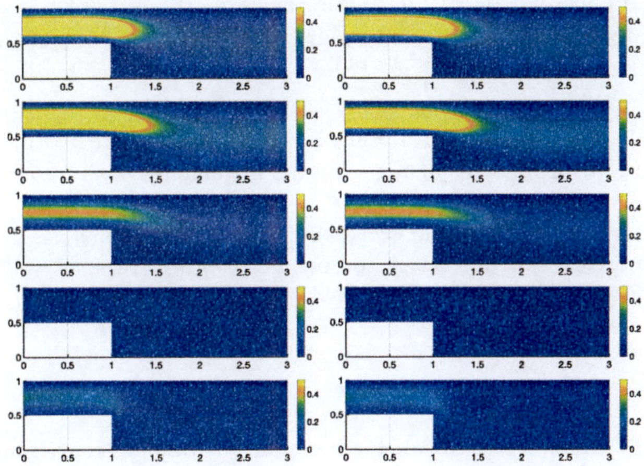

Fig. 6. Magnitudes of the velocity computed with the FOM (left column) and with the Loewner Galerkin ROMs generated from $k = 100$ (right column) frequency samples. Magnitudes of the velocities at times $t = 1.19$, $t = 1.82$, $t = 2.45$, $t = 3.08$, $t = 3.71$ are shown top to bottom.

4 Conclusions and Future Work

We have presented an extension of the Loewner framework to compute ROMs of quadratic-bilinear systems arising from semi-discretized incompressible Navier-Stokes equations.

The application to the Navier-Stokes equations showed the potential of the Loewner framework, but also raises a number of important questions that still need to be addressed. One question is data generation. Currently, transfer function component measurements are obtained via computational simulations; obtaining them from experiments is an open issue. The most important issue is stability. Our numerics have shown that the standard Loewner framework, which generates a Petrov-Galerkin reduced order model, $\widetilde{\mathbf{W}} \neq \widetilde{\mathbf{V}}$, may not be stable. We note that [8, page B255] also report instability of their ROM based on Petrov-Galerkin and interpolation when applied to Burgers' equation with smaller viscosity. In our experiments, combining the projection matrices $\widetilde{\mathbf{W}}, \widetilde{\mathbf{V}}$ generated by the standard Loewner framework and applying a Galerkin projection with the larger projection matrix $[\widetilde{\mathbf{W}}, \widetilde{\mathbf{V}}]$ gave good results. However, using this Galerkin projection destroys the purely data driven aspect, since this Galerkin projection ROM is computed by explicitly projecting the FOM, while standard Loewner Petrov-Galerkin ROM can be computed from data alone.

The specific incompressible Navier-Stokes system (1) is somewhat limiting. First, the output (1c) does not depend on pressure. Extending the output to $\mathbf{y}(t) = \mathbf{C}_1\mathbf{v}(t) + \mathbf{C}_2\mathbf{p}(t)$ leads to an output of the type $\mathbf{y}(t) = \mathbf{C}_3\boldsymbol{\Theta}_r\widetilde{\mathbf{v}}(t) + \mathbf{C}_4\mathbf{Q}(\boldsymbol{\Theta}_r\widetilde{\mathbf{v}}(t) \otimes \boldsymbol{\Theta}_r\widetilde{\mathbf{v}}(t))$ in the resulting projected system corresponding to (6). Thus in addition to the quadratic term in (6a) another quadratic term appears in the output. Incorporation of quadratic terms in the output equation is under investigation. Moreover, Dirichlet boundary condition inputs lead to inputs given by $\mathbf{g}$ as well as its derivative $\frac{d}{dt}\mathbf{g}$. These derivative terms leads to additional terms in projected equations and in the transfer function. They also impact the behavior of transfer function components at infinity, which has to be addressed by extending the approach in [4] for systems governed by the Oseen equation.

Acknowledgements. The authors gratefully acknowledge beneficial discussions with Drs. Antoulas and Gosea, constructive comments of two referees, as well as support by NSF grants CCF-1816219 and DMS-1819144.

References

1. Ahmad, M.I., Benner, P., Goyal, P., Heiland, J.: Moment-matching based model reduction for Navier-Stokes type quadratic-bilinear descriptor systems. ZAMM Z. Angew. Math. Mech. **97**(10), 1252–1267 (2017)
2. Antoulas, A.C., Beattie, C.A., Gugercin, S.: Interpolatory model reduction, Computational Science & Engineering, vol. 21. Society for Industrial and Applied Mathematics (SIAM), Philadelphia, PA (2020). https://doi.org/10.1137/1.9781611976083, https://doi.org/10.1137/1.9781611976083
3. Antoulas, A.C., Gosea, I.V., Heinkenschloss, M.: On the Loewner framework for model reduction of Burgers' equation. In: King, R. (ed.) Active Flow and Combustion Control 2018, pp. 255–270. Springer-Verlag, Berlin, Heidelberg, New York (2018). https://doi.org/10.1007/978-3-319-98177-2_16, https://doi.org/10.1007/978-3-319-98177-2_16
4. Antoulas, A.C., Gosea, I.V., Heinkenschloss, M.: Data-driven model reduction for a class of semi-explicit DAEs using the Loewner framework. In: Reis, T., Grundel, S., Schöps, S. (eds.) Progress in Differential-Algebraic Equations II, pp. 185–210. Springer International Publishing, Cham (2020). https://doi.org/10.1007/978-3-030-53905-4_7, https://doi.org/10.1007/978-3-030-53905-4_7
5. Antoulas, A.C., Gosea, I.V., Ionita, A.C.: Model reduction of bilinear systems in the Loewner framework. SIAM J. Sci. Comput. **38**(5), B889–B916 (2016). https://doi.org/10.1137/15M1041432

6. Antoulas, A.C., Lefteriu, S., Ionita, A.C.: Chapter 8: a tutorial introduction to the Loewner framework for model reduction. In: Benner, P., Cohen, A., Ohlberger, M., Willcox, K. (eds.) Model Reduction and Approximation: Theory and Algorithms, pp. 335–376. SIAM, Philadelphia (2017). https://doi.org/10.1137/1.9781611974829.ch8, https://doi.org/10.1137/1.9781611974829.ch8
7. Bänsch, E., Benner, P., Saak, J., Weichelt, H.K.: Riccati-based boundary feedback stabilization of incompressible Navier-Stokes flows. SIAM J. Sci. Comput. **37**(2), A832–A858 (2015)
8. Benner, P., Breiten, T.: Two-sided projection methods for nonlinear model order reduction. SIAM J. Sci. Comput. **37**(2), B239–B260 (2015)
9. Benner, P., Goyal, P.: An iterative model reduction scheme for quadratic-bilinear descriptor systems with an application to Navier-Stokes equations. In: Keiper, W., Milde, A., Volkwein, S. (eds.) Reduced-Order Modeling (ROM) for Simulation and Optimization, pp. 1–19. Springer Verlag (2018). DOI: DOIurl10.1007/978-3-319-75319-5_1 https://doi.org/10.1007/978-3-319-75319-5_1
10. Benner, P., Goyal, P., Gugercin, S.: $\mathcal{H}_2$-Quasi-optimal model order reduction for quadratic-bilinear control systems. SIAM J. Matrix Anal. Appl. **39**(2), 983–1032 (2018)
11. Elman, H.C., Silvester, D.J., Wathen, A.J.: Finite Elements and Fast Iterative Solvers with Applications in Incompressible Fluid Dynamics, 2nd edn. Numerical Mathematics and Scientific Computation. Oxford University Press, Oxford (2014). https://doi.org/10.1093/acprof:oso/9780199678792.001.0001, http://dx.doi.org/10.1093/acprof:oso/9780199678792.001.0001
12. Gosea, I.V., Antoulas, A.C.: Data-driven model order reduction of quadratic-bilinear systems. Numer. Linear Algebra Appl. **25**(6), e2200 (2018)
13. Gu, C.: QLMOR: a projection-based nonlinear model order reduction approach using quadratic-linear representation of nonlinear systems. Computer-Aided Design of Integrated Circuits and Systems, IEEE Trans. **30**(9), 1307–1320 (2011)
14. Heinkenschloss, M., Sorensen, D.C., Sun, K.: Balanced truncation model reduction for a class of descriptor systems with application to the Oseen equations. SIAM J. Sci. Comput. **30**(2), 1038–1063 (2008)
15. Layton, W.: Introduction to the numerical analysis of incompressible viscous flows. In: Computational Science & Engineering, vol. 6. Society for Industrial and Applied Mathematics (SIAM), Philadelphia, PA (2008).https://doi.org/10.1137/1.9780898718904. https://doi.org/10.1137/1.9780898718904
16. Mayo, A.J., Antoulas, A.C.: A framework for the solution of the generalized realization problem. Linear Algebra Appl. **425**(2-3), 634–662 (2007). https://doi.org/10.1016/j.laa.2007.03.008, https://doi.org/10.1016/j.laa.2007.03.008
17. Rugh, W.J.: Nonlinear System Theory. The Volterra/Wiener approach. Johns Hopkins Series in Information Sciences and Systems. Johns Hopkins University Press, Baltimore (1981). PDF version with corrections (2002). https://sites.google.com/site/wilsonjrugh. Accessed 13 Feb 2018

Part V: Active Flow Control and Performance of Turbomachinery

The Effect of Periodic Disturbance Patterns on the Efficiency of Active Flow Control in a Linear Stator Cascade

Karl Neuhäuser(✉) and Rudibert King

Chair of Measurement and Control, Technische Universität Berlin,
10623 Berlin, Germany
karl.neuhaeuser@tu-berlin.de

Abstract. Constant volume combustion (CVC) is a promising way to substantially increase gas turbine efficiency. Shockless explosion combustion and pulsed detonation combustion are potential candidates for CVC integration in gas turbines. Both these schemes operate with sequentially ignited tubes that introduce highly transient flow characteristics in neighboring machine components. One of the most critical parts of turbomachinery is the compressor. In this work, the impact of the firing pattern on the flow field in a compressor stator located upstream is studied. In particular, five combustion patterns are investigated in terms of local blockage and compressor performance. As wake measurements indicate, patterns with low fluctuations of local blockage exhibit less flow deflection, reduced losses, and higher pressure increases compared to those with large fluctuations. Furthermore, the effect of active flow control (AFC) utilizing end-wall blowing is investigated. Experiments show that the impact of AFC varies depending on the pattern, favoring patterns with low fluctuations of local blockage. The results indicate that closed-loop AFC is favorable when a disturbance rejection is of interest but does not significantly improve performance compared to steady blowing.

Keywords: Constant volume combustion · Compressor · Closed-loop control · Total pressure loss · Static pressure increase · Steady blowing

1 Introduction

Progress in increasing the efficiency of gas turbines by conventional means has reached a plateau. Over the last century, improvements have been achieved mainly with enhanced materials, better cooling, and optimized aerodynamic blade design [1]. However, these improvements have exhausted their potential. New approaches are needed, such as enhanced process design exploiting recuperation and intercooling [2].

Constant volume combustion (CVC) is another promising way to substantially increase the efficiency of gas turbines by replacing the conventional Brayton cycle with the thermodynamically more efficient Humphrey cycle [3].

R. King and D. Peitsch (Eds.): AFCC 2021, NNFM 152, pp. 243–257, 2022.
https://doi.org/10.1007/978-3-030-90727-3_15

The Humphrey cycle promises increased fuel efficiency [4] and thus reduced costs and environmental impact. In contrast to previous advancements, the potential of CVC goes beyond a low single-digit increase.

Two possible implementations of CVC are shockless explosion combustion, in which simultaneous auto-ignition in the combustion tubes is achieved via gradual stratification of the fuel-air mixture [5], and pulsed detonation combustion, in which a pulsating detonation driven by deflagration and detonation waves is performed [6]. Both concepts require execution in single, distinct combustion tubes and result in a highly transient operation that affects the compressor and the turbine; see Fig. 1 (left).

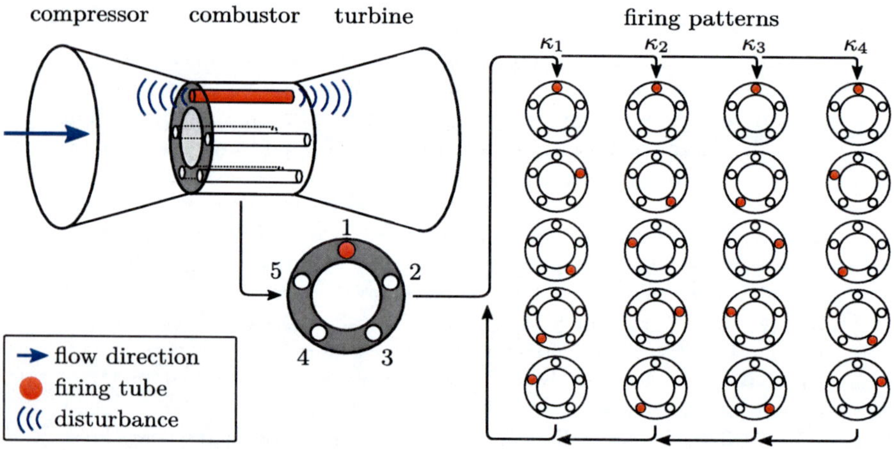

Fig. 1. Schematics of a gas turbine with a CVC combustor consisting of five tubes. Disturbances originating in the combustor propagate into adjoining turbomachinery. The right-hand side shows possible periodic firing sequences/patterns that always skip a fixed number of tubes between firing events.

Neither component, compressor nor turbine, is designed for transient operation and therefore must be carefully evaluated for negative effects, such as flow separation, stall, and performance degradation. However, the extent to which turbomachinery will be affected can be expected to vary significantly when using different firing patterns of a discrete set of tubes. Some possible patterns for an exemplary combustor consisting of five combustion tubes are depicted in Fig. 1 (right). In an actual CVC gas turbine combustor, the number of tubes is potentially larger than in the example from Fig. 1. For more combustion tubes, simultaneous firing of tubes will be advisable, yielding several copies of such patterns for every time instant.

Of the adjoining turbomachinery components, the compressor is considered critical for operation in CVC because it maintains a positive pressure gradient and experiences the CVC-related disturbance coming from the downstream flow direction. To counteract the disturbance, the potential of active flow control

(AFC) in the form of air injection at the sidewall or on the blade surface has been studied in previous work [7–9]. It has been shown that AFC can be effectively used to increase the static pressure. However, when the energy used for AFC is taken into account, there is hardly any improvement in efficiency and usually degradation. Closed-loop AFC has proven to be more effective than open-loop AFC in terms of disturbance suppression but has no significant advantage in terms of losses or increased static pressure.

In the previous studies mentioned above, a disturbance generator simulated the CVC-related disturbance by a local and temporal cross-section reduction, for example, the opening and closing of a flap [7]. However, only one firing pattern[1] was used as reference. The pattern is similar to pattern κ_1 from Fig. 1, in which the tubes fire consecutively in a row. It is the most intuitive among the many patterns imaginable, but some significant drawbacks can be expected in terms of increased static pressure and total pressure loss. In this work, five disturbance patterns, including the κ_1-pattern, are investigated and compared.

The remainder of the paper is organized as follows. First, due to a large number of potential patterns, a particularly promising uniform type of pattern is defined in Sect. 2. Because each of the derived patterns is unique, we introduce a general property called "local blockage" to evaluate the pattern performance. In Sect. 3, the experimental setup consisting of a linear stator cascade with a variable disturbance generator and AFC devices is described. Disturbed base flow measurements without AFC are presented in Sect. 4. Both open- and closed-loop AFC implementations and the results are described in Sect. 5. The results are summarized and discussed in Sect. 6.

2 κ-Patterns

The potential patterns of a CVC process depend, among other things, on the number of tubes and ignition times. We denote the number of tubes by $n_{\text{tubes}} \in \mathbb{N}^+$ and the period duration in which the firing sequence repeats by $\mathrm{T} \in \mathbb{R}$. Using this notation, the so-called κ-pattern construction rule in a cyclic combustion chamber or in a set of tubes is introduced as follows.

1. The tubes ignite at equidistantly distributed times so that the sequence completes after T and starts again from the beginning.
2. The number of the next firing tube is the number of the previous tube plus $j \in \mathbb{N}^+$.

It is also assumed that the first firing tube is tube no. 1, as in Fig. 1, and that a set of tubes always contains a prime number of tubes. The latter guarantees that no tube is excluded and all tubes fire exactly once during one sequence. Using, for example, $n_{\text{tubes}} = 5$, these rules lead to four different patterns, which are

[1] There are no real firing tubes in any of the compressor test rigs, including the one used in the current contribution. Whenever a "firing process" or "firing pattern" is mentioned in what follows, we refer to a local and temporal cross-section reduction by the moving flaps.

depicted in Fig. 1 (right). Note that the κ_1-pattern and the κ_2-pattern mirror the κ_4-pattern and κ_3-pattern, respectively, but produce a different rotation orientation with respect to the suction/pressure side of the stator vanes. The ignition times t_j are obtained in the general case via

$$t_j = (j-1)\,\frac{\mathrm{T}}{n_{\mathrm{tubes}}} \quad \forall j \in \{1, 2, ..., n_{\mathrm{tubes}}\}. \tag{1}$$

They indicate the beginning of the firing process of a single tube with the duration t_{f}. For a real application, it seems plausible that this process, during which ignition, combustion, and ejection of the hot gas into the turbine occurs, takes up a considerable part of the period because the firing frequency will be high. It is unlikely that tubes remain unfired and therefore open for a long time because the pressure gain from the combustion process would otherwise not occur. During the firing process, the flow upstream of a tube is temporarily blocked because the combustion chamber is closed. Otherwise, hot gas would emerge into the compressor because the firing tubes have a higher pressure regime.

Due to the span of the firing process, it is inevitable that several tubes block the flow simultaneously despite their different ignition times. If simultaneously blocked tubes are located in close proximity to each other, as in the κ_1-pattern for example, their effect on the local flow field is superimposed. However, for patterns that skip one, two, or more tubes, a weaker effect can be expected. This circumstance can be quantified using the concept of total and local blockage, which is introduced next. We define total blockage, that is, the normalized passage area blocked by all pipes simultaneously, as

$$\text{total blockage} := \frac{\text{area currently blocked by all tubes}}{\text{maximum blockable area}}. \tag{2}$$

For the κ-patterns, the total blockage is constant at all times with exactly the same value for all patterns. In contrast, the local blockage of a specific passage or any other local part of the compressor, defined as

$$\text{local blockage} := \frac{\text{area currently blocked from tubes in a specific local part}}{\text{maximum blockable area in that part}}, \tag{3}$$

is highly transient and dependent on the pattern. For the κ-patterns, the κ_1-pattern and its mirrored variant exhibit the strongest local blockage, as immediately adjacent tubes fire subsequently. It can be assumed that the greater the local blockage, the greater the risk of flow separation and thus increased losses in the upstream compressor passages. In the next section, the experimental setup is outlined, as is how the concept of local blockage under given geometric parameters can be applied.

3 Experimental Setup

The test rig consists of a linear stator cascade, a disturbance generator, and AFC devices. The setup can be seen in Fig. 2.

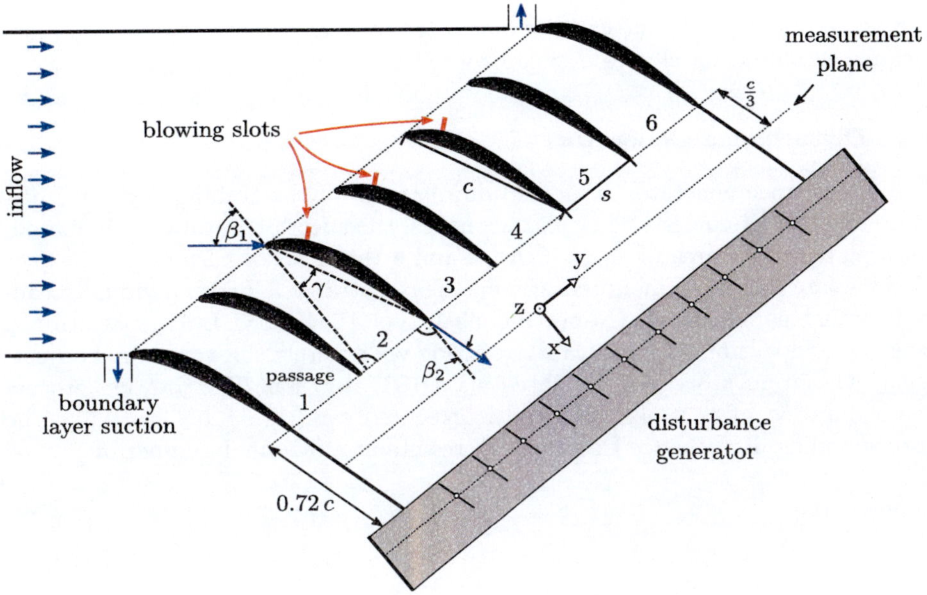

Fig. 2. Linear stator cascade with disturbance generator

Table 1. Geometric cascade parameters

Parameter	Symbol	Value	Unit	Parameter	Symbol	Value	Unit
Chord length	c	0.375	m	Outflow angle	β_2	17.5	Degrees
Height	h	0.3	m	Inflow angle	β_1	51.5	Degrees
Spacing	s	0.183	m	Stagger angle	γ	25.9	Degrees

3.1 Linear Stator Cascade

The stator cascade has seven linearly arranged, high-turning compressor blades. Their geometry was provided by Rolls-Royce and has a De Haller number (DH) of DH = 0.653 and a diffusion factor (DF) of DF = 0.48. Further geometrical parameters are listed in Table 1. The cascade is mounted to an open wind tunnel and operated at a phase-averaged Reynolds number of

$$Re = \frac{\mathrm{u}\,c}{\nu} = 6 \cdot 10^5. \tag{4}$$

The inflow velocity u is measured using a Prandtl tube (Electro-Mation GmbH, type L). Wake measurements are conducted using a five-hole probe (Aeroprobe, type PC5-TIP-2-5-C240-152-025), which is located in the measurement plane at c/3 downstream of the trailing edge. The probe is traversable in the y-z-plane. Both devices are equipped with differential pressure sensors (First Sensor AG, type HCLA12X5B). The measurements were recorded on a dSpace, type DS1005, and an additional real-time capable device (Speedgoat Performance with IO332-6 cards) was used for closed-loop control.

3.2 Disturbance Generator

The disturbance generator is located downstream of the trailing edge at a distance of $0.72\,c$. It consists of 11 plates; that is, the effect of 11 tubes is emulated. Each plate has an area of $A_{\text{plate}} = h\,s/2$ and a thickness of 1.5 mm.

They are rotatably mounted around their central axis and individually controlled via stepper motors (Nanotec GmbH, type PD4C-USB). Any pattern can be generated without mechanical retrofitting via a purely software-based adjustment. The firing process is simulated via a 180° rotation. The rotation process has a duration of $t_f \approx 42/60\,\text{T}$ and is executed every $\text{T} = 1/f_d = 0.5\,\text{s}$. The corresponding disturbance frequency f_d results in a Strouhal number of

$$Sr = \frac{f_d\,c}{\text{u}} \approx 0.03. \tag{5}$$

3.3 Active Flow Control Setup

AFC is used in the form of pressurized air that is fed from a supply source via a flow meter (Festo, SFAB-200U) into a pressure tank, where a proportional valve (Hoerbiger, type PRE-U) regulates the desired pressure level. The six blowing slots are located at a distance of $0.3\,c$ from the leading edge at both end walls on the suction side and are connected to the tank via six solenoid switching valves (Festo, type MHJ-10). The pressurized air is injected into the passage under an angle of 15° to flow direction. Solenoid valves are used because they have faster switching characteristics, take up less space, and are cheaper compared to proportional valves. However, they are subject to the limitation of a binary control variable because they can only be fully opened or closed. The implications for closed-loop control are explained in a later section.

3.4 Local Blockage at the Test Rig

Based on definition (3), knowing the geometrical parameters, the rotation time span t_f, and assuming a constant maximum acceleration/deceleration of the plates rotation angle, the local blockage B_{L} can be calculated as

$$B_{\text{L}}(t) = \frac{1}{4\,A_{\text{plate}}} A'_{\text{proj}}(t). \tag{6}$$

$A'_{\text{proj}}(t)$ is the sum of the projected area of the plates located behind a specific passage in the disturbance plane under the outflow angle at time $t \in \mathbb{R}^+$. Because there are 11 plates, five distinct patterns are possible and will be investigated. The time progression of their local blockage is shown in Fig. 3. The κ_1-pattern exhibits huge local blockage at one point in time due to subsequently

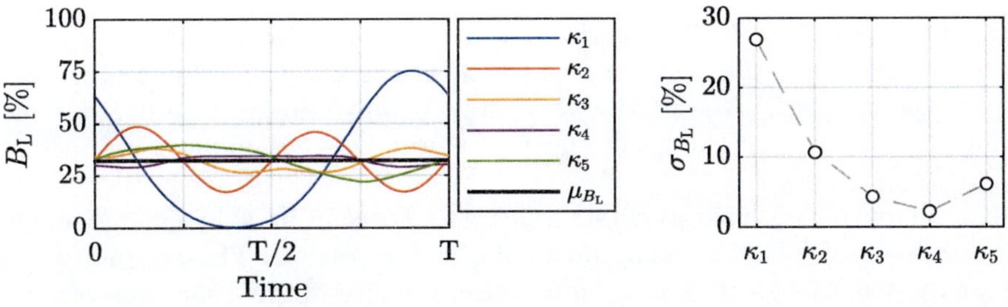

Fig. 3. Local blockage of passage 4 for all κ-patterns at the test rig. Left: time progression of the local blockage B_L over one period. μ_{B_L} is the average value of all patterns. Right: standard deviation σ_{B_L} of B_L for that period.

blocked tubes, as was already suggested in the previous section. Up to 75 % of the passage is blocked, whereas the maximum value for the other patterns is only around 50 %. The local blockage of the κ_2-pattern adheres to a double-periodic structure that arises from the skipping of a single plate. For κ_3 and κ_4, no distinct structure is recognizable. Over time, they deviate slightly from the mean value μ_{B_L}, which is approximately 32 %. It is not until κ_5 that a sine-like curve with a small amplitude arises again. The mean value is the same for all patterns, even for κ_1 and κ_2. However, the standard deviation σ_{B_L} over one period, shown in Fig. 3 (right), varies and has a minimum for κ_4. It becomes clear that κ_1 and κ_2 exhibit large fluctuations of local blockage and that skipping the maximum number of plates as in the κ_5-pattern does not lead to a better temporal distribution of local blockage compared with κ_4 in a setup with 11 plates.

4 Disturbed Base Flow Measurements

The following section presents experimental results of the disturbed base flow wake measurements behind passage 4 without AFC. Because the local blockage as introduced in the previous section allows for quantification of the impact that a pattern has, the question remains to what extent the pattern affects the compressor flow. To answer this question, wake measurements behind passage 4 were conducted at 154 positions over 6 s for all 5 patterns. The total pressure loss ζ and static pressure increase c_p were calculated for every time instant k by

$$\zeta(y, z, k) = \frac{p_{t,1} - p_{t,2}(y, z, k)}{q_1}, \quad c_p(y, z, k) = \frac{p_2 - p_1(y, z, k)}{q_1}. \tag{7}$$

p_t, p, and q are the total, static, and dynamic pressure, respectively, with index 1 representing the cascade inlet and index 2 representing the measurement plane. Inflow measurements are phase-averaged. The results for κ_1, κ_2, and κ_4 are plotted in Fig. 4. Because κ_3 and κ_5 showed similar results to κ_4, they are not explicitly shown here. The coordinates are normalized to the spacing s and half the passage height $h/2$. The origin of the radial coordinate ($z/(h/2) = 0$) is located at the end wall, whereas the origin of the axial coordinate ($y/s = 0$) can be seen in Fig. 2. Due to the symmetric passage design, the contour plots contain data only from the lower end wall to the passage mid-section, meaning $z/(h/2) \in [0, 1]$.

For all patterns, basic flow structures that occur in a linear cascade are recognizable in the field measurements of ζ in Fig. 4 (top). Those structures are the area of reduced total pressure due to the velocity deficit that runs vertically over the entire height of the measuring plane at $y/s \approx 0.25$ and the corner vortex whose center is located at $y/s \approx 0.4$ and $z/(h/2) \approx 0.3$. However, the positioning of these structures and thus the entire flow field changes under the influence of the disturbance, which is particularly evident for the κ_1-pattern and the κ_2-pattern. The time progression of the field measurement of ζ for κ_1 shows a deflection of the flow structures from the left to the right side and vice versa. Almost the same observations can also be made for c_p, with the difference that the flow structures are less recognizable.

Around 3/4 T, the velocity deficit of κ_1 is enlarged near the mid-section. This is an indication of increased suction side flow separation but is not definitive proof. Flow separation can lead to huge losses and thus decreased performance. Potential reasons for the occurrence of this phenomenon are the elevated passage flow velocity, and, foremost, the large deflection due to the immense local blockage of κ_1 at that moment of the period.

For κ_2, the deflection follows a double-periodic progression, meaning that the displacement from the left to the right happens twice over the course of one period. It is less evident compared to κ_1 but is still recognizable. Only the structure of the κ_4-pattern does not undergo any displacement. Nevertheless, a change does occur even for that pattern, that is, a change in the amplitude. Looking at the map averaged values $\overline{\zeta}$ and $\overline{c}_p$ in Fig. 4 (middle), we notice a reduced but not disappearing amplitude compared to the other patterns.

Finally, the cycle and map averaged values $\widetilde{\zeta}$ and $\widetilde{c}_p$ are calculated and plotted in Fig. 4 (bottom). Each red dot represents a full map measurement over one period. In total, 29 measurements were conducted. The standard deviations of $\widetilde{\zeta}$ and $\widetilde{c}_p$ were estimated based on the repeated measurement of one pattern. It is assumed that the other patterns all exhibit the same standard deviation. Both metrics show remarkable agreement with the standard deviation of local blockage from Fig. 3. The pattern with the best values for $\widetilde{\zeta}$ and $\widetilde{c}_p$ is the one with the most constant local blockage, κ_4. The worst performance is observed for κ_1. The large gap between κ_1 and the other patterns may be due to the aforementioned flow separation on the blades' suction side, but further investigations are necessary to shed light on this phenomenon.

Although the metrics fit remarkably well with the standard deviation of local blockage in Fig. 3, an important point must be noted. Even though the means of the patterns vary, they do not differ by more than twice the standard deviation, except for κ_1. Therefore, they cannot be considered significantly distinguishable.

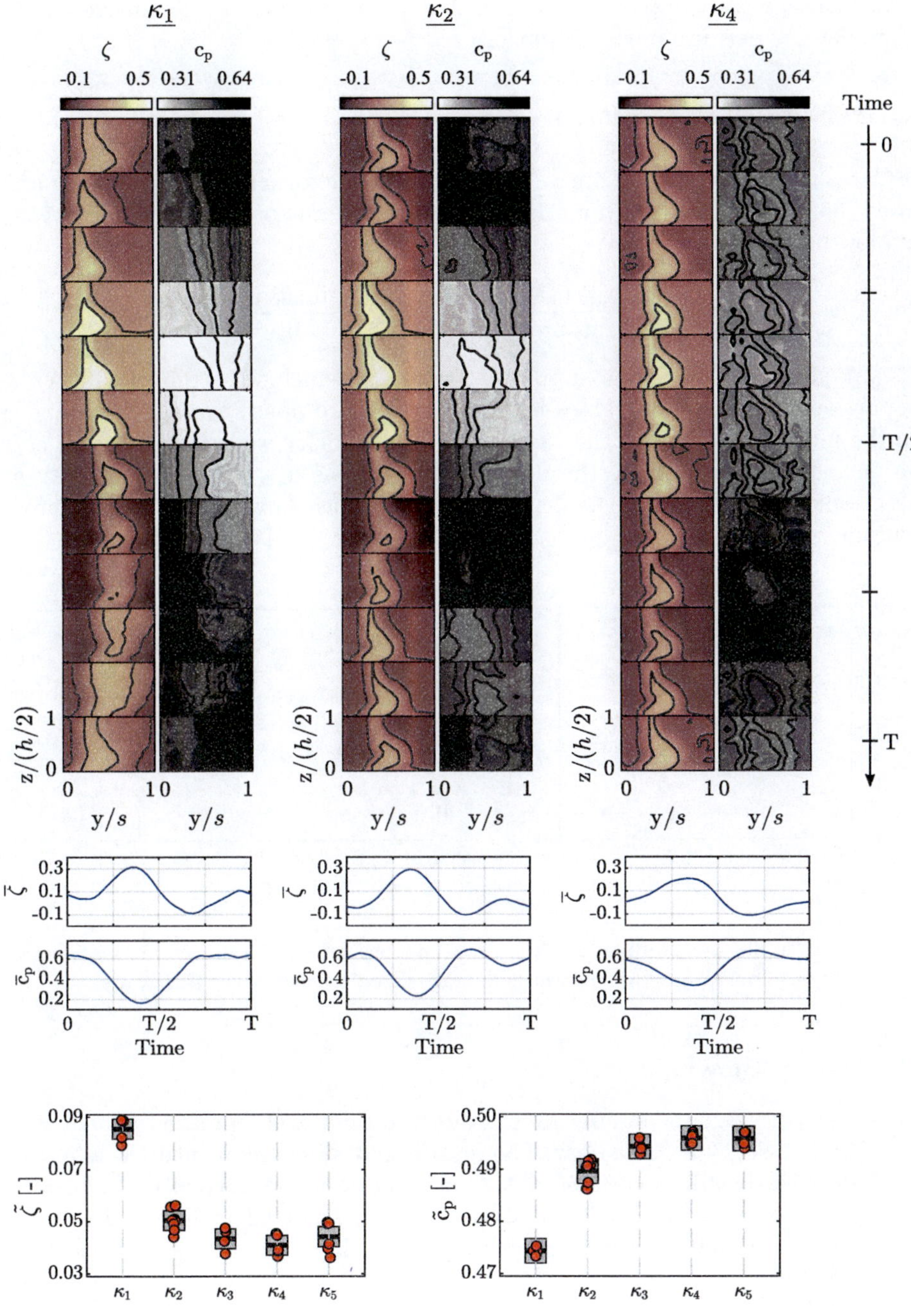

Fig. 4. Top: wake measurements for the κ_1-, κ_2-, and κ_4-patterns. Middle: time progression of the map averaged values. Bottom: cycle averaged values of all patterns. (Red dots: data points, black lines: average value, grey bars: standard deviation)

Local blockage can hence be used as an indicator but not as a definitive metric to predict the performance of a pattern.

5 Active Flow Control

AFC was used in two ways: open-loop control in the form of steady blowing and closed-loop control. For both methods, the momentum coefficient for each of the six blowing slots,

$$\widetilde{c}_\mu = \frac{\dot{m}_{act}\, u_{act}}{\dot{m}_1\, u} \quad \text{with} \quad \dot{m}_{act} = \frac{\dot{m}_{supply}}{6}, \tag{8}$$

was varied by adjusting the supply pressure p_V, which was controlled via the proportional valve. All variables in this equation are phase-averaged. $\dot{m}_1$ and u are the mass flow entering the cascade and its velocity, respectively. $\dot{m}_{supply}$ is the mass flow measured by the flow meter, whereas u_{act} is the jet velocity at the blowing slot outlet that is derived from the mass flow using the continuity equation.

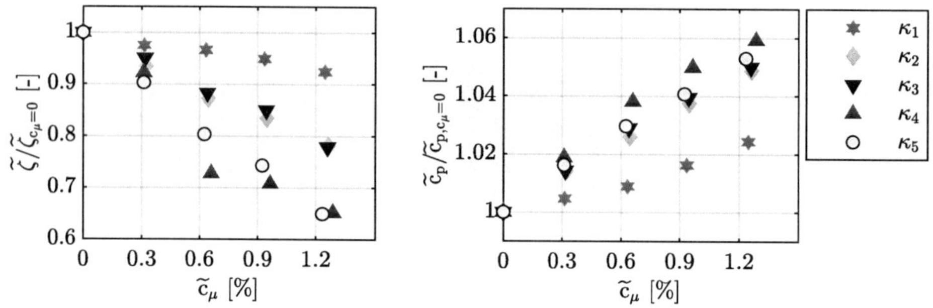

Fig. 5. Map- and cycle-averaged metrics for all patterns. Left: total pressure loss. Right: Static pressure increase. Both metrics are normalized to their base flow value.

5.1 Steady Blowing

For steady blowing, the supply pressure in the tank was regulated to values of $p_V \in \{2.04,\ 2.75,\ 3.33,\ 3.88\}$ bar. The resulting averaged momentum coefficients $\widetilde{c}_\mu$ varied slightly due to randomness. The results of the normalized field- and phase-averaged measurements for all patterns are plotted in Fig. 5. The measurements show that a larger $\widetilde{c}_\mu$ results in a reduced total pressure loss and improved static pressure increase, but there are notable differences between the patterns. Again, patterns with a constant local blockage tend to be favorable, meaning that better performance can be reached by applying the same $\widetilde{c}_\mu$. The κ_1-pattern drops off in comparison, as the metrics are only marginally improved by the AFC. Regarding the other patterns, there is a slight advantage for κ_4 in terms of $\widetilde{c}_p$ and a more significant one for both κ_4 and κ_5 in $\widetilde{\zeta}$.

5.2 Closed-Loop Control

Closed-loop control was carried out using the solenoid valve position in the lower half of passage 4 as control input. The other valves receive the same control input, but with a time shift corresponding to the time shift of the pressure maxima in each passage. Because the total pressure $p_{t,2}$ showed a higher sensitivity towards end wall blowing, it is used as a control variable $y \in \mathbb{R}$. The measurement position for control is chosen as the position where the control variable shows the greatest deviation from the base flow when the valve is open and the maximum supply pressure is applied. Next, pseudo random binary sequence (PRBS) identification experiments were conducted and the parameters of a linear, time-invariant, asymptotically stable second-order transfer function with time delay were identified by minimization of the squared error sum. The model was then transformed to its state-space representation

$$\underline{x}(k_i + 1) = \mathbf{A}\,\underline{x}(k_i) + \underline{b}\,u(k_i) \quad , \; \underline{x}(0) = \underline{x}_{0,i} \tag{9a}$$

$$y(k_i + n_d) = \underline{c}^T\,\underline{x}(k_i). \tag{9b}$$

Here, the control input $u \in \{0, 1\}$ is the solenoid valve position. $\underline{x} \in \mathbb{R}^{n_x}$ is the state with the model order n_x, $t_i = k_i\,\Delta t$ is the time, $\Delta t = \mathrm{T}/n_p$ is the sampling time, and $k_i \in \{1, 2, .., n_p\}$ is the sampling index in the i-th iteration. The number of time steps in a period is n_p, the initial condition is $\underline{x}_{0,i} \in \mathbb{R}^{n_x}$, and $n_d \in \mathbb{N}^+$ is the time delay, which is omitted in what follows. Furthermore, $\mathbf{A} \in \mathbb{R}^{n_x \mathrm{x} n_x}$, $\underline{b} \in \mathbb{R}^{n_x}$, and $\underline{c} \in \mathbb{R}^{n_x}$. A control error $e \in \mathbb{R}$ is defined as $e(k_i) = r(k_i) - y(k_i)$, with the reference $r \in \mathbb{R}$.

The periodic nature of the disturbance allows for the application of an iterative learning control (ILC) algorithm that exploits the periodicity of the combustion process. An overview of ILC algorithms is given in [10]. The approach used here is optimization-based due to the binary solution domain. To that end, an augmented model for the i-th iteration is written as

$$\underline{e}_i = \underline{r}_i - \mathbf{G}\underline{u}_i - \mathbf{F}\underline{x}_{0,i}, \tag{10a}$$

with system matrices

$$\mathbf{F} = \begin{bmatrix} \underline{c}^T \\ \underline{c}^T\mathbf{A} \\ \underline{c}^T\mathbf{A}^2 \\ \vdots \\ \underline{c}^T\mathbf{A}^{n_p} \end{bmatrix}, \quad \mathbf{G} = \begin{bmatrix} 0 & 0 & \dots & 0 & 0 \\ \underline{c}^T\underline{b} & 0 & \dots & 0 & 0 \\ \underline{c}^T\mathbf{A}\underline{b} & \underline{c}^T\underline{b} & \dots & 0 & 0 \\ \vdots & \vdots & \ddots & \vdots & \vdots \\ \underline{c}^T\mathbf{A}^{n_p-1}\underline{b} & \underline{c}^T\mathbf{A}^{n_p-2}\underline{b} & \dots & \underline{c}^T\underline{b} & 0 \end{bmatrix} \tag{10b}$$

and lifted vectors of appropriate dimensions.

$$\underline{e}_i = \left[e(k_i)\; e(k_i + 1) \dots e(k_i + n_p - 1)\right]^{\mathrm{T}}, \tag{10c}$$

$$\underline{u}_i = \left[u(k_i)\; u(k_i + 1) \dots u(k_i + n_p - 1)\right]^{\mathrm{T}}, \tag{10d}$$

$$\underline{r} = \left[r(k_i)\; r(k_i + 1) \dots r(k_i + n_p - 1)\right]^{\mathrm{T}} \tag{10e}$$

Now, at the beginning of each iteration, the quadratic program

$$\underline{u}_i^* = \arg\min_{\underline{u}_i} \frac{1}{2} \left(\underline{e}_i^T \mathbf{W}_e \, \underline{e}_i + \underline{u}_i^T \mathbf{W}_u \, \underline{u}_i + \Delta \underline{u}_i^T \mathbf{W}_\Delta \, \Delta \underline{u}_i \right) \tag{11a}$$

$$\text{s.t.} \quad \Delta \underline{u}_i = \underline{u}_i - \underline{u}_{i-1} \tag{11b}$$

$$\Delta \underline{x}_{0,i} = \underline{x}_{0,i} - \underline{x}_{0,i-1} \tag{11c}$$

$$\underline{e}_i = \underline{e}_{i-1} - \mathbf{G} \Delta \underline{u}_i - \mathbf{F} \Delta \underline{x}_{0,i} \tag{11d}$$

$$\underline{u}_i \in \{0, 1\}^{n_p} \tag{11e}$$

is solved. Because $\underline{u}$ is in a binary domain, the solution can only be obtained with an appropriate solver. Here, a branch-and-bound algorithm [11] is used after a reformulation of Eq. (11), following the implementation from [12]. The rounding strategy from [12] provided an improved initial guess for the branch-and-bound algorithm, which was then used to conduct a depth-first search.

The weighting matrices $\mathbf{W}_e$, $\mathbf{W}_u$, and $\mathbf{W}_\Delta \in \mathbb{R}^{n_p x n_p}$ are chosen in a way that the Hessian $\mathbf{H} = \left(\mathbf{G}^T \mathbf{W}_e \mathbf{G} + \mathbf{W}_u + \mathbf{W}_\Delta \right)$ is positive definite. When designing the weighting matrices, special attention was paid to the convergence behavior of the controller. Among other improvements of the algorithm, a gradual adjustment of $\mathbf{W}_\Delta$ after convergence was implemented. This approach was proposed by [13] to prevent the controller from getting stuck in local minima. Because the solution domain is integer-valued, this is likely to happen. The gradual adjustment allows further integer-valued steps when the algorithm is stuck in such a local minimum.

5.3 Open vs. Closed Loop: Disturbance Rejection

As proven by previous work [7], closed-loop AFC is effective in supressing periodic disturbances in a stator cascade. A typical task is to follow a reference trajectory representing an undisturbed case. The ILC algorithm finds a binary control trajectory that allows for the best reference tracking with respect to Eq. (11). Figure 6 shows the result of the controller for the κ_5-pattern compared to the open-loop case. Both open- and closed-loop AFC consume the same mass flow and therefore have the same phase-averaged $\widetilde{c}_\mu$. The blue line shows the base flow, and the black dashed line is the reference. Note that the reference does not represent the average total pressure at that position but rather a total pressure that corresponds to the total pressure loss coefficient for the undisturbed case. The red and yellow lines are the measurements for open loop and closed loop after convergence, respectively. In Fig. 6 (right), the normalized 2-norm of the control error of each iteration is plotted. In the case of base flow and open loop, the values remain constant apart from random noise, while for the closed-loop case the learning behavior of the algorithm becomes apparent. After approximately 15 iterations, a converged state is reached.

In case of steady blowing, the disturbance is partially rejected in the first half of the period but is amplified in the second half. However, with the previously described ILC algorithm, the valves are open only at certain times so that the amplification of the disturbance in the second half is avoided or a little

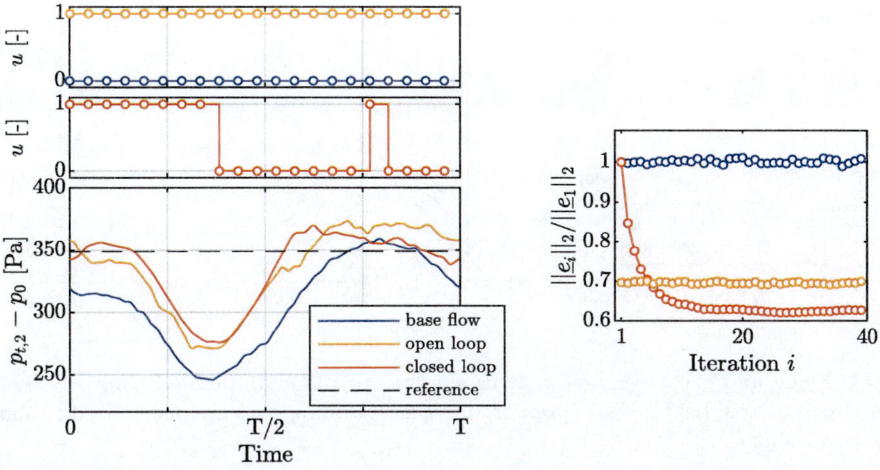

Fig. 6. Reference tracking results for open loop and closed loop. Left, top, and left, middle: applied control trajectories. Left, bottom: Measured control variables over one period. Right: convergence behavior. The closed-loop control trajectory and measurement are from the $i = 39$-th iteration.

reduced. The control authority of compressed air does not allow to always bring the controlled variable towards the reference. It becomes clear that closed-loop control is better for disturbance rejection because the converged error norm is approximately 7% points lower than that of steady blowing.

5.4 Open vs. Closed Loop: Performance

The effective disturbance rejection does not necessarily lead to greater efficiency in terms of $\widetilde{\zeta}$ and $\widetilde{c}_p$. To compare the performances, field measurements were conducted for the closed-loop controller after convergence by holding the converged control input trajectory. The averaged metrics are plotted together with the open-loop results in Fig. 7. To achieve the same values of $\widetilde{c}_\mu$, the supply pressure for the closed loop was increased. Because the valves are open for approximately half of the period, the same mass flow was injected.

The results show that the controller achieves a slightly better static pressure increase at the same total pressure loss. Although the advantage lies at 1% and is thus far below the margin of the 7% better disturbance rejection, the controller was able to achieve the same and even slightly better performance in terms of averaged total pressure loss and static pressure increase.

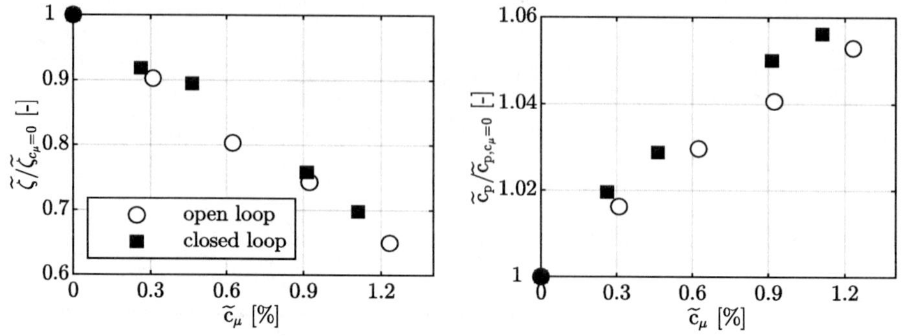

Fig. 7. Field- and cycle-averaged metrics for the open loop and closed loop (κ_5). Left: total pressure loss. Right: static pressure increase. Both metrics are normalized to their base flow value.

6 Conclusion

In this work, a linear stator compressor cascade under the influence of different firing patterns of CVC was investigated. A disturbance generator allowed for the application of arbitrary firing patterns without mechanical retrofitting. Because the number of possible patterns to investigate is potentially large, the analysis was limited to a specific type of pattern, denoted as κ_j. In this type of pattern, the ignition times are equally distributed over the course of a period, and the number of the next firing tube is the number of the previous tube plus $j \in \mathbb{N}^+$. Furthermore, the concept of local blockage was introduced, which makes the impact of patterns quantifiable. The analysis showed that patterns with successive firings of adjacent tubes experience large fluctuations of local blockage over the course of one period. In contrast, patterns where multiple tubes are skipped between two firing events experience a more constant local blockage.

In the subsequent wake measurements, several observations could be made. First, the flow field for patterns with large fluctuations of local blockage experiences a strong deflection of the flow structure over the course of one period. The deflection is observed for κ_1 and κ_2. For κ_2, the deflection follows a double-periodic progression. The other investigated patterns, which have a more constant local blockage, show no deflection. Additionally, indications for suction side flow separation can be observed for the κ_1-pattern, which possibly leads to major performance reduction. However, the performance proves to be better for patterns with constant local blockage.

Second, AFC in the form of end wall blowing has a greater positive effect on the performance for patterns with constant local blockage. Closed-loop AFC in the form of an iterative learning control approach has the same effect as open-loop AFC in terms of performance improvement but is more effective in terms of disturbance rejection.

It can be concluded that the firing pattern of CVC has a significant impact on compressor performance in the base flow and for AFC applications. Local blockage can be used as an indicator to identify patterns that are well-suited in terms of compressor performance. Nevertheless, the generalizabilty of this concept needs to be verified in further studies.

Acknowledgment. The authors acknowledge the support for this research by the Deutsche Forschungsgemeinschaft (DFG) in the context of the Collaborative Research Center CRC 1029 'Substantial Efficiency Increase in Gas Turbines through Direct Use of Coupled Unsteady Combustion and Flow' through sub-project B06.

References

1. Jansohn, P.: Modern gas turbine systems: High efficiency, low emission, fuel flexible power generation. Woodhead Publishing Limited (2013)
2. Gülen, S.C.: Etude on gas turbine combined cycle power plant - Next 20 years. J. Eng. Gas Turbines Power **138**(5), 051701 (2016)
3. Kaemming, T.A., Paxson, D.E.: Determining the pressure gain of pressure gain combustion. In: 2018 Joint Propulsion Conference, pp. 4567–4581 (2018)
4. Stathopoulos, P.: Comprehensive thermodynamic analysis of the Humphrey cycle for gas turbines with pressure gain combustion. Energies **11**(12), 3521–3541 (2018)
5. Bobush, B.C., Berndt, P., Paschereit, C.O., Klein, R.: Shockless explosion combustion: an innovative way of efficient constant volume combustion in gas turbines. Combust. Sci. Technol. **186**(10–11), 1680–1689 (2014)
6. Roy, G.D., Frolov, S., Sergei, M., Netzer, D., Borisov, A.A.: High-Speed deflagration and detonation: fundamentals and control. In: International Colloquium on Control and Detonation Processes (2001)
7. Steinberg, S.: Learning control algorithms for an unsteady vane flow field. Ph.D. thesis, TU Berlin (2018). https://doi.org/10.14279/depositonce-6695
8. Mihalyovics, J., Brück, C., Peitsch, D., Vasilopoulos, I., Meyer, M.: Numerical and experimental investigations on optimized 3D compressor airfoils. In: Turbo Expo: Power for Land, Sea and Air, vol. 50992, p. V02AT39A038 (2018)
9. Fietzke, B., Mihalyovics, J., King, R., Peitsch, D.: Binary repetitive model predictive active flow control applied to an annular compressor stator cascade with periodic disturbances. In: ASME Turbo Expo 2021: Turbomachinery Technical Conference and Exposition. ASME Digital Collection (2021)
10. Bristow, D.A., Tharayil, M., Alleyne, A.: A survey of iterative learning control. IEEE Control Syst. Mag. **26**(3), 96–114 (2006)
11. Morrison, D.R., Jacobson, S.H., Sauppe, J.J., Sewell, E.C.: Branch-and-Bound algorithms: a survey of recent advances in searching, branching, and pruning. Discrete Optim. **19**, 79–102 (2016)
12. Arnold, F., Neuhäuser, K., King, R.: Experimental comparison of two integer valued iterative learning control approaches at a stator cascade. ASME J. Eng. Gas Turbines Power **142**(1), 011020 (2020)
13. Schäpel, J.S., King, R., Yücel, F., Völzke, F., Paschereit, C.O., Klein, R.: Fuel injection control for a valve array in a shockless explosion combustor. In: Proceedings of the ASME Turbo Expo 2018: Turbomachinery Technical Conference and Exposition, vol. 6, p. V006T05A007 (2018)

DBD Plasma Actuation on the Blades of Axial-Flow Turbomachinery

David Greenblatt(✉), Omer Pfeffermann, and David Keisar

Faculty of Mechanical Engineering, Technion – Israel Institute of Technology, 320003 Haifa, Israel
davidg@technion.ac.il

Abstract. Flow separation, or stall, in axial flow turbomachinery results in a loss of pressure or compression in the case of fans and compressors, or the loss of power or thrust generation in the case of turbines. Wave-power-based Wells turbines, in particular, suffer so acutely from blade stall during normal operation, that it compromises their viability as a major renewable energy resource. In this research, pulsed dielectric barrier discharge (DBD) plasma actuators were implemented on the blades of a mono-plane Wells turbine impeller and its full-bandwidth performance was evaluated. An initial parametric study indicated that blade-tip reduced frequencies ≥2.5 produced the greatest impeller acceleration from rest. The corresponding physical pulsation frequency was then used as a basis for conducting nominally steady-state experiments as well as experiments involving acceleration and deceleration of the impeller. Data so acquired, corresponding to a reduced frequency range of 0.9 to 2.5, was compiled to construct an impeller performance map. Plasma pulsations dramatically increased the effective impeller bandwidth by producing useful net power well beyond flow ratios where mono-plane impellers spin down to a standstill. In fact, the shaft power at a 17° blade-tip angle of attack exceeded the plasma input power by a factor of 33. These findings are potentially game-changing for wave energy generation and axial flow turbomachinery in general.

Keywords: Turbomachinery · Wave energy · Wells turbine · DBD plasma actuators · OWC turbine · Deep stall · Separation control

1 Introduction

Flow separation on the blades of rotating machinery, referred to as stall, severely limits their operational envelope, and can lead to high levels of vibration and in some cases, to damage or failure This is true for both incompressible-flow fans and compressors where the pressure or compressibility drop severely compromises the operation and safety of the machine [1–4]. Flow separation is also encountered in the low-pressure turbine stage of high-altitude unmanned air vehicles resulting in propulsion losses [5, 6]. Moreover, wave-power-based oscillating water column (OWC) turbines, eponymously named Wells turbines, in particular, suffer so acutely from blade stall throughout their normal operation, that it compromises their viability as a major renewable energy resource [7].

R. King and D. Peitsch (Eds.): AFCC 2021, NNFM 152, pp. 258–271, 2022.
https://doi.org/10.1007/978-3-030-90727-3_16

The simplifying assumptions associated with blade-element momentum theory [8] for incompressible or mildly compressible flows, allows the flow over a blade element (or section) to be viewed as analogous to that over an airfoil. While this is a significant simplification, it is nevertheless reasonable to posit that the inception and mechanism of stall over a blade element and airfoil are similar. By extension, active separation control methods that are effective on airfoils will also be effective on the blades of these axial-flow machines. This supposition can be used as a guiding principle for selecting and applying separation control methods.

The active separation control method selected for this research, is the single dielectric barrier discharge (DBD) plasma actuator [9]. There are three primary reasons for this selection. First, they effectively control separation on airfoils and wings throughout a large Reynolds and Mach number range (see references in [10]). Second, they have very low inertia and can easily withstand the centrifugal forces associated with high-speed rotation. Third, when pulse-modulated at low duty cycles, their electrical power consumption is low (typically watts/meter), which means that they can produce a net system benefit [11]. The axial-flow configuration selected presently for evaluating the efficacy of DBD plasma actuators is the Wells turbine impeller referred to above [12, 13]. Wells turbines convert hydraulic energy associated with undulating wave motion into an oscillating air column within an axial flow duct that houses the impeller. They present a particularly extreme and challenging environment because the conditions on the blades vary both temporally and spatially. Fixed-speed and fixed-geometry impellers cannot run efficiently in varying amplitude oscillatory duct flows and thus, during normal operation, the blade angles-of-attack vary between large positive and negative values leading to massive stall [13]. Indeed, Wells turbine stall produces such enormous power and bandwidth losses, noise and mechanical vibration, that it compromises their viability [14]. It therefore comes as no surprise that many techniques have been proposed to reduce or eliminate Wells turbine stall [15–24].

A proof-of-concept study performed by the present authors determined that pulse-modulated perturbations at dimensionless frequencies near unity produced enormous increases in post-stall Wells turbine performance and were effective at angles of attack up to 50° [25]. This reinforced our hypothesis that the excitation mechanism of separation control observed on airfoils and wings is also active on the rotating impeller blades. Furthermore, DBD plasma actuators can be configured to operate on both sides of the blades [26] and this renders them ideal for the oscillatory conditions experienced by the impeller.

The objective of this investigation is to characterize the performance of a Wells turbine impeller, throughout its operational range, when equipped with DBD plasma actuators. The axial-flow test facility described in [25], designed to accommodate blade-mounted DBD plasma actuators, was used to achieve this objective. The experimental facility, detailing the actuator implementation, measurement techniques and controlling parameters are described in Sect. 2. This is followed by a description of the methods employed and a discussion of the results in Sect. 3. Conclusions and future research are presented in Sect. 4.

2 Experimental Facility

The experimental facility (see Fig. 1) comprised a transparent air duct with an upstream bell-mouth contraction, an impeller assembly, and a 1.5 kW downstream suction fan (Soler and Palau HRT/4–500). The duct was made from three flanged 750 mm long acrylic tubes, with inner and outer radii of $R_i = 245$ mm and $R_o = 250$ mm respectively. The mono-plane impeller was comprised of a hub and six symmetric NACA 0015 blade profiles ($b = 140$ mm and $c = 100$ mm), that were 3D printed using ABS (Acrylonitrile Butadiene Styrene), resulting in a nominal solidity $\sigma = 0.5$. The hub diameter was 200 mm (thus tip and hub radii were $R_h = 100$ mm and $R_t = 240$ mm) with a half-ellipsoid nose fairing on the air inlet side. The blades were secured within slots at a 90° stagger angle between two (non-conducting Delrin and ABS) fore and aft rotating hub elements mounted on a Delrin shaft. The shaft was mounted on two bearings and the upstream part of the assembly was secured in place by three support rods, covered with a NACA 0015 fairings, at stagger angles of 0°.

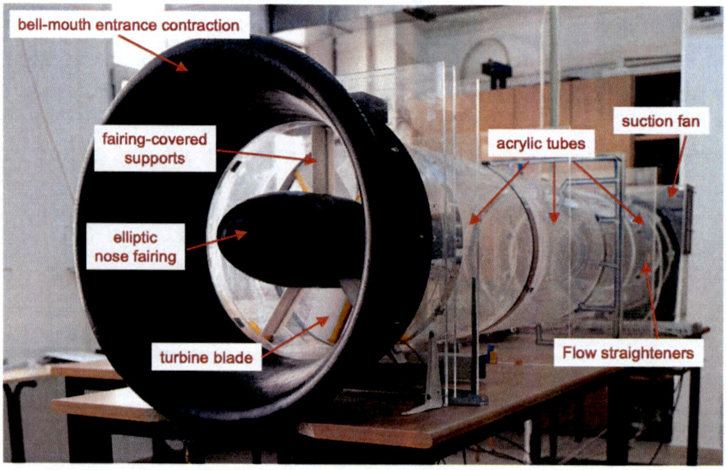

Fig. 1. Photograph of the axial flow facility from the air inlet side, showing the main externally visible components.

A photograph of the upstream part of the impeller assembly with the nose fairing removed (Fig. 2, left) shows the support-rod fairings and impeller blades, equipped with DBD plasma actuators. Also visible is an end-type slip-ring (Mercotac 215-2K) used to connect high voltage pulses to the DBD actuators. A photograph of the downstream part of the impeller assembly with the hub fairing removed (Fig. 2, right) shows the custom dynamometer, comprising a 1.0 Nm torque transducer (Kistler 4502A), a 1.0 Nm hysteresis brake (Magtrol HB-140), an optical speed (ω) sensor, and flexible couplings. A through-bore-type slip-ring (Prosper SRH3899) was mounted downstream of the impeller to serve as an earth for the plasma actuators.

The suction fan used to drive the system was accelerated to its design speed, namely 1,500 rpm, where the static pressure p_s versus volumetric flowrate Q characteristic curve

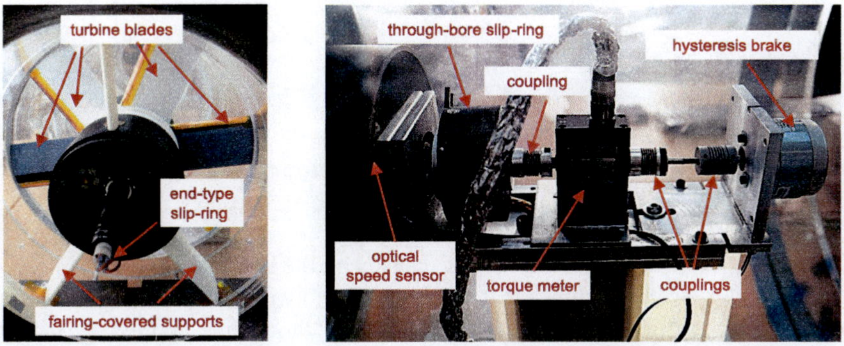

Fig. 2. Annotated photographs of: (left) the upstream part of the impeller with the nose faring removed; and (right) the downstream part of the impeller with the fairing removed.

was calibrated independently [25]. The axial velocity upstream of the impeller was thus calculated by simple conservation of mass, namely $U_x = 4Q/\pi(R_i^2 - R_h^2)$.

Each blade was printed with a leading-edge recess to accommodate the actuators that extended over the full span. The actuators were composed of a 20 mm wide × 20 μm thick copper electrode, encapsulated by a 300 μm thick translucent silicone rubber dielectric, and a 10 mm wide × 20μm thick exposed copper electrode. The actuators were driven in parallel by a modified Minipuls 2® high-voltage generator (GBS Elektronik GmbH) at $f_{\text{ion}} = 11,900\,\text{Hz}$, using a CPX400D–Dual 420-W power supply. For the experiments presented herein, the actuators were pulse modulated at low duty cycles (d.c.) in the frequency range $60\,\text{Hz} \leq f_p \leq 400\,\text{Hz}$. Current ($I$) supplied to the Minipuls 2 was monitored by means of a Fluke 115 Digital Multimeter, thereby facilitating the gross power supplied to the actuator: $P_{\text{in}} = V_{DC}I$ (watts) [10]. Based on direct calibration [25], the net contribution of the pulsed plasma perturbations to the impeller torque was estimated to be $O(10^{-4})$ Nm.

3 Discussion of Results

3.1 Relevant Definitions

For the present implementation, the actuators were mounted along the entire span of the blades and therefore their pulse-modulation frequency f_p was fixed. Under the assumption of negligible radial and azimuthal velocity components, the magnitude of velocity relative to the blade can be written as:

$$W(r) = (U_x^2 + \omega^2 r^2)^{1/2} \tag{1}$$

and therefore, the nominal reduced frequency $F^+ = f_p c/W(r)$ varies along the span.

This should be contrasted with airfoils and H-bladed vertical axis wind turbines where the reduced frequency is constant along the span [27, 28]. For convenience, we define the nominal reduced frequencies at the tip ($U_t = \omega R_t$) and hub ($U_h = \omega R_h$), namely:

$$F_t^+ = \frac{f_p c}{U_t}(\varphi^2 + 1)^{-1/2} \tag{2}$$

and

$$F_h^+ = \frac{f_p c}{U_t} \frac{\zeta}{(\zeta^2 \varphi^2 + 1)^{1/2}} \tag{3}$$

for $U_t \neq 0$, where $\varphi \equiv U_x/U_t$ is the flow (or speed) ratio and $\zeta \equiv R_t/R_h$ is the tip-to-hub ratio. Similarly, the angles of attack at the tip and hub are expressed as $\alpha_t = \tan^{-1}(\varphi)$ and $\alpha_h = \tan^{-1}(\varphi\zeta)$.

Due to relatively large frictional torque caused by the bearings and slip-rings (common in laboratory-scale machines that produce $O(10^1)$ watts), we distinguish between the aerodynamic torque applied to the impeller T_A, the torque applied by the brake (torque meter reading) T_B and the frictional torque T_F. In this research, because T_F is non-negligible, the central objective is to determine T_A or, more specifically, the aerodynamic shaft power $P_A = T_A\omega$ throughout the turbine's operational range.

3.2 Effect of Modulation Frequency

In experiments performed previously [25], the suction fan was accelerated linearly to 1,500 rpm at an arbitrarily assigned time t_0, and the impeller was allowed to spin up and settle to a steady-state baseline condition, typically between 240 and 260 rpm. Under these conditions, the aerodynamic torque produced by the massively stalled blades was balanced by the friction in the slip-rings and bearings (see next section). The introduction of perturbations at $f_p \geq 200$ Hz caused the impeller to gradually speed up and continuously accelerate ($\dot{\omega} \equiv d\omega/dt > 0$). When the impeller speed reached 500 rpm, the perturbations were terminated, and the impeller spun down to its baseline values. This was done in order to remain within the design limits of the facility and avoid damage (see next section).

Due to small variations in initial conditions described above, for the present experiments, perturbations were initiated prior to accelerating the suction fan at t_0. The results of the impeller response and suction pressure, following a linear acceleration of the suction fan over 10 s, are shown in Fig. 3 and Fig. 4, respectively. For all cases, during spin-up of the suction fan (Fig. 4), the impeller is initially stationary (Fig. 3) with $F_t^+ = F_h^+ = F^+ = f_p c/U_x$ and starts rotating after 10–15 s. Over the next 20 s, all perturbation frequencies produce an improvement over the baseline steady-state case, but only frequencies of $f_p \geq 200$ Hz produced continuous acceleration of the impeller. In instances where the impeller reached 500 rpm, namely for the $f_p \geq 200$ Hz and 400 Hz cases, the fan was decelerated linearly to zero (indicated on the figures). In these experiments as well as those reported in [25], $f_p = 200$ Hz corresponds to $1.4 \leq F_t^+ \leq 2.5$ and this is well within the optimal reduced frequency range observed on airfoils [29, 30]. This further supports our conjecture that the instability mechanism operating on stationary airfoils is also active in the rotating frame. A new observation from the present data set is that perturbations are effective even at angles of attack as high as 90°.

The system pressure losses (Fig. 4) closely track the increases in rpm (or power) and facilitate the instantaneous calculation of Q from the fan curve, and hence fan power $P_{\text{fan}} = -p_s Q$ and net impeller aerodynamic efficiency $\eta = (P_A - P_{\text{in}})/P_{\text{fan}}$. An unexpected result was that following deceleration of the fan under $f_p = 200$ Hz and

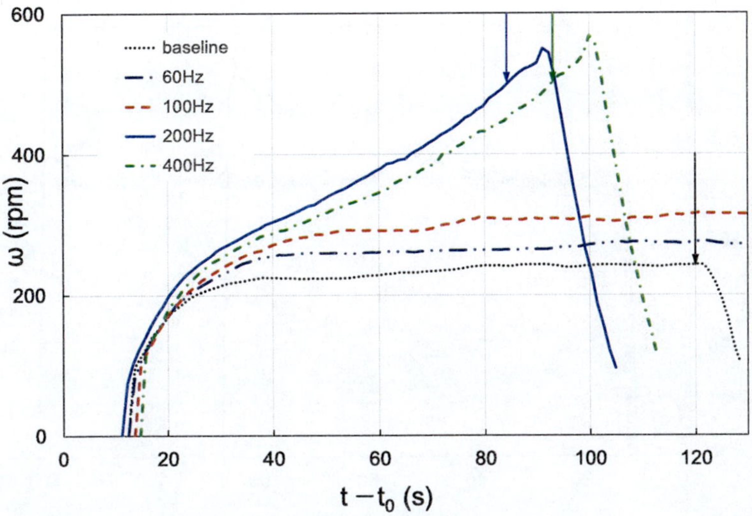

Fig. 3. Impeller rotational speed-up resulting from acceleration (from $t_0 = 0$) and deceleration of the suction fan for the uncontrolled (baseline) and different perturbation frequencies (see corresponding suction pressure in Fig. 4). The fan accelerates and decelerates linearly over 10 s. Vertical arrows indicate the initiation of fan deceleration.

400 Hz perturbations (see indications in Fig. 3 and Fig. 4), the impeller *accelerated* for approximately 7 s. In contrast, following fan deceleration in the baseline case, the speed remained constant for 1.5 s following deceleration of the fan and then spun down. This intriguing observation can be explained as follows. At $\omega = 500$ rpm, the flow ratio $\varphi = 0.56$, the tip angle of attack $\alpha_t = 29°$ and hence the impeller is deeply stalled. Under these conditions, the reduced frequencies at the tip are $F_t^+ = 1.4$ and 2.8. As the fan decelerates, U_x decreases and therefore, so do φ and α. For example, when U_x halves its value, the blades experience light (or shallow) stall conditions, corresponding to $\varphi = 0.28$ and $\alpha_t = 16°$. And because pulsed perturbations are particularly effective within this regime on airfoils at $F_t^+ = 0.7$ and 1.4, the aerodynamic efficiency (C_L/C_D) of the blades increases dramatically. We again implicitly assert that even though the conditions vary spatially and temporally on the blades, the separated shear layer excitation mechanism active on airfoils is also active on the rotating blades. Indeed, this improvement is so dramatic that the impeller accelerates even though the dynamic pressure relative to the blade, namely $1/2\rho(U_t^2 + U_x^2)$, decreases.

3.3 Overall Turbine Performance

One difficulty associated with controlling the massive stall conditions described in the previous section was that the turbine speed could not easily be maintained constant. This is because the perturbations increased impeller speed, corresponding to lower angles of attack and hence greater flow control effectiveness [10]. Hence, the impeller not only sped up, but the rate of speed-up also increased. Lightly loading the impeller with the hysteresis brake (<100 Nmm) led to a more gradual acceleration of the impeller, while

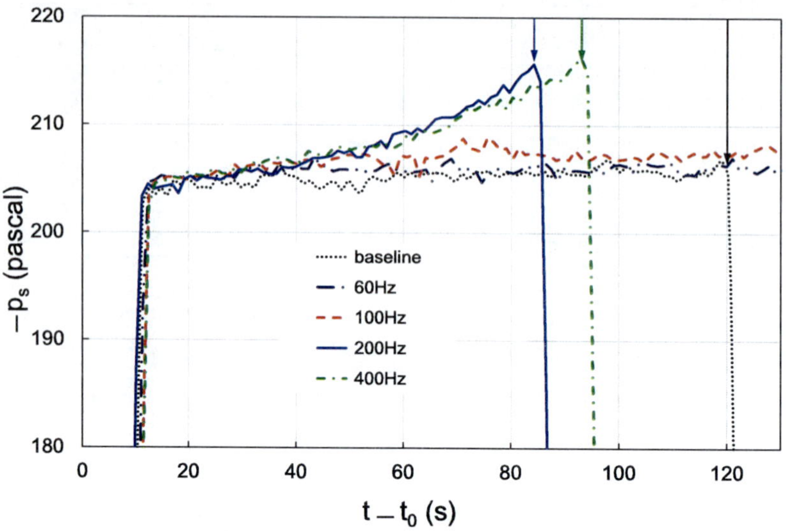

Fig. 4. Suction pressure produced in the tube following acceleration (from $t_0 = 0$) and deceleration of the fan for the uncontrolled (baseline) and different perturbation frequencies (see corresponding impeller speed in Fig. 3). The fan accelerates and decelerates linearly over 10 s. Vertical arrows indicate the initiation of fan deceleration.

slightly higher loading (>100 Nmm), produced deceleration and a lower settling speed. Although closed-loop control can be used to attain a set-point for the machine, this was considered to be outside of the objectives of this paper. Instead, we conducted a series of experiments under accelerating and deceleration conditions. We then used a composite of these experiments to construct a performance map.

To account for all of the combinations mentioned above, the unsteady system equation must be considered, namely:

$$J_0 \dot{\omega} = T_A - T_B - T_F \tag{4}$$

Hence with the shaft nominally unloaded and rotating at a constant speed, the aerodynamic torque is balanced by the frictional torque as described in the previous section. Our objective here is to quantify T_A and hence determine the aerodynamic shaft power $P_A = T_A \omega$ under all operational conditions.

In order to determine the non-negligible frictional torque due to the slip-rings and bearings, the simplified form of Eq. (4), namely $\dot{\omega} = T_F / J_0$ was employed. To determine $\dot{\omega}$ under frictional loading alone, namely under the conditions $U_x = 0$, the impeller was spun up to 500 rpm, after which the fan was decelerated to zero, and the perturbations were terminated. The impeller was then allowed to spin down to zero under quiescent conditions ($U_x = 0$) to obtain $\omega = \omega(t)$. These experiments were performed periodically throughout the experimental campaign and a data sample is shown in Fig. 5, where the deceleration is referenced to arbitrary time t'. Linear least-squares curves were fitted to the data and an average was used to determine $\dot{\omega} = -3.66$ rad/s^2. In order to determine J_0, the individual components of the impeller (hub, bearings, slip-rings and blades) were weighed and then $J_0 = 0.050\ \mathrm{kg\,m^2}$ was calculated based on the geometric configuration of the impeller assembly. Finally, the frictional torque, namely $T_F = J_0\dot{\omega} = -0.183$ Nm, was determined and considering that the maximum brake torque was 1 Nm, friction comprised a sizeable fraction of the overall load.

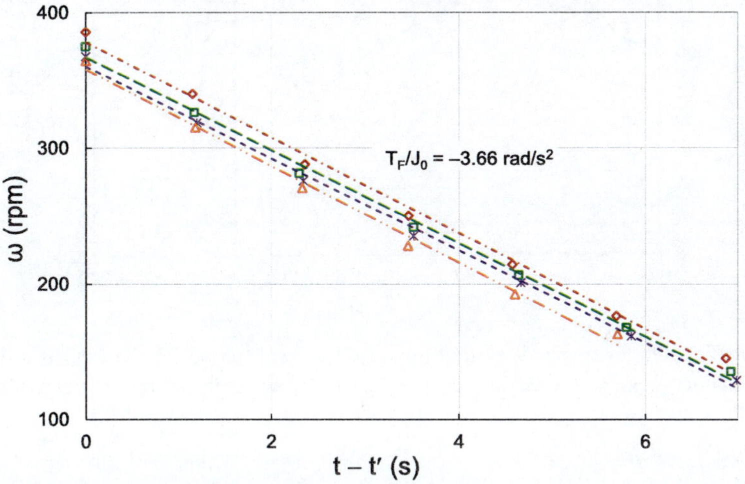

Fig. 5. Representative data sets acquired throughout the experimental campaign, illustrating spin-down of the impeller under frictional loading alone.

In an attempt to attain constant impeller speeds, two separate experiments were performed. The first illustrates the effect of spinning up the turbine and then terminating the plasma perturbations, while the second shows the effect of braking the turbine while the plasma perturbations are active. In the first experiment, the impeller was allowed to speed-up, and then the brake was set close to its maximum setting (Fig. 6). This resulted in a gradual deceleration until the pulsations were terminated, following which the impeller spun down to a standstill. The clearest evidence that the perturbations are responsible for driving the turbine is that the turbine immediately and rapidly spins down when they are terminated. For this experiment, the aerodynamic torque was determined in the region of constant torque with perturbations activated, where $\dot{\omega}$ was estimated by a least-squares curve fit.

In the second experiment, a lower brake setting was employed, resulting in a near constant rotational speed (Fig. 7). A subsequent increase in loading brought the impeller to a standstill. In this experiment, two relevant data were extracted. The first was based on average values between 120 and 150 s under the assumption that $\dot{\omega} = 0$. The aerodynamic shaft power, namely 82 W (Eq. 4), is slightly higher than the estimate in [25] and is achieved with a plasma input power of 2.5 W, representing an astonishing ratio $P_A/P_{in} = 33$. An accumulation of previous experience on airfoils [10, 29] shows that as the post stall angle is reduced, the separation control effect increases. Nevertheless, this remains an extraordinary result in light of the fact that the angle-of-attack varies from 17° to 30° from tip to hub. The second relevant data were extracted from the deceleration phase where $\omega(t)$ was determined using a fourth-order polynomial curve-fit.

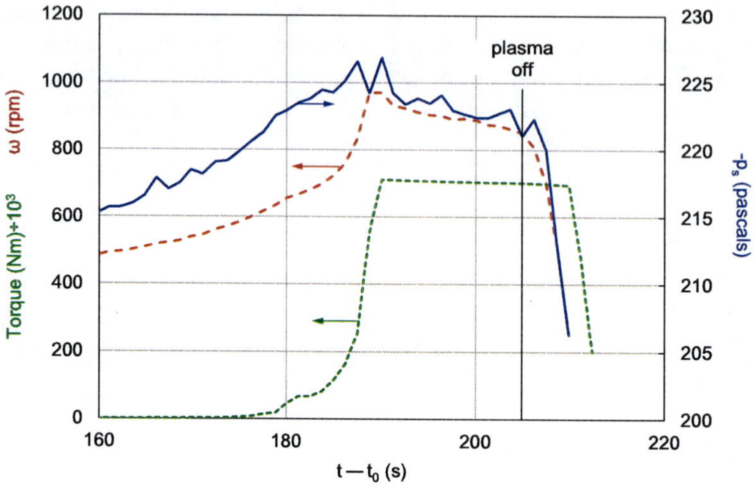

Fig. 6. Impeller rotational speed, system pressure loss and torque for the loaded turbine with plasma pulsations at $f_P = 200\,\text{Hz}$ and d.c. $= 1.7\%$. Arrows indicate axis corresponding to the data set.

The final experiment in this series involved unloading the turbine and allowing it to accelerate under the braking effect of friction alone. Although this experiment exceeded the design load of the blades, it was deemed critical to obtaining the turbine performance over a larger bandwidth. The results from this experiment (Fig. 8) show both ω and $\dot{\omega}$ initially increasing. At approximately 96 s, corresponding to 80 rad/s or 760 rpm, the impeller accelerated dramatically, reached inflection 99 s, and continued to accelerate ($\dot{\omega} > 0$) until a blade failed. Under these conditions, the $\dot{\omega}$ term dominates as shown in the figure and the aerodynamic torque is calculated according to $T_A = J_O\dot{\omega} + T_F$.

The data extracted from Fig. 6, Fig. 7 and Fig. 8 are summarized on the performance map shown in Fig. 9. The steady-state and mild deceleration data are indicated as solid symbols and the acceleration and deceleration phases are shown by open symbols. Error bars indicate uncertainty resulting from differentiation of the experimental data. Despite the disparate methods employed for accumulating these data, they amalgamate to produce a consistent map that clearly reveals the effects of pulsed plasma perturbations. In the relatively low flow ratio range, namely $\varphi \leq 0.22$ range ($\alpha_t \leq 13°$), corresponding to values below which the peak power occurs, it is not clear if perturbations produce a positive gross net effect because no baseline data is available. Here the peak aerodynamic efficiency $\eta = (P_A - P_{\text{in}})/P_{\text{fan}}$ is 48.5%. This is somewhat lower than values around 65% cited in the literature [13] and is most likely due to the relatively low tip Reynolds numbers that peak at 2.4×10^5. At flow ratios greater than some critical value, conventional impellers produce zero power. For example, comparable solidity ($\sigma = 0.51$) mono-plane impellers [13], stall at $\varphi \geq 0.24$, resulting in a total shut-down of power generation with accompanying noise and potentially destructive vibrations. In our experiments, the critical value falls between 0.22 and 0.29 above which plasma pulsations always produce net positive aerodynamic efficiency $\eta = (P_A - P_{\text{in}})/P_{\text{fan}}$. This has a dramatic effect on the turbine bandwidth and this factor alone has the potential

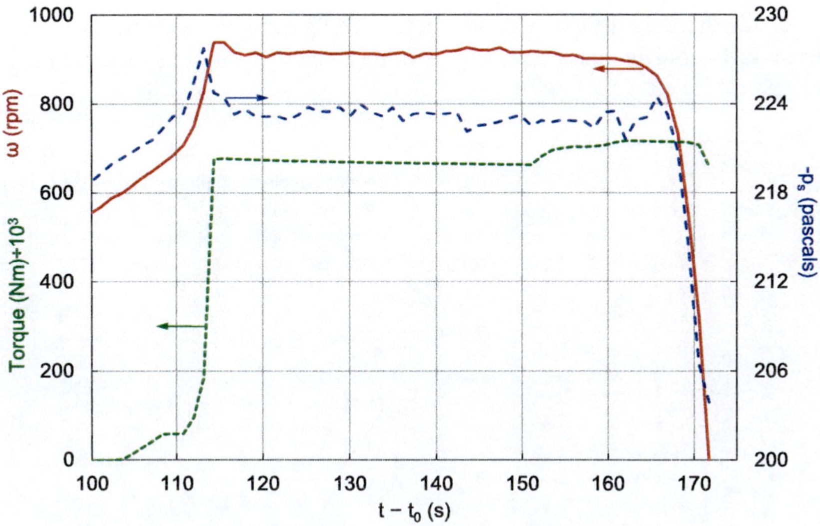

Fig. 7. Impeller rotational speed under increasing load until spin-down with plasma pulsations at $f_p = 200$ Hz and d.c. $= 1.7\%$. Arrows indicate axis corresponding to the data set.

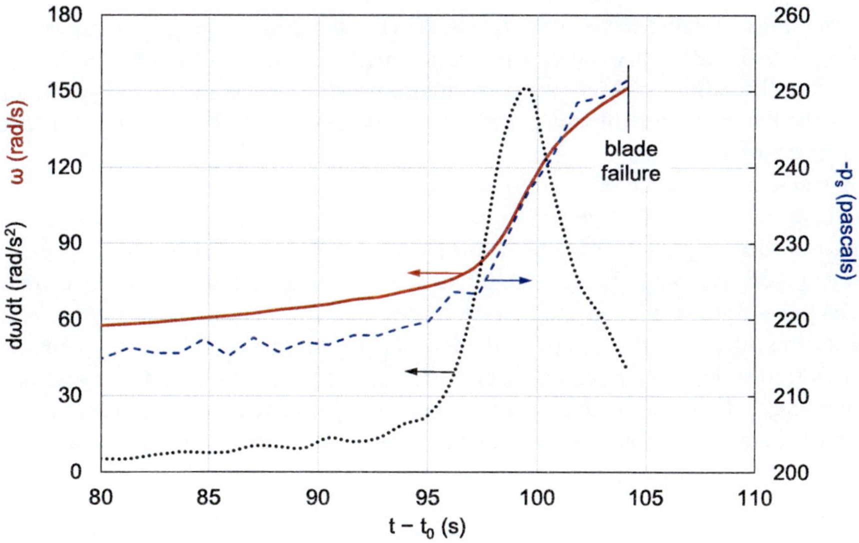

Fig. 8. Impeller speed and acceleration to failure, together with fan pressure, under frictional loading with pulsations at $f_p = 200$ Hz and d.c. $= 1.7\%$. Arrows indicate axis corresponding to the data set.

to dramatically increase the overall efficiency of Wells turbines. A question still open is whether perturbations increase peak power values, and this should be a topic for future research.

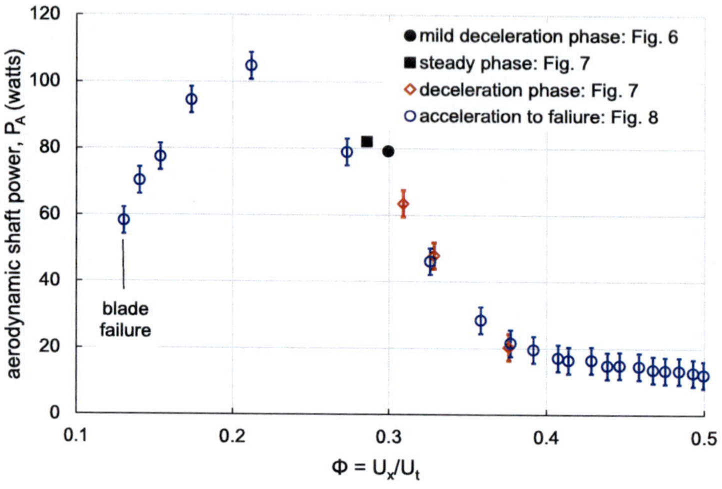

Fig. 9. Amalgamation of data acquired to construct a Wells turbine performance map in the presence of plasma pulsations at $f_p = 200$ Hz.

In the present experiments with $f_p = 200$ Hz, the reduced frequency range varies from $F_t^+ = 2.5$ under fully stalled stationary conditions ($\alpha_t = 90°$) to $F_t^+ = 0.9$ under incipient stall conditions ($\alpha_t = 12°$) corresponding to the peak in Fig. 9. As hypothesized previously, the most probably reason why $f_p = 200$ Hz pulsations consistently produce the most favorable results is that the corresponding reduced frequency range falls well within range known to be effective on airfoils, namely $0.5 \leq F^+ \leq 3$ [29, 30]. These frequencies are effective at generating nearly two-dimensional spanwise vortices that deflect the shear layer towards the airfoil surface, thereby increasing lift. Despite the loss of two-dimensionality along the blade span of the impeller, this mechanism still appears to be highly effective.

The present proof-of-concept study focused on performance measurements and, although tremendous potential was demonstrated, a number of topics are suggested for future research. First, although it was hypothesized that the inception and mechanism of stall over blades and airfoils are similar, this was not directly validated. In future studies, blade surface pressure measurements, as well as optical flow-field measurements, must be performed to gain an understanding of the blade aerodynamics. Second, our experiments were performed under laboratory-scale conditions, where the maximum tip Reynolds numbers ($\leq 2.4 \times 10^5$) were an order of magnitude smaller than those on in-service machines. Thus, a large test facility should be developed to evaluate the viability of plasma actuators under full-scale conditions. If the blade-airfoil analogy is indeed valid, then effective airfoil separation control shown by [31] at $Re > 2 \times 10^6$ (Mach number of 0.4) bodes well for full-scale applications. The main caveat is that the plasma body force per unit width must scale with the tip dynamic pressure and chord, namely $\frac{1}{2}\rho(U_x^2 + U_t^2)c$ and this may require different high voltage generators [32] to those employed here. Third, another DBD technology that may also have significant potential for controlling blade separation due to its effectiveness on airfoils is pulsed-periodic nanosecond excitation [33]. Pulses of $O(10^1)$ nanoseconds can produce O(1) microsecond overheating of several hundred degrees within the discharge region. Periodic pulses thus produce a series of compression waves and in this sense the mechanism of pulsed-periodic nanosecond excitation is different from that employed presently.

4 Conclusions and Future Research

The present investigation studied the effect of pulsed DBD plasma actuators on Wells turbine impeller performance throughout its operational range. Experiments were conducted initially to isolate the most effective pulsation frequency without biasing the data with different initial conditions. A pulsation frequency of $f_p = 200$ Hz, corresponding to $F^+ = 2.5$ under massively stalled conditions produced the greatest acceleration of the impeller. This physical frequency was then used as a basis for conducting nominally steady-state experiments as well as experiments involving acceleration and deceleration of the impeller. The unsteady equation of rotary motion was used, in conjunction with the experimental data, to construct the impeller performance map.

The most significant steady-state result indicated an aerodynamic shaft power of 82 W at a tip angle of attack of 17°, which, remarkably, was 33 times higher than the plasma input power. Decelerating the impeller under these conditions, in conjunction with the equation of motion, revealed the deep-stall benefits of plasma perturbations. On the other hand, accelerating the impeller to destruction under frictional loads alone, again in conjunction with the equation motion, allowed characterization of virtually the entire operational range. The dramatic increase in the operational bandwidth shows that DBD plasma actuators have the potential to be a game-changing OWC technology. Positive implications for axial flow fans, compressors and low-pressure turbine stages can also clearly be extrapolated. The reduced frequency at the tip varied between 0.9 and 2.5 in the post-stall regime, which fell well within the effective excitation range on two-dimensional airfoils. Thus, the excitation mechanism still appears to be highly effective on the blades despite the loss of flow two-dimensionality.

Future research should focus on understanding the aerodynamics of the problem and evaluating DBD plasma effectiveness at higher Reynolds numbers. Furthermore, an oscillatory bi-directional-flow facility, representative of Wells turbine operational conditions, should be constructed. Application of DBD plasma actuators under these oscillatory conditions will provide a clearer evaluation of overall system benefits. Such a system should also include some form of closed-loop control with sensors for flow direction and incipient stall, with the ability to initiate and terminate perturbations [11, 34] on the appropriate stalling surface [26].

Acknowledgments. This research was supported by the Carl E. Schustak Energy Research and Development Fund. The authors wish to thank Mark Epshtein, Ahmad Zidane and Dor Polonsky for assisting with the design and construction of the facility as well as assisting with data acquisition.

References

1. de Jager, B.: Rotating stall and surge control: a survey. In: Proceedings of the IEEE Conference on Decision and Control. pp. 1857–1862. IEEE (1995). https://doi.org/10.1109/cdc.1995.480612
2. Sheard, A.G., Corsini, A.: The mechanical impact of aerodynamic stall on tunnel ventilation fans. Int. J. Rotating Mach. **2012**, 1–12 (2012). https://doi.org/10.1155/2012/402763
3. Bianchi, S., Corsini, A., Sheard, A.G., Tortora, C.: A critical review of stall control techniques in industrial fans (2013). https://doi.org/10.1155/2013/526192
4. Greitzer, E.M.: Axial compressor stall phenomena. J. Fluids Eng. Trans. ASME. **102**, 134–151 (1980). https://doi.org/10.1115/1.3240634
5. Huang, J., Corke, T.C., Thomas, F.O.: Plasma actuators for separation control of low-pressure turbine blades. AIAA J. **44**, 51–57 (2006). https://doi.org/10.2514/1.2903
6. Hultgren, L.S., Ashpis, D.E.: Demonstration of separation control using dielectric barrier discharge plasma actuators. AIAA J. **56**, 4614–4620 (2018). https://doi.org/10.2514/1.J056976
7. QUB: Islay LIMPET wave power plant. Contract JOR3-CT98–0312 (2002)
8. Dixon, S.L.: Fluid mechanics, thermodynamics of turbomachinery (1975). https://doi.org/10.1016/0017-9310(77)90019-9
9. Corke, T.C., Post, M.L., Orlov, D.M.: SDBD plasma enhanced aerodynamics: concepts, optimization and applications (2007). https://doi.org/10.1016/j.paerosci.2007.06.001.
10. Keisar, D., Hasin, D., Greenblatt, D.: Plasma actuator application on a full-scale aircraft tail. AIAA J. **57**, 616–627 (2019). https://doi.org/10.2514/1.J057233
11. Ben-Harav, A., Greenblatt, D.: Plasma-based feed-forward dynamic stall control on a vertical axis wind turbine. Wind Energy **19**, 3–16 (2016). https://doi.org/10.1002/we.1814
12. Antonio, F.D.O.: Wave energy utilization: A review of the technologies (2010). https://doi.org/10.1016/j.rser.2009.11.003
13. Raghunathan, S.: The wells air turbine for wave energy conversion (1995). https://doi.org/10.1016/0376-0421(95)00001-F
14. Boake, C.B., Whittaker, T.J.T., Folley, M., Ellen, H.: Overview and initial operational experience of the LIMPET wave energy plant. In: Proceedings of the International Offshore and Polar Engineering Conference, pp. 586–594 (2002)
15. Setoguchi, T., Santhakumar, S., Takao, M., Kim, T.H., Kaneko, K.: Effect of guide vane shape on the performance of a Wells turbine. Renew. Energy. **23**, 1–15 (2001). https://doi.org/10.1016/S0960-1481(00)00163-4
16. Amundarain, M., Alberdi, M., Garrido, A.J., Garrido, I., Maseda, J.: Wave energy plants: Control strategies for avoiding the stalling behaviour in the Wells turbine. Renew. Energy. **35**, 2639–2648 (2010). https://doi.org/10.1016/j.renene.2010.04.009
17. Lekube, J., Garrido, A.J., Garrido, I., Otaola, E., Maseda, J.: Flow control in wells turbines for harnessing maximum wave power. Sensors (Switzerland) **18**, 535 (2018). https://doi.org/10.3390/s18020535
18. Mishra, S., Purwar, S., Kishor, N.: Maximizing output power in oscillating water column wave power plants: an optimization based MPPT algorithm. Technologies. **6**, 15 (2018). https://doi.org/10.3390/technologies6010015
19. Halder, P., Samad, A., Thévenin, D.: Improved design of a Wells turbine for higher operating range. Renew. Energy. **106**, 122–134 (2017). https://doi.org/10.1016/j.renene.2017.01.012
20. Gato, L.M.C., Webster, M.: An experimental investigation into the effect of rotor blade sweep on the performance of the variable-pitch Wells turbine. Proc. Inst. Mech. Eng. Part A J. Power Energy **215**, 611–622 (2001). https://doi.org/10.1243/0957650011538848

21. M'zoughi, F., Bouallègue, S., Ayadi, M., M'zoughi, F., Garrido, A.J., Garrido, I.: Modelling and airflow control of an oscillating water column for wave power generation. In: 2017 4th International Conference on Control, Decision and Information Technologies, CoDIT 2017, pp. 938–943. Institute of Electrical and Electronics Engineers Inc. (2017). https://doi.org/10.1109/CoDIT.2017.8102718
22. Gato, L.M.C., Eça, L.R.C., De, A.F.: Performance of the wells turbine with variable pitch rotor blades. J. Energy Resour. Technol. Trans. ASME. **113**, 141–146 (1991). https://doi.org/10.1115/1.2905794
23. Shehata, A.S., Xiao, Q., Selim, M.M., Elbatran, A.H., Alexander, D.: Enhancement of performance of wave turbine during stall using passive flow control: First and second law analysis. Renew. Energy. **113**, 369–392 (2017). https://doi.org/10.1016/j.renene.2017.06.008
24. Das, T.K., Samad, A.: Influence of stall fences on the performance of Wells turbine. Energy **194**, 116864 (2020). https://doi.org/10.1016/j.energy.2019.116864
25. Greenblatt, D., Pfeffermann, O., Keisar, D., Göksel, B.: Wells turbine stall control using plasma actuators. AIAA J. **59**, 765–772 (2021). https://doi.org/10.2514/1.J060278
26. Greenblatt, D., Lautman, R.: Inboard/outboard plasma actuation on a vertical-axis wind turbine. Renew. Energy. **83**, 1147–1156 (2015). https://doi.org/10.1016/j.renene.2015.05.020
27. Greenblatt, D., Schulman, M., Ben-Harav, A.: Vertical axis wind turbine performance enhancement using plasma actuators. Renew. Energy. **37**, 345–354 (2012). https://doi.org/10.1016/j.renene.2011.06.040
28. Greenblatt, D., Ben-Harav, A., Mueller-Vahl, H.: Dynamic stall control on a vertical-axis wind turbine using plasma actuators. AIAA J. **52**, 456–462 (2014). https://doi.org/10.2514/1.J052776
29. Greenblatt, D., Wygnanski, I.: The control of flow seperation by periodic excitation. Prog. Aerosp. Sci. **36**, 487–545 (2000). https://doi.org/10.1016/S0376-0421(00)00008-7
30. Greenblatt, D., Göksel, B., Rechenberg, I., Schüle, C.Y., Romann, D., Paschereit, C.O.: Dielectric barrier discharge flow control at very low flight Reynolds numbers. AIAA J. **46**, 1528–1541 (2008). https://doi.org/10.2514/1.33388
31. Kelley, C.L., et al.: High Mach number leading-edge flow separation control using AC DBD plasma actuators. In: 50th AIAA Aerospace Sciences Meeting Including the New Horizons Forum and Aerospace Exposition (2012). https://doi.org/10.2514/6.2012-906
32. Thomas, F.O., Corke, T.C., Iqbal, M., Kozlov, A., Schatzman, D.: Optimization of dielectric barrier discharge plasma actuators for active aerodynamic flow control. AIAA J. **47**, 2169–2178 (2009). https://doi.org/10.2514/1.41588
33. Roupassov, D.V., Nikipelov, A.A., Nudnova, M.M., Starikovskii, A.Y.: Flow separation control by plasma actuator with nanosecond pulsed-periodic discharge. AIAA J. **47**, 168–185 (2009). https://doi.org/10.2514/1.38113
34. Greenblatt, D., Neuburger, D., Wygnanski, I.: Dynamic stall control by intermittent periodic excitation. J. Aircr. **38**, 188–189 (2001). https://doi.org/10.2514/2.2751

Numerical Analysis of Unsteady Compressor Performance Under Boundary Conditions Caused by Pulsed Detonation Combustion

Nicolai Neumann[1](✉), Tim Rähse[2], Panagiotis Stathopoulos[2], and Dieter Peitsch[1]

[1] Institute of Aeronautics and Astronautics, Chair for Aero Engines, Technische Universität Berlin, Marchstraße 12–14, 10587 Berlin, Germany
{nicolai.neumann,dieter.peitsch}@tu-berlin.de
[2] Institute of Fluid Dynamics and Technical Acoustics, Chair for Unsteady Thermodynamics in Gas Turbine Processes, Technische Universität Berlin, Müller-Breslau-Straße 8, 10623 Berlin, Germany
{raehse,stathopoulos}@tu-berlin.de

Abstract. Pressure gain combustion is a revolutionary concept to increase gas turbine efficiency and thus potentially reduces the environmental footprint of power generation and aviation. Pressure gain combustion can be realized through pulsed detonation combustion. However, this unsteady combustion process has detrimental effects on adjacent turbomachines. This paper identifies realistic time-variant compressor outlet conditions, which could potentially stem from pulsed detonation combustion. Furthermore, a low fidelity approach based on the 1D-Euler method is applied to investigate the performance of a compressor exposed to these outlet boundary conditions. The simulation results indicate that the efficiency penalty due to unsteady compressor operation remains below 1% point. Furthermore, between 80% and 95% of the fluctuations' amplitudes are damped till the inlet of the 4-stage compressor.

Keywords: Pressure gain combustion · Unsteady 1D-Euler simulation · Compressor performance

Abbreviations and Nomenclature

E^3	Energy efficient engine
PDC	Pulsed detonation combustion
PGC	Pressure gain combustion
A	Cross sectional area
c	Absolute velocity
E	Internal energy
$F_{x,Outlet}$	Outlet surface force
$F_{x,Inlet}$	Inlet surface force

R. King and D. Peitsch (Eds.): AFCC 2021, NNFM 152, pp. 272–287, 2022.
https://doi.org/10.1007/978-3-030-90727-3_17

$F_{x,Endwall}$	Surface force caused by a change in area
$F_{x,Blade}$	Blade force
h	Specific enthalpy
$\dot{m}$	Mass flow
p	Pressure
t	Time
t_{close}	Time during which the combustor is closed
T	Temperature
V_p	Volume of the plenum
W	Work input
$\Delta\Phi$	Relative amplitude of static pressure
ϵ	Unsteady damping
η	Isentropic efficiency
ρ	Density
γ	Ratio of specific heats
$_{ax}$	Quantity in axial direction
$_{in}$	Inlet of component
$_{out}$	Outlet of component
$_{p}$	Quantity in plenum
$_{ma}$	Mass-averaged
$_{wa}$	Work-averaged

1 Introduction

Despite the current pandemic, aviation is seen as an increasing market in the years to come [1,2]. To reduce aviation's emission footprint, the industry committed to the ACARE flightpath 2050, which e.g. aims for a 75% reduction of CO_2 emission compared to a reference aircraft from 2000 [3]. Due to the high energy density requirement both gravimetric and volumetric for long haul flights, which are responsible for 20% of aviation's CO_2 emissions [4], there is currently no alternative to hydrocarbon fuels. Hydrogen appears to be an option for short- and medium-range aircraft when burnt in modified gas turbines [4]. Other propulsion concepts like pure electric or distributed propulsion are still in their infancy and limited to small commuter and regional aircraft. Hence, conventional gas turbine based engines are still indispensable and therefore require further development to meet aforementioned goals. However, the conventional gas turbine engine is a very mature technology, making efficiency improvements increasingly difficult to attain. That is why revolutionary concepts such as pressure gain combustion (PGC) attract a lot of attention. Instead of a constant pressure combustion, which in fact occurs at a pressure loss, the fuel is burned at quasi-constant volume conditions leading to a total pressure increase. In their simplified ideal thermodynamic studies, Heiser and Pratt [5] determined thermal efficiencies of up to 80% using PGC. More recent thermodynamic studies [6,7] also account for reduced turbine efficiency and losses associated with the secondary air system for blade cooling. The PGC cycles, indeed, outpace their Joule counterparts by up to 10% points in thermal efficiency, especially at low cycle pressure ratios. However, the disadvantage of pressure gain combustion lies in its unsteadiness, which results in time-variant boundary conditions for adjacent turbomachines. Hence, it is very likely that the PGC device will be surrounded by two plena, which dampen the fluctuations. The turbine is most severely affected, when hot gas pressure waves strike onto the turbine blades. That is object of research both experimentally and numerically for several years [8–12]. Less attention is directed towards the compressor, which will be the focus of this paper.

The unsteady periodic combustion leads to a likewise unsteady throttling of the upstream compressor, because the inlet of a combustion tube is closed after ignition, until combustor pressure level allows for refilling. This leads to a mass accumulation in the upstream plenum rising compressor outlet pressure associated with detrimental effects onto compressor performance as the literature survey confirms. Hoke and Bradley [13] connected a turbocharger to two pulsed detonation combustion (PDC) tubes. Even when applying a simultaneous firing pattern, the feasibility could be demonstrated. Sakurai et al. [14] successfully connected two combustion tubes to a radial compressor. One tube operated in PGC mode, whereas the other ran in conventional combustion mode, achieving sustained operation. However, they found cycle performance to be lower compared to full conventional combustion, which they attributed to a decrease in compressor efficiency. When switching the second tube to PGC mode as well, the gas turbine operation could not be sustained. Lu and Zheng [15], in turn, conducted a third experimental study of the PDC-compressor interaction by connecting an arrangement of four PDC tubes to a centrifugal compressor. After examining the sensitivity to the firing pattern and frequency, they concluded that the compressor operates closer to the surge line. At Technische Universität Berlin, various 2D and 3D compressor cascades are exposed to throttling devices that simulate PGC boundary conditions. Active flow control is applied successfully in order to stabilise the flow [16–19].

Numerical studies in this particular field do not always allow a direct comparison. That is due to the way the compressor outlet boundary, which simulates the presence of a PGC device, is specified. Because of the resolution also in circumferential direction, 3D-CFD allows to simulate a sequential firing pattern by a rotating section of elevated pressure. Thus, only part of the compressor outlet domain is throttled. Examples are de Almeida and Peitsch [20–22], who performed 3D-CFD simulations of the rear part of the Rolls Royce E3E and NASA GE E^3 high pressure compressors at varying back pressure frequencies (12%–300% of blade passing frequency) and amplitudes (5%–20%). At low amplitudes, they found the efficiency penalty to be less than 1%, independent of frequency. At high amplitudes, penalties went up to 15%. The authors further claim, that most of the fluctuations are damped across a single stage. Numerical investigations based on a 1D-method usually use boundary conditions with much lower pressure amplitudes. This is because the complete compressor outlet is throttled in 1D. Neumann et al. [23] used a coupled numerical scheme to compute compressor boundary conditions that would result from a PDC device with three and five combustion tubes. They showed that results of a unsteady quasi 1D-Euler simulation match those computed by 3D-CFD, where also the complete outlet domain is throttled. Furthermore, the efficiency penalty was estimated to be below 1% point for these specific cases and the results suggest that 90% of the fluctuations are damped over the last four stages.

This paper builds upon these findings [23] and substantially extends it by answering the following research questions:

1. What range of compressor boundary conditions can be expected when operated together with a PDC?
2. What is the impact of these boundary conditions onto compressor performance in terms of efficiency? Which part of the compressor is most affected by a certain set of boundary condition?

For that purpose, the paper presents a simplified approach to estimate the amplitude and frequency of pressure fluctuations caused by different plenum volumes and PDC specifications. The results will serve as an input for boundary conditions of 1D-compressor simulations. The compressor performance is evaluated for a wide range of throttle frequencies and amplitudes using a validated 1D-Euler method. In doing so, a comprehensive evaluation of multi-stage compressor performance is presented, which combines the derivation of the compressor boundary conditions and their influence on compressor operation. To the authors' knowledge, no such study has been carried out so far.

2 Method and Compressor Model

The compressor performance is simulated primarily by an unsteady quasi 1D-Euler approach due to its resolution in time and space. However, the momentum and energy equations of the 1D-Euler approach require source terms, in order to account for blade force and work input within the compressor. These source terms are calculated by a mean line method. 1D-Euler and mean line method are briefly introduced next.

The 1D-Euler method was first developed for the simulation of a shockless explosion combustion by Berndt [24] and then extended for the simulation of turbomachines by Dittmar and Stathopoulos [25] and Neumann et al. [23]. For this purpose, the force of the compressor blades ($F_{x,blade}$) and the work of the rotor blades (W_{Rotor}) are integrated into the momentum and energy equations, respectively, as seen in Eq. 1.

$$\frac{\partial}{\partial t}\begin{pmatrix} \rho A \\ \rho c A \\ \rho E A \end{pmatrix} + \frac{\partial}{\partial x}\begin{pmatrix} \rho c A \\ \rho c^2 A + pA \\ cA\left(\rho E + p\right) \end{pmatrix} = \frac{\partial}{\partial x}\begin{pmatrix} 0 \\ F_{x,Blade} + pA \\ W_{Rotor} \end{pmatrix} \tag{1}$$

The solver uses a second-order finite volume scheme. The inlet conditions are fixed total temperature and total pressure. Static pressure is prescribed at the outlet using a ghost cell. The compressor domain consists of 511 cells each having a length of 8×10^{-4} m. The mean line method used to compute the source terms in the 1D-Euler equations has been recently developed at the Chair for Aero Engines. A comprehensive introduction is given in [26]. The source terms are computed as follows. The force in axial direction $F_{x,Blade}$ of Eq. 1 is computed from Newton's 2. law of motion.

$$\dot{m} \cdot (c_{x,out} - c_{x,in}) = \sum F_i \tag{2}$$

The sum of forces $\sum F_i$ constitutes of four forces as defined in Eq. 3.

$$\sum F_i = F_{x,Inlet} - F_{x,Outlet} - F_{x,Endwalls} + F_{x,Blade} \tag{3}$$

For a blade row depicted in Fig. 1, $F_{x,Inlet}$ and $F_{x,Outlet}$ act on the inlet and outlet of the control volume with the respective pressure. The mean line method computes entry and exit of a blade row, hence $F_{x,Inlet}$ and $F_{x,Outlet}$. Furthermore, whenever there is a change in area along the x-axis, a force $F_{x,Endwalls}$ acts onto the control volume. Figure 1 depicts this force only a the hub but it could likewise act at the casing. $F_{x,Endwalls}$ is computed by the 1D-Euler code as it is part of the momentum equation (compare right-hand side of Eq. 1). Lastly, there is the blade force $F_{x,Blade}$ in axial direction created by the blades and required for Eq. 1. Combining Eq. 2 with Eq. 3 and solving for $F_{x,Blade}$ gives the required quantity:

$$F_{x,Blade} = \dot{m}(c_{x,out} - c_{x,in}) + F_{x,Outlet} - F_{x,Inlet} + F_{x,Endwalls}, \tag{4}$$

where $F_{x,Outlet} = p_{out} \cdot A_{out}$ and $F_{x,Inlet} = p_{in} \cdot A_{in}$.

The work W_{Rotor} introduced by the rotor blades is calculated from the first law of thermodynamics, Eq. 5. Since the blade row is assumed adiabatic no heat is exchanged.

$$W_{Rotor} = \dot{m}(h_{t,out} - h_{t,in}) \tag{5}$$

The mean line method is executed to compute the source terms for several operating points on a constant speed line of the compressor. The source terms are stored

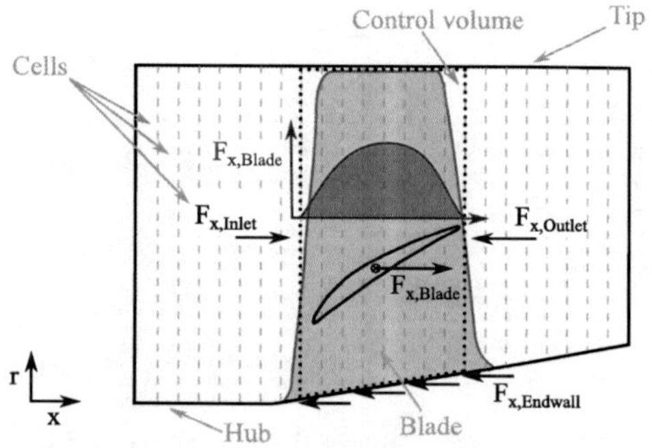

Fig. 1. Sketch of rotor blade with control volume and definition of force source terms, cells size is exaggerated for visibility

in a look-up table, which is used by the 1D-Euler code. Depending on the current axial velocity at the blade leading edge, the 1D-Euler code interpolates the corresponding axial force source term and work source term. These two are distributed among all cells of the blade row in the 1D-Euler code. As sketched in Fig. 1, a parabolic distribution is chosen, which resembles the typical blade loading.

The last four stages of the well-known NASA/GE E^3 high-pressure compressor are modeled for this study [27,28]. The compressor originally consists of ten stages. The analysis is limited to the last four stages starting with rotor 7 because they are mainly affected by the relevant range of boundary conditions. The compressor inlet conditions are selected as if the full ten-stage compressor were operating. The subsequent simulations and analyses use the operating point defined in Table 1.

Table 1. Specifications of the NASA/GE Energy Efficient Engine high pressure compressor operating point of the last four stages

Reduced corrected air mass flow	8.95 kg/s
Pressure ratio	2.36:1
Rotational speed	12669.5 rpm
Inlet total temperature	590.8 K
Inlet total pressure	990.5 kPa
Outlet static pressure	2169 kPa

3 Validation

The compressor model and the presented methods have been validated in a recent publication [23]. Principally, the steady-state results of the mean line method require validation to verify the correctness of the source terms and unsteady results of the 1D-Euler code are compared to results of unsteady 3D-CFD (URANS) to prove the accuracy of the general approach.

3.1 Steady-State Conditions

Results of the steady-state CFD, 1D-Euler and mean line method simulations of the E^3 compressor are compared with published experimental data.

The experimental data of the last four stages is taken from [28] and depicted in Fig. 2. Since the last four stages were not tested separately but always together with the front block of the compressor, throttling at a constant mechanical speed intersects different reduced speed lines on the characteristic as the temperature at the inlet of rotor 7 changes. Hence, experimental data points shown in Fig. 2 are not on a single reduced speed line. Still, the 3D-CFD and the mean line simulations were performed at a constant mechanical speed of 12669.5 rpm and at constant inlet total temperature of 590 K and total pressure of 990 kPa at rotor 7. These inlet condition were reported in [27] for the predicted design point and selected here, even though they lead to an offset in isentropic efficiency. However, the performance in terms of pressure ratio is captured accurately by 3D-CFD. Since inter blade values of the CFD simulations are used to tune the loss models of the mean line method, also mean line data matches those of the experiment. Regarding isentropic efficiency, mean line results agree with 3D-CFD results but less agreement with experimental data is achieved as explained above. Another shortcoming is that both mean line pressure ratio and isentropic efficiency deviate close to surge from 3D-CFD data. A likely explanation is the assumption of constant flow blockage for all operating points in the mean line simulations. The source terms for the 1D-Euler code will be derived from mean line data of the depicted speed line as outlined in Sect. 2.

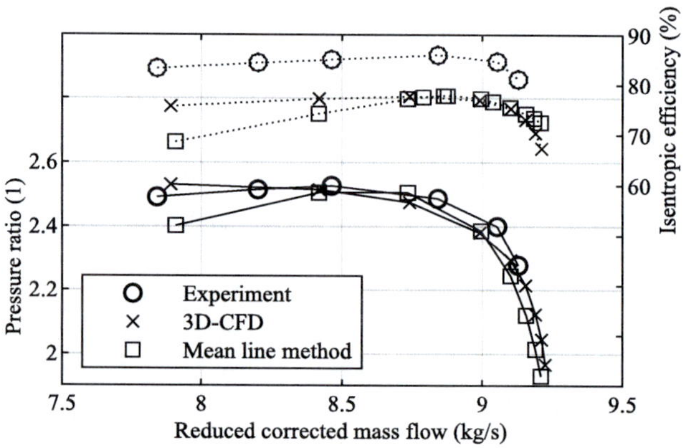

Fig. 2. Comparison of experimental [28], 3D-CFD [23] and mean line method compressor map

3.2 Unsteady Conditions

The unsteady 1D-Euler results are validated against unsteady 3D-CFD results because of their presumably higher fidelity. The complete validation can be found in [23]. Here, only the comparison for a fluctuation with a frequency of 60 Hz and a relative pressure amplitude of 3.6% at the compressor outlet is presented. The relative amplitude is defined in Eq. 6 and is used for the comparison of the results of both numerical methods.

$$\Delta\Phi = \frac{\Phi_{max} - \Phi_{min}}{2 \cdot \Phi_{mean}} \tag{6}$$

$\Delta\Phi$ represents in Eq. 6 the relative amplitude of the periodic variation of static pressure Φ in time. Figure 3 depicts relative amplitudes along the compressor length from 1D-Euler and 3D-CFD simulations. The latter are computed only between rows using a spatial mass flow averaged static pressure. Overall, a good agreement is found between 1D-Euler and 3D-CFD simulations. 3D-CFD tend to give smaller relative amplitudes which might be explained by viscous effects. They are ignored in the 1D-Euler method between blade rows. In absolute terms, the relative amplitude is reduced for both 3D-CFD and 1D-Euler from 3.6% to 0.3%. Thus, the application of the lower fidelity 1D-Euler method is justified as it gives similar results compared to unsteady 3D-CFD for this representative case.

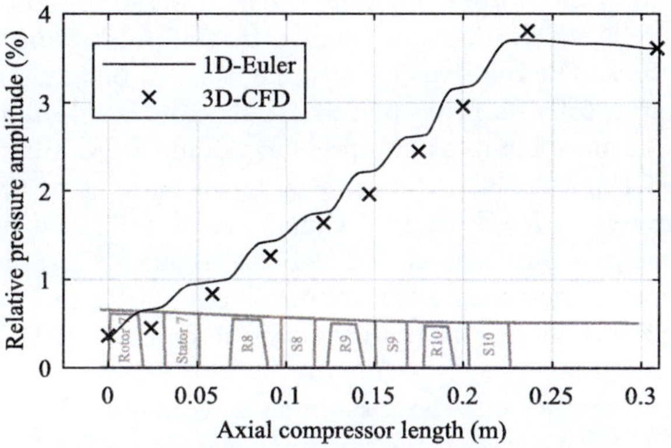

Fig. 3. Comparison of relative pressure amplitude computed by 1D-Euler and 3D-CFD

4 Calculation of Boundary Condition

The 1D-Euler method requires a time-dependent static pressure outlet boundary condition. This outlet pressure simulates the presence of a plenum and PDC tubes downstream of the compressor. Simulations using a 0D-compressor characteristic, 0D-plenum and multiple 1D-PDC tubes show that the resulting pressure trace inside the plenum has a sinusoidal shape [23]. Thus, a compressor outlet boundary condition might be described by a sinusoidal signal of a certain frequency and relative amplitude. In this section realistic ranges for frequencies and relative amplitudes are identified. This is accomplished from two perspectives: Firstly, depending on both the PDC and plenum specifications and state of the fluid, resulting pressure fluctuation are computed. Secondly, maximum tolerable relative amplitudes are identified based on the available surge margin of the compressor using its steady-state map.

The relative amplitude of the pressure fluctuation is derived by an approach by Wintenberger [29]. According to Wintenberger, the relative pressure amplitude can be estimated by Eq. 7.

$$\Delta\Phi = \frac{\dot{m}_0 t_{\text{close}}}{V_P \overline{\rho_P}} \left(\frac{\gamma}{2} + \frac{\gamma - 1}{4} \cdot \frac{\dot{m}_0 t_{\text{close}}}{V_P \overline{\rho_P}} \right) \tag{7}$$

That equations is derived from the unsteady mass and energy equations for a control volume, which comprises the plenum from the compressor exit till the PDC valve plane. This system of equations has been solved separately for the closed part of the cycle and for the open part of the cycle. Details and assumptions used in the derivation can be found in [30]. The amplitude is controlled by a single non-dimensional parameter $\frac{\dot{m}_0 t_{\mathrm{close}}}{V_P \bar{\rho}_P}$, which represents the ratio of the amount of mass added to the plenum during the closed part of the cycle to the average mass in the plenum. In Eq. 7 only the volume of the plenum V_P and t_{close} are variables. $\dot{m}_0$, $\bar{\rho}_P$ and γ are defined by the compressor operating point. t_{close} is defined as the relative closing time divided by the product of firing frequency and number of tubes. The relative closing time is set to 0.05, i.e. the combustion tube is closed during 5% of a firing period. That value is selected as it results in a good match with more complex compressor-plenum-PDC simulations in [23].

The results of Eq. 7 are depicted in Fig. 4, where lines of constant relative amplitude are presented for different frequencies (X-axis) and plenum volumes (Y-axis). The plenum volume is varied between 0.005 m^3–0.1 m^3, which corresponds to 3%–60% of the compressor volume. Frequencies are selected between 20 Hz and 1000 Hz resulting from the literature research for PDCs [31]. The frequency of the sinusoidal wave is a function of the firing frequency of a single tube and the total number of tubes assuming a sequential firing pattern. A firing frequency of 20 Hz is selected for this analysis, since that has been achieved in experiments [32]. Thus, the frequency range between 20 Hz and 1000 Hz corresponds to a number of PDC tubes between 2 and 50. Theoretical approaches such as in [33] tend to suggest higher firing frequencies in the order of 200 Hz. This is a result of the assumption of a completely formed detonation once the fuel air mixture ignites. In reality, a deflagration to detonation transition is required, which reduces wave velocity and results in lower cycle frequencies.

Three main observations can be made from Fig. 4:

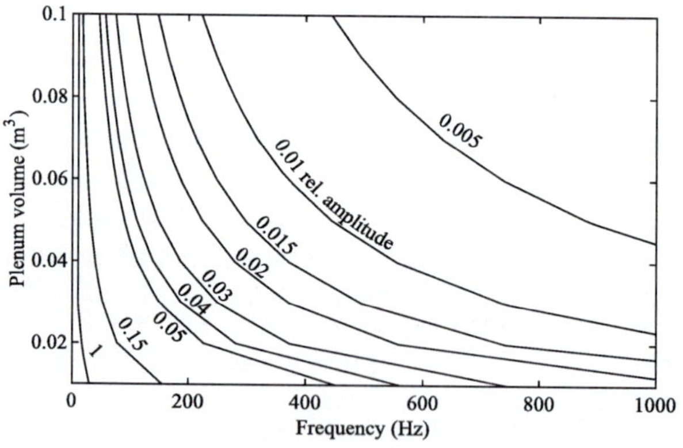

Fig. 4. Isolines of constant relative amplitude for varying frequencies and plenum volumes

- At high plenum volumes and low frequencies (top left), relative amplitudes are very sensitive to frequency but little to plenum volume. Even a small increase in frequency substantially reduces relative amplitude.
- Vice versa is true at high frequencies and low plenum volumes. Relative amplitude is very sensitive to plenum volume but little sensitive to frequency. Here, even a small increase in plenum volume significantly reduces the relative amplitude.
- Around an imaginary diagonal line on the plot, increasing both plenum volume and frequency results in lower relative amplitudes.

According to Fig. 4, relative amplitudes reach values as high as their mean value (relative amplitude = 1). Obviously, that is intolerable from a compressor stability perspective. Therefore, a maximum permissible relative amplitude is inferred from the compressor map, which is a simplification, since the compressor map depicts steady-state operation. The time-averaged operating point is positioned at a pressure ratio of 2.36 for this study, which is below the intended design point of the compressor. However, that operating point allows for relative amplitudes of 5% without exceeding the surge line on the steady-state map and eve provides some safety margin. As a result, the compressor model will be simulated under boundary conditions with frequencies in the range of 50 Hz–1000 Hz and amplitudes between 0.5%–5% around a mean static outlet pressure of 2169 kPa, which corresponds to a pressure ratio of 2.36 of the last four stages.

5 Compressor Simulation Results

Previously identified boundary conditions were simulated for the operating point defined in Table 1. The simulations were run for multiple periods to allow for an unsteady periodic convergence. Eventually, convergence was verified by a method, which relies on fundamental concepts from digital signal processing including the discrete Fourier transform and cross correlation [34].

5.1 Unsteady Damping

Any excitation is damped in a viscous and compressible medium. In case of a compressor exposed to a fluctuating outlet pressure profile, the pressure waves lose strength and eventually vanish while travelling upstream. This rate of dissipation is quantified by the unsteady damping ϵ of the relative amplitude (compare Eq. 6).

$$\epsilon = \frac{\Delta\Phi_2 - \Delta\Phi_1}{\Delta\Phi_2} \tag{8}$$

$\Delta\Phi_2$ is the relative amplitude at the compressor exit and $\Delta\Phi_1$ is located arbitrarily further upstream.

Unsteady damping is plotted over frequency and relative amplitude. By evaluating different planes within the compressor, the damping characteristic can be inferred. Figure 5 depicts unsteady damping at the inlet of the last stage and Fig. 6 shows unsteady damping upstream of the first rotor.

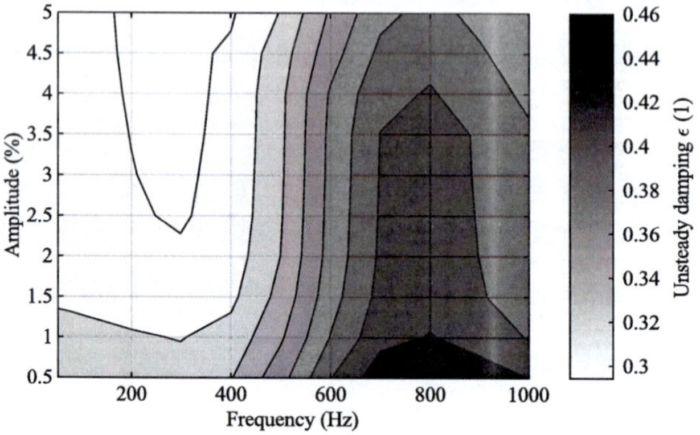

Fig. 5. Unsteady damping at the inlet of the last stage

White areas indicate less damping whereas black stands for a higher degree of damping. From Fig. 5 is can be inferred that low frequencies are less damped than high frequencies. According to the data, roughly 30% of the fluctuation's pressure amplitude is attenuated at 200 Hz whereas 46% are damped at 800 Hz. Furthermore, the results suggest that relative amplitude does not have a strong effect on unsteady damping. That is especially true for frequencies between 400 Hz–700 Hz.

Taking into account Fig. 6, which depicts unsteady damping taken at the inlet, a different picture is drawn. Low frequencies are more damped than high frequencies. Frequencies below 200 Hz have an unsteady damping of as high as 94%, whereas only 82% of the outlet fluctuation is attenuated at high frequencies. It can be concluded that fluctuations with a high frequency show a high amount of damping within the last stage, but as soon as their amplitudes are lower the rate of damping reduces. Hence, at the inlet higher relative amplitudes are observed for fluctuations with a high frequency compared to a fluctuation with a low frequency.

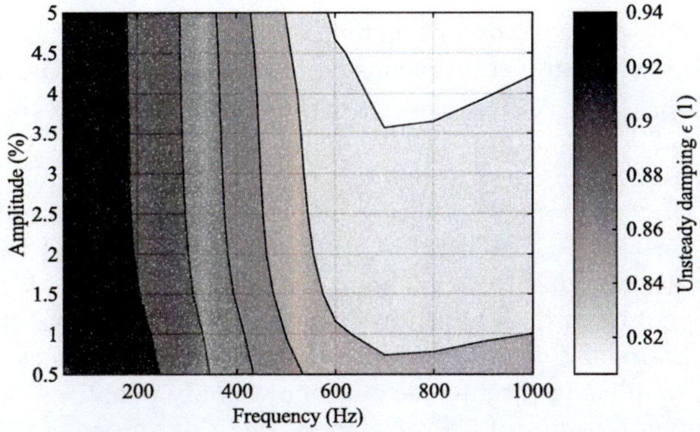

Fig. 6. Unsteady damping at the inlet of the four-stage compressor

5.2 Isentropic Efficiency

Next, compressor isentropic efficiency is presented for the range of selected boundary conditions. Isentropic efficiency is a suitable metric as it relates the ideal energy input with the real energy input required to achieve a certain pressure ratio. The definition is given in Eq. 9.

$$\eta_{is} = \frac{\Delta h_{ideal}}{\Delta h_{real}} = \frac{(p_{t,out}/p_{t,in})^{\overline{\gamma}/(\overline{\gamma}-1)} - 1}{T_{t,out}/T_{t,in} - 1} \tag{9}$$

Temperatures and pressure are averaged to obtain a representative value for a complete firing period. For that, an averaging procedure presented by [35] and successfully applied in [10,23] to calculate isentropic efficiency will be used. Instead of using instantaneous values in Eq. 9 to yield a time resolved efficiency, temperatures are mass-averaged and pressures are work-averaged over one period. A mass-averaged temperature is computed by Eq. 10 for both inlet and outlet:

$$T_t{}^{ma} = \frac{\int_{t_1}^{t_2} \rho c T_t d\xi}{\int_{t_1}^{t_2} \rho c d\xi} \tag{10}$$

The work-averaged pressures at inlet and outlet are averaged differently according to [35]. Work-averaged total inlet pressure is defined as:

$$\left(p_{t,in}^{\text{wa}}\right)^{(\gamma-1)/\gamma} = \frac{\int_{t_1}^{t_2} \rho_{in} c_{in} T_{t,in}(\xi) d\xi}{\int_{t_1}^{t_2} \rho_{in} c_{in} \left(\frac{T_{t,in}(\xi)}{p_{t,in}(\xi)^{(\gamma-1)/\gamma}}\right) d\xi} \tag{11}$$

and work-averaged outlet pressure as:

$$\left(p_{t,out}^{\text{wa}}\right)^{(\gamma-1)/\gamma} = \frac{1}{\Delta t} \int_{t_1}^{t_2} p_{t,out}^{(\gamma-1)/\gamma}(\xi) d\xi. \tag{12}$$

The mass and work-averaged definition for isentropic efficiency has the advantage that the assumption of an instantaneous compression through the compressor is not required [35].

Again, data is depicted over relative amplitude and frequency in Fig. 7. The isentropic efficiency is presented as delta efficiency. That is the difference between a steady-state efficiency of an operating point having the same minimum surge margin as the unsteady throttled operation. It can be inferred that isentropic efficiency penalties remain below 1% point for the selected boundary conditions. The efficiency penalty is dominantly influenced by relative amplitude. The frequency of the boundary condition does not have a strong effect on efficiency. At low relative amplitudes, isentropic efficiency is only penalized by 0.1% point. At high relative amplitudes of 5%, efficiency penalties amount to 1% point. For every percentage point increase in relative amplitude, the efficiency penalty is increases by roughly 0.2% point. The penalty can be decreased by going for a lower relative amplitude of the boundary condition. That can be achieved by a larger plenum, as Fig. 4 suggests.

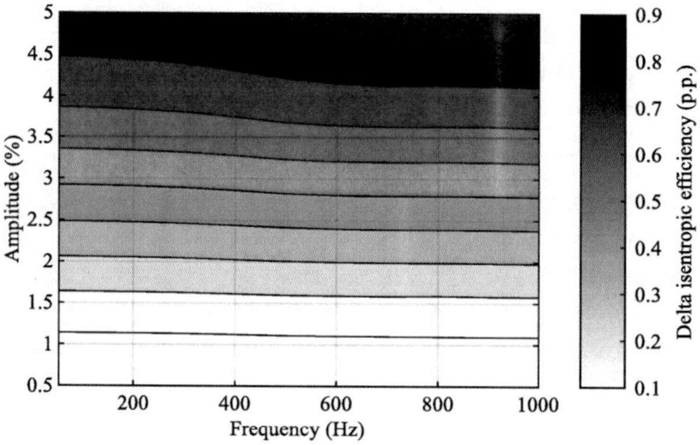

Fig. 7. Isentropic efficiency penalty caused by unsteady boundary condition

6 Conclusion

This paper presents the application of a low fidelity numerical method, in order to investigate compressor behaviour at unsteady boundary conditions similar to those caused by pulsed detonation combustion. For this purpose, a range of reasonable unsteady compressor boundary conditions is derived for different pulsed detonation combustion specifications and plenum sizes. That time-variant static pressure trace is further prescribed at the outlet of a 4-stage compressor model. A fast quasi 1D-Euler method simulates the compressor performance for multiple periods. The results are evaluated with regards to unsteady damping and isentropic efficiency. The main conclusions are:

– relative amplitude and frequency range from 0.5%–5% and 50 Hz–1000 Hz, respectively, for a realistic pulsed detonation combustor and the underlying 1D-approach according to the simplified model.
– The unsteady damping depends on fluctuation frequencies. The last stage dampens between 30%–46% of the fluctuation's amplitude. Between 84% and 92% are attenuated till the inlet of the 4-stage compressor model.
– Compressor isentropic efficiency is influenced by the relative amplitude of the fluctuations. A penalty due to unsteady operation of up to 1% point compared to a comparable steady-state operating point is observed.

The penalty in compressor efficiency is minor, which is likely specific to this compressor, as the modelled compressor does not suffer from a strong decrease in efficiency when moving closer to the surge line (compare Fig. 2). However, the authors believe that the fluctuations' amplitudes are more restricted by stability concerns than by the expected penalty on compressor efficiency. Especially as the last stages are critical in terms of stability at high reduced speeds encountered at a possible cruise operating point. Further research using the presented method is in order that adequately evaluates the influence of pressure gain combustion on compressor stability.

Acknowledgement. Funding: The authors gratefully acknowledge the support by the Deutsche For-schungsgemeinschaft (DFG) as part of the Collaborative Research Center SFB 1029 "Substantial efficiency increase in gas turbines through direct use of coupled unsteady combustion and flow dynamics" in project D01.

References

1. International Civil Aviation Organization - ICAO: Annual report of the council, May 2018
2. International Air Transport Association - IATA: 20 year passenger forecast, 2020
3. Krein, A., Williams, G.: Flightpath 2050: Europe's vision for aeronautics. In: Innovation for Sustainable Aviation in a Global Environment: Proceedings of the Sixth European Aeronautics Days, p. 63 (2012). https://doi.org/10.2777/50266
4. Fuel Cell and Hydrogen Joint Undertaking: Hydrogen powered aviation: a fact-based study of hydrogen technology, economics, and climate impact by 2050 (2020)
5. Heiser, W.H., Pratt, D.T.: Thermodynamic cycle analysis of pulse detonation engines. J. Propul. Power **18**(1), 68–76 (2002). https://doi.org/10.2514/2.5899
6. Neumann, N., Woelki, D., Peitsch, D.: A comparison of steady-state models for pressure gain combustion in gas turbine performance simulation. In: Proceedings of GPPS Beijing 2019, Beijing, China, 16–18 September 2019 (2019)
7. Stathopoulos, P.: Comprehensive thermodynamic analysis of the humphrey cycle for gas turbines with pressure gain combustion. Energies **11**(12), 3521 (2018). https://doi.org/10.3390/en11123521
8. Glaser, A.J., Caldwell, N., Gutmark, E.: Performance measurements of a pulse detonation combustor array integrated with an axial flow turbine. In: 44th AIAA Aerospace Sciences Meeting and Exhibit, pp. 1–12, January 2006. https://doi.org/10.2514/6.2006-1232

9. Schliwka, T., Tiedemann, C., Peitsch, D.: Interaction of main flow and sealing air across a turbine cavity under unsteady conditions. In: ISABE - 22th International Symposium on Air Breathing Engines, Phoenix, Arizona, USA, October 2015. ISABE-2015-21288
10. Xisto, C., Petit, O., Grönstedt, T., Rolt, A., Lundbladh, A., Paniagua, G.: The efficiency of a pulsed detonation combustor-axial turbine integration. Aerosp. Sci. Technol. **82–83**, 80–91 (2018). https://doi.org/10.1016/j.ast.2018.08.038. ISSN: 1270-9638
11. Heinrich, A., Herbig, M., Peitsch, D., Topalovic, D., King, R.: A testrig to evaluate turbine performance and operational strategies under pulsating inflow conditions. In: AIAA - Propulsion and Energy 2019 Forum, Indianapolis, Indiana, USA. American Institute of Aeronautics and Astronautics, August 2019. Ch. AIAA 2019-4039, ISBN: 978-1-62410-590-6. https://doi.org/10.2514/6.2019-4039
12. Liu, Z., Braun, J., Paniagua, G.: Integration of a transonic highpressure turbine with a rotating detonation combustor and a diffuser. Int. J. Turbo Jet-Engines (2020). ISSN: 0020-7403. https://doi.org/10.1515/tjeng-2020-0016
13. Hoke, J., Bradley, R., Stutrud, J., Schauer, F.: Integration of a pulsed detonation engine with an ejector pump and with a turbo-charger as methods to self-aspirate. In: 40th AIAA Aerospace Sciences Meeting & Exhibit, Reno, USA, p. 615 (2002). https://doi.org/10.2514/6.2002-615
14. Sakurai, T., Nakamura, S.: Performance and operating characteristics of micro gas turbine driven by pulse, pressure gain combustor. In: Proceedings of the ASME Turbo Expo 2020: Turbomachinery Technical Conference and Exposition, 21–25 September 2020 (2020). https://doi.org/10.1115/GT2020-15000
15. Lu, J., Zheng, L., Wang, Z., Wang, L., Yan, C.: Experimental investigation on interactions between a two-phase multi-tube pulse detonation combustor and a centrifugal compressor. Appl. Therm. Eng. **113**, 426–434 (2017). https://doi.org/10.1016/j.applthermaleng.2016.10.188
16. Staats, M., Nitsche, W.: Experimental investigations on the efficiency of active flow control in a compressor cascade with periodic non-steady outow conditions In: Volume 2A: Turbomachinery. American Society of Mechanical Engineers, June 2017. https://doi.org/10.1115/gt2017-63246
17. Brück, C., Mihalyovics, J., Peitsch, D.: Experimental investigations on highly loaded compressor airfoils with different active ow control parameters under unsteady ow conditions. In: Proceedings of GPPS Montreal, Montreal, Canada, May 2018. GPPS-2018-0054. https://doi.org/10.5281/zenodo.1343489
18. Werder, T., Liebich, R., Neuhäuser, K., Behnsen, C., King, R.: Active flow control utilizing an adaptive blade geometry and an extremum seeking algorithm at periodically transient boundary conditions. J. Turbomach. **143**(2) (2020). ISSN: 0889-504X. https://doi.org/10.1115/1.4049787
19. Fietzke, B., Mihalyovics, J., King, R., Peitsch, D.: Binary repetitive model predictive active flow control applied to an annular compressor stator cascade with periodic disturbances. In: Proceedings of the ASME Turbo Expo 2021: Turbomachinery Technical Conference and Exposition, 07–11 June 2021. American Society of Mechanical Engineers (2021). GT2021-58 744
20. de Almeida, V.B.C., Peitsch, D.: Aeroelastic assessment of a highly loaded high pressure compressor exposed to pressure gain combustion disturbances. J. Glob. Power Propul. Soc. **2**, 477–492 (2018). https://doi.org/10.22261/jgpps.f72ouu
21. de Almeida, V.B.C., Peitsch, D.: Multirow performance and aeroelastic analyses of a compressor subjected to disturbances from pressure gain combustion. In: Proceedings of the 15th ISUAAAT, Oxford, UK, September 2018. ISUAAAT15-031

22. de Almeida, V.B.C., Motta, V., Peitsch, D.: Unsteady aerodynamics of a high pressure compressor working under pressure gain combustion disturbances. In: IGTC - International Gas Turbine Congress, Tokyo, Japan, November 2019. IGTC2019-0070. http://igtc2019.org/IGTC19_ContentListWeb_4.html/#wepm14_01
23. Neumann, N., Asli, M., Garan, N., Peitsch, D., Stathopoulos, P.: A fast approach for unsteady compressor performance simulation under boundary condition caused by pressure gain combustion. Appl. Therm. Eng. **196**, 117223 (2021). https://doi.org/10.1016/j.applthermaleng.2021.117223. ISSN: 1359-4311
24. Berndt, P., Klein, R., Paschereit, C.O.: A kinetics model for the shockless explosion combustion. In: Turbo Expo: Power for Land, Sea, and Air, vol. 49767. American Society of Mechanical Engineers (2016). V04BT04A034
25. Dittmar, L., Stathopoulos, P.: Numerical analysis of the stability and operation of an axial compressor connected to an array of pulsed detonation combustors. In: Proceedings of the ASME Turbo Expo 2020: Turbomachinery Technical Conference and Exposition, 21–25 September 2020. American Society of Mechanical Engineers (2020)
26. Neumann, N., Peitsch, D.: Introduction and validation of a mean line solver for present and future turbomachines. In: Proceedings of ISABE, Canberra, Australia, 22–26 September 2019 (2019). ISABE-2019-24441
27. Holloway, P., Koch, C., Knight, G., Shaffer, S.: Energy efficient engine - high pressure compressor detail design report (1982). NASA-CR-165558
28. Cline, S., Fesler, W., Liu, H., Lovell, R., Shaffer, S.: High pressure compressor component performance report (1983). NASA-CR-168245
29. Wintenberger, E., Shepherd, J.E.: Thermodynamic cycle analysis for propagating detonations. J. Propul. Power **22**(3), 694–698 (2006). https://doi.org/10.2514/1.12775
30. Wintenberger, E.: Application of steady and unsteady detonation waves to propulsion. Ph.D. thesis, California Institute of Technology (2004)
31. Perkins, H.D., Paxson, D.E.: Summary of pressure gain combustion research at NASA, April 2018. NASA TM-2018-219874
32. Völzke, F.E., Yücel, F.C., Gray, J.A.T., Hanraths, N., Paschereit, C.O., Moeck, J.P.: The influence of the initial temperature on DDT characteristics in a valveless PDC. In: King, R. (ed.) Active Flow and Combustion Control 2018. NNFMMD, vol. 141, pp. 185–196. Springer, Cham (2019). https://doi.org/10.1007/978-3-319-98177-2_12 ISBN: 978-3-319-98177-2
33. Mitrofanov, V., Zhdan, S.: Thrust performance of an ideal pulse detonation engine. Combust. Explos. Shock Waves **40**(4), 380–385 (2004). https://doi.org/10.1023/B:CESW.0000033559.75292.8e
34. Clark, J.P., Grover, E.A.: Assessing convergence in predictions of periodic-unsteady flowfields. J. Turbomach. **129**(4), 740–749 (2006). https://doi.org/10.1115/1.2720504
35. Suresh, A., Hofer, D.C., Tangirala, V.E.: Turbine efficiency for unsteady, periodic flows. J. Turbomach. **134**(3) (2011). https://doi.org/10.1115/1.4003246

Efficiency Increase and Start-Up Strategy of an Axial Turbine Stage Under Periodic Inflow Conditions Using Extremum Seeking Control

Daniel Topalovic[1](✉), Rudibert King[1], Markus Herbig[2], Alexander Heinrich[2], and Dieter Peitsch[2]

[1] Chair of Measurement and Control, Technische Universität Berlin, 10623 Berlin, Germany
{daniel.topalovic,rudibert.king}@tu-berlin.de

[2] Chair for Aero Engines, Technische Universität Berlin, 10587 Berlin, Germany
{markus.herbig,alexander.heinrich,dieter.peitsch}@tu-berlin.de

Abstract. Since the efficiency increase of state-of-the-art gas turbines has become incrementally smaller, a significant improvement seems unlikely using the established concepts. Replacing the underlying constant pressure combustion with a constant volume combustion may change this and may yield significant efficiency increases. A possible realization of such a machine can be accomplished using firing tubes that utilize pulsed detonation combustion. This results in unsteady, periodic boundary conditions, generating challenges to the operation of the machine. A possible solution to operational difficulties is presented here using extremum seeking control (ESC). This model-free, easy to parameterize control approach is tested using a mock-up test rig designed specifically to mimic the pulsed flow conditions in front of an axial turbine. The ESC is defined in such a way that it maximizes either the calculated turbine efficiency or the specific work that is converted by the turbine by changing the synchronization between firing tubes only, leaving the parameter space of single firing events unaffected. Further, a concept for the start-up of such a gas turbine using ESC is introduced.

Keywords: Extremum seeking control · Pulsed detonation combustion · Axial turbine · Efficiency increase · Closed-loop control

1 Introduction

The accomplished efficiency gains of state-of-the-art gas turbines have been decreasing over time [1] due to the fact that the efficiency increase was mainly driven by the evolutionary improvements of single turbine components. Therefore, it seems unlikely that such an approach will yield significant efficiency improvements in the future.

R. King and D. Peitsch (Eds.): AFCC 2021, NNFM 152, pp. 288–302, 2022.
https://doi.org/10.1007/978-3-030-90727-3_18

One possible option to push beyond the estimated limits of state-of-the-art gas turbines is through a change of the underlying thermodynamic cycle. Current gas turbines utilize the Brayton cycle. It has been theoretically shown that if this cycle were replaced by the Humphrey cycle, an additional increase in thermal efficiency would be attainable [2], possibly allowing an expansion of current efficiency limits of gas turbines. The necessary Humphrey cycle must be realized by a change from the currently used isobaric combustion process to an isochoric one.

Different technical approximation approaches to an isochoric combustion process have been published, e.g., rotating detonation combustion (RDC) [3], shockless explosion combustion (SEC) [4], and pulsed detonation combustion (PDC) [5]. SEC and PDC applications are considered here. However, for reasons of space, explanations are only given in reference to the PDC case.

A PDC-based gas turbine features several detonation tubes. These are driven in a cyclic manner, where each cycle is divided into three stages: actual combustion of a combustible mixture, purging of the detonation tube, and refilling of the tube for the next cycle. Due to the pulsating nature of detonation tubes, it can be expected that other machine components will experience non-constant boundary conditions. A turbine positioned downstream will therefore experience not only an inhomogeneous periodic inflow over time, but also strong pressure and velocity fluctuations. The non-constant inflow to the turbine might decrease the overall efficiency of the machine. This should be mitigated for obvious reasons, passively by the coupling of machine components, e.g., the detonation tubes and the turbine, using buffering elements such as plena, or actively, e.g., through control.

A possible reduction of pressure fluctuations in front of a turbine stage under pulsating inflow conditions has been investigated in [6], using a mock-up turbine test rig for the analysis of the turbine behavior under pulsating inflow conditions [7]. In [6], a suitable operation mode minimizing the pressure fluctuations in front of a turbine stage using only the synchronization of firing tubes has been shown, based on a combination of an optimal open-loop control and a closed-loop control using extremum seeking. The applied ESC was chosen as a closed-loop control approach because it is model free, is easy to parameterize, and allows for the compensation of model uncertainties. Furthermore, disturbances are rejected that are not addressed by open-loop control alone. While a successful operation regarding the minimization of pressure fluctuations was already shown, the versatility of ESC regarding possible different operational objectives was not investigated. In this contribution, it is shown how such an ESC approach can be used to maximize not only the specific work, but also the efficiency of a turbine stage using only the synchronization between the firing events. Further, a possible start-up concept for a multi-tube PDC gas turbine is introduced and applied using the so-called TU Pulse test rig.

The remainder of this contribution is organized as follows. Section 2 describes the turbine test rig to which the proposed concepts were experimentally applied. In Sect. 3, the basic concept of ESC and the implemented algorithm are presented. Section 4 defines the different applications of ESC-supported operation that were addressed, and results are shown and discussed in Sect. 5. A conclusion is drawn in Sect. 6.

2 Experimental Setup

The TU Pulse test rig was designed for preliminary studies of the behavior of a PDC-based gas turbine. It represents the link between combustion tubes and the first stage of the actual turbine. The setup was specifically designed for research focusing on the turbine behavior under pulsating inflow conditions. Therefore, no actual combustion was needed in this test rig, but mock-up tubes mimicked the impulse-like character from fluid discharge of an actual unsteady combustion system. Detailed information characterizing the rig can be found in [7]. Here, a short overview defines the main characteristics. The most important elements that affect the flow of the test rig are exemplified by the block diagram in Fig. 1a), and an additional impression of the realization is given in Fig. 1b).

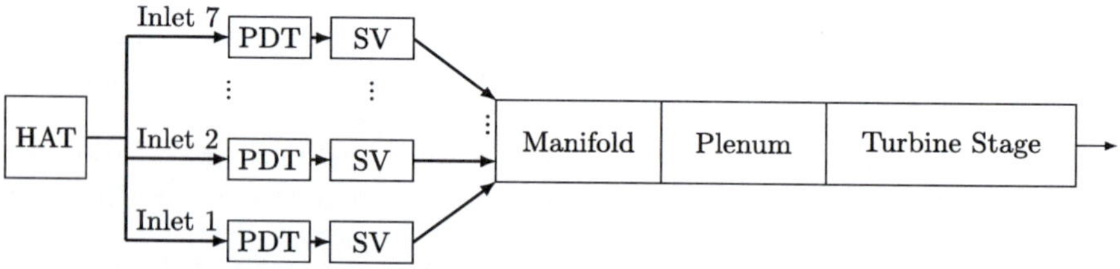

a) Block diagram of components affecting flow.

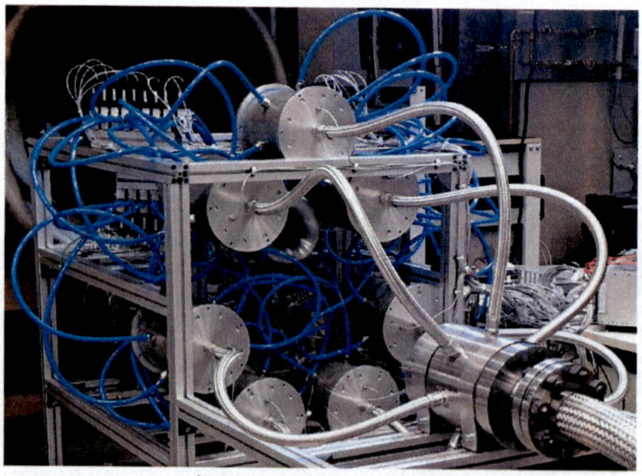

b) Actual test rig.

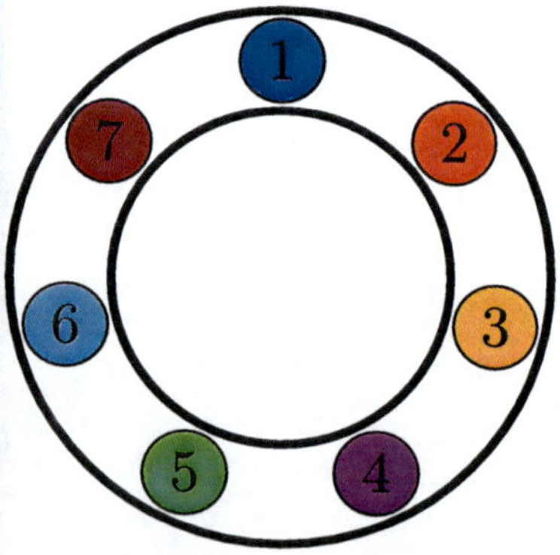

c) Inlet positions.

Fig. 1. TU pulse.

The setup was operated with air from the so-called Hot-Acoustic-Test-Rig (HAT), a joint facility of German Aerospace Center (DLR) and Technische Universität Berlin [8], which supplied it with an air mass flow of up to $0.78\,\mathrm{kg\,s^{-1}}$, with a maximum flow temperature of 823 K. The supplied air mass flow was distributed into seven inlets, where each represents a PDC combustion chamber. Figure 1c) exemplifies the inlet positions. In each inlet, a pressure damping tank (PDT) serves as a hot air reservoir. It decouples upstream effects of the different inlets. Additionally, a battery of six parallel connected solenoid valves (SVs) was installed at the end of each inlet.

The SVs were packed to a battery of five synchronously controlled valves and a single permanently opened one. From this point, each battery is referred to as one SV. The permanently opened SV in each inlet was used to create a steady base mass flow to the components downstream to reduce the strain on the turbine stage. The ends of the inlets were linked to a manifold, which was connected to a turbine stage via a small plenum. The turbine stage was mechanically coupled to a compressor to allow for load adjustment. For the experiments published here, only a subset of available sensors of the TU Pulse were used. Temperature measurements were done with type-K thermocouples. The sensor measuring the upstream turbine temperature, T_4, was mounted into the 12 o'clock position (first inlet) in the manifold, while the sensor measuring the downstream temperature, T_5, was mounted into the 3 o'clock position behind the turbine. The total pressure upstream of the turbine, p_{T4}, was obtained using a pitot tube, which was mounted in the line of sight of the fourth inlet position in the plenum in front of the turbine. The total pressure, p_{T5}, was determined using a five-hole probe downstream of the turbine. The corresponding pressure and temperature values were time-averaged over two actuation periods. Further, six pressure sensors with a high bandwidth (Kulite XTEH-10L-190) were flush-mounted in the circumference of the plenum. These sensors were used to characterize the pressure fluctuations in front of the turbine stage. In all performed experiments, the supplied air mass flow was $0.27\,\mathrm{kg\,s^{-1}}$, resulting in an operation point of 50% of the calculated design rotation speed of the turbine.

The rotation speed was measured using a laser and a photodiode, which generated a TTL-signal that was evaluated by an integrated counter. In all performed experiments, an inflow temperature of around 293 K was used. For the actuation of the test rig, impulse-like flows coming from the inlets, so-called firing events, were defined and realized through the application of control signals to the battery of SVs in each inlet. A firing event was defined by periodically opening and afterwards closing an SV with a defined firing frequency f and duration, given as duty cycle. During the experiments, the duty cycle was set to 50%, which described the part of an actuation period during which a valve was opened. The associated firing frequency was fixed at 20 Hz in all experiments. The beginning of a firing event within a period was defined as the firing instant or trigger. The synchronization between the firing instants in a common period defined the firing pattern, which was later changed by the controller. The upper part of Fig. 2 shows, for example, a generic firing pattern, which was used as a starting point for each experiment. The peaks indicate the firing instants of this pattern. In the lower part of Fig. 2, the resulting control signals of the corresponding SVs can be seen, where 1 indicates an open SV and 0 a closed one.

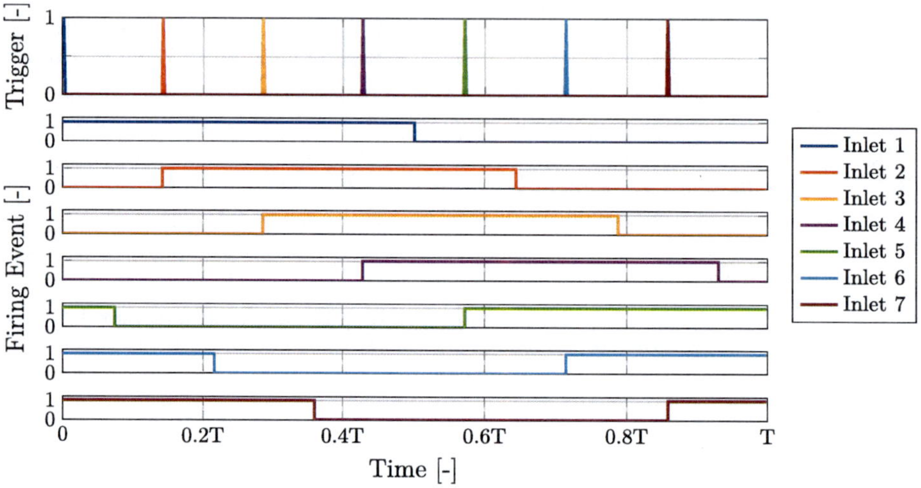

Fig. 2. Reference firing pattern defined by triggers and corresponding firing events over a period with $T = 1/f$.

2.1 Characterization of the Turbine Behavior

For the application of an ESC, an objective function must be defined that describes the system behavior in such a way that the control can either maximize or minimize this objective function to achieve a desired effect. Here, three figures of merit based on the sensor values of the TU Pulse were used as objective functions: specific work, turbine efficiency, and variance of the pressure sensor signals in front of the turbine. The specific work $\tilde{w}$ can be characterized by Eq. (1) [9]

$$\tilde{w} = c_p(T_5 - T_4) \ , \tag{1}$$

with c_p representing the specific heat capacity in kJ/(kgK), which is assumed to be constant here. The temperatures T_4 and T_5 are given in K. While T_4 and T_5 can be measured directly in the setup, such a direct calculation should be treated carefully because of measurement noise and sensor dynamics. The first can be compensated for online through filtering, but this induces additional delays that must be accounted for in further examination. In the evaluations performed here, we have used a moving average filter of the obtained signals over two full actuation periods to compensate for measurement noise. The dynamic of the sensors and the influence of the filters are addressed by a delayed analysis within the ESC. While the specific work can be a useful characterization of the turbine behavior, one of the main parameters to describe the performance of a turbine is the isentropic efficiency η, which is described by [9]

$$\eta = \frac{1 - \frac{T_4}{T_5}}{1 - (\frac{p_{T5}}{p_{T4}})^{(\frac{\kappa-1}{\kappa})}}, \tag{2}$$

where κ represents the heat capacity ratio, which is assumed to be constant here. Similar to the specific work, a direct calculation of the efficiency is omitted and the signals that are used in this calculation are filtered by a moving average over two actuation periods. Notice that this averaging approach will result in a higher boundary estimation of the real efficiency, as is discussed in [10]. Since a high resolution in time of the efficiency would require temperature sensors with a very high bandwidth, which were not used in the setup, this estimation had to be used.

As discussed in [6], a possible operation strategy for a turbine under pulsating inflow conditions could be defined by a homogenization of the pressure field in front of the turbine. The homogeneity of the flow can be characterized by the variance σ^2 of all $n_y = 6$ pressure sensors y_i, based on two full actuation periods in the circumference of the plenum upstream of the turbine. With $2n_p$ discrete measurement points,

$$\sigma^2 = \frac{1}{2n_p n_y} \sum_{k=0}^{2n_p-1} (\sum_{i=1}^{ny} y_i^2(k)) - (\frac{1}{2n_p n_y} \sum_{k=0}^{2n_p-1} (\sum_{i=1}^{ny} y_i(k)))^2 \tag{3}$$

follows.

3 Extremum Seeking Control

ESC is a model-free control approach that can be applied if the control target is defined as a maximization or minimization of a static objective function. The basic idea behind ESC is an online, gradient-based optimization. To that end, the gradient of the objective function with respect to the control inputs is estimated. Depending on the system and the problem formulation, the gradient can be estimated in a continuous or discrete fashion. More specifically, a gradient approximation is obtained by applying perturbations to the control input. As a model is not required in ESC, it can be used as well in systems too complex or too uncertain to be modeled. However, this advantage comes at the cost of time-consuming gradient estimations, which additionally scale with the number of control variables. As a result, a rather slow convergence toward an extremum occurs. Moreover, a gradient-based search does not guarantee finding the global optimum. Because ESC does not stop the perturbations of the control input, even when the desired extremum or its neighborhood is reached, the value of the objective function is not held constant using ESC, but oscillates near the extremum.

The control input that is investigated in this contribution is the firing pattern, which consists of seven firing instants in total. Since all firing tubes use the same frequency, the number of required control inputs to define a firing pattern can be reduced to six by relating them to a fixed firing instant of a first tube.

As objective functions, the efficiency, the specific work of the turbine, and the pressure variance in the plenum are examined here.

Since the consequences of a change of the firing pattern appear only when a full firing period has passed, no continuous actuation of the system is possible. Therefore, no standard ESC approach as described in the literature can be used here. Instead, an approach called 'iterative learning control based on extremum seeking' [11] will be exploited. This was already applied for the minimization of the pressure variance in front of a turbine stage under pulsating inflow conditions in [6]. The basic idea behind this ESC approach is that a discrete optimization of the objective function performed subsequently over several iterations. Given an iteration step, an open-loop control is applied to the system. The system response at each time step in the iteration is recorded and evaluated in the cost function at the end of this iteration step. Using this value of the objective function, a new control trajectory for the next iteration is calculated and applied to the system, resulting in a closed-loop control over the iterations. As an easy to parameterize approach for the minimization of the objective functions, a gradient descent approach with a simple finite difference gradient estimation was used. The applied algorithm can be summarized as follows [6]:

Adapted Gradient Descent ESC Algorithm:

1. Define iteration counter as $j = 0$ and perturbation counter as $n = 0$.
2. An initial open-loop control $\underline{u}_{j,n}$ is applied to the system.
3. After a fixed amount of time is passed and a steady state of the system has been attained, the cost $Q(\underline{u}_{j,n})$ is evaluated.
4. The perturbation counter is increased to $n = n + 1$ and the $(n + 1)$st entry in the open-loop control $\underline{u}_{j,n}$ is perturbed with a small, fixed value d.
5. The previous two steps are repeated until all possible perturbations with a perturbation size of $\pm d$ were realized.
6. The attained objective function values are used to estimate the gradient $\underline{g}$.
7. $n = 0$ and $j = j + 1$ are set, and $\underline{u}_{j,0} = \underline{u}_{j-1,0} + \lambda \underline{g}$ is calculated. Afterwards, the algorithm starts with step 2 again.

The parameter λ represents a step-size and d the perturbation amplitude. Both represent the only tuning parameters in this approach and were chosen through manual tuning in this contribution, starting with low values respectively. $d = 0.01$ was applied in all experiments; the varying values of λ are given below. The first initial open-loop control was defined in all experiments as the reference firing pattern, which was shown in Fig. 2. The fixed amount of time that is defined in the algorithm before a steady state is reached and evaluated was defined very conservatively with a length of 40 times the actuation period. This was done to eliminate the delaying influence of the used filtering of the objective functions and to guarantee that no transient behavior was falsely accounted for in the evaluation. For simplicity reasons, the ESC was always performed as a minimization. Therefore, in the experiments maximizing the efficiency, the actual objective function was defined as the efficiency times -1, but for a more intuitive understanding, the efficiency will be shown in the results. Further, in the performed experiments, the applied firing instances were normalized over the period length T using $\underline{u}^*_{j,n} = \underline{u}_{j,n}/T$ to represent the firing pattern in an interval between 0 and 1.

4 Applications Using ESC

In this contribution, multiple possible ways for the operation of a PDC-based gas turbine using ESC are presented. While simple maximization or minimization of system states, e.g., the specific work and the efficiency of the turbine, can be done using ESC, an ESC-supported start-up strategy for a PDC-based gas turbine is also introduced.

4.1 Maximization of Efficiency and Specific Work

The simplest operational goal can be defined as a maximization or minimization of possible objective functions characterizing the machine. Here, two possible targets are considered: the absolute value of the specific work and the efficiency of the turbine. As will be shown in the results, these characteristics depend on the applied firing patterns used in the actuation of the test rig. Hence, the introduced ESC approach can be applied directly to the test rig. Note that the specific work of the turbine has a negative magnitude due to its definition and is therefore minimized.

4.2 Start-Up Strategy of a PDC-Based Gas Turbine

A starting process of a multitube PDC-based gas turbine may be challenging due to the intrinsic nature of such a machine. One possible starting strategy can be defined by beginning to fire with the PDC tubes with no evaluation of feedback and readjustments. After a start-up is completed successfully, the parameters defining the firing events, e.g., fuel mass per firing event, firing pattern, and firing frequency, could be modified steadily until a desired operation point is reached. While such a start-up approach would be easy to define, possible drawbacks may occur in actual operation. For instance, the interaction of multiple tubes within the starting process may affect the start-up. Since all firing tubes are connected via a plenum with the turbine and each firing event excites the pressure field in front of the turbine, induced pressure fluctuations resulting from one firing event may affect other firing tubes. This could result, e.g., in a suboptimal purging and/or refilling of firing tubes that would therefore affect new firing events, thus possibly cascading in an unstable operation mode.

As a possible alternative start-up strategy, an ESC-supported starting process is introduced here that considers the possible interactions of multiple firing tubes. This approach can be subdivided into two phases, i.e., continuous operation of all tubes as in a conventional gas turbine to start the rotation and then changing of the firing mode of individual tubes to an unsteady operation. The instant at which each combustion tube starts to fire in the PDC mode can then be defined manually, e.g., by an operator, or by a pre-defined function.

To determine when an additional firing tube should be added to the unsteady operating tubes, the output signals coming from the machine could be used. One possible signal to determine whether the unsteady operation should start in the next combustion tube could be the variance of the pressure sensor signals in the plenum that characterizes the corresponding pressure fluctuation. In the approach described here, the pressure variance is minimized through a firing pattern altering ESC, ensuring that a desired state is reached. Further, after each added combustion tube, the power of the conventionally working tubes is reduced to reach the desired operation point of the complete system. This procedure is repeated until all combustion tubes fire in PDC mode.

In the experiments performed in the cold test rig, we used an operator-based determination to decide when an additional firing tube should be added. This decision was made using the variance of the pressure sensor signals, as defined in Eq. (3), that was minimized through the ESC over the start-up process. Additionally, a cold bypass stream was used to start the turbine rotating. Further, since no extra inlets exist at the test rig, which allowed for the supply of a bypass mass flow, the mass flow was channeled through the firing inlets that were not yet added to the operation by keeping all the corresponding SVs open. In the first phase of the start-up and after each added firing inlet, the supply mass flow through the HAT was manually adapted by the operator in such a way that a mass flow of $0.27\,\mathrm{kg\,s^{-1}}$ was maintained.

5 Results

5.1 Efficiency and Specific Work Increase

An experiment to maximize the turbine efficiency, as calculated in Eq. (2), was performed using an ESC that adapts the firing pattern, named η-ESC. In Fig. 3, the applied firing pattern and the corresponding online calculated efficiency are shown. The latter is represented as a percentage of the efficiency that was measured in an experiment using the same inflow conditions, e.g., mass flow and design rotation speed, but no pulsation (relative efficiency).

Starting from a relative base efficiency slightly above 85% compared to the steady flow case, using the reference pattern, the conservatively parameterized ESC with $\lambda = 0.05$ adapted the firing pattern in such a way that a relative turbine efficiency of around 105% was reached at the end of the experiment. This resulted in a significant increase of 20% points in relative efficiency compared to the reference case with no changes to the inflow conditions other than the firing pattern. A convergence into three clusters of the firing instants that define the pattern can be seen in subfigure b), where the firing instants are shown as normalized to an interval of [0, 1]. This interval represents the length of an actuation period of 0.05 s.

Note that the upper limit of 1 indicates the end of an actuation period, which is followed by the starting instant of the next period, which is represented by the lower limit of 0. Therefore, both limits represent the same instant in time, and inlets 1 and 7 belong to the same cluster. Further, no change of the firing instant of the first inlet can be seen, since it has been fixed at 0 s in each period to decrease the amount of control inputs for the ESC.

Since the maximum efficiency of the turbine stage under pulsating operation was not known beforehand, the experiment was performed using the fixed inflow parameters and the variable firing pattern until no further increase of efficiency was realized by the ESC. This experiment indicates a strong sensitivity of the turbine behavior and efficiency under pulsating inflow conditions regarding the firing pattern, similar to the sensitivity of the pressure fluctuation in front of the turbine, as shown in [6]. Further, it consolidates the assumption that the firing pattern should not be chosen arbitrarily for the operation of a turbine under pulsating inflow conditions. The obtained pattern of clustering was not predictable beforehand and could be replicated in repeated experiments, always resulting in a cluster of the first and seventh inlets, the second and third inlets into another cluster, and the remaining inlets into a third cluster. However, the positive effect of clustering with respect to the turbine is less favorable for the last compressor stage, see [12].

In a second experiment, the cost function for the so-called $\tilde{w}$-ESC was changed to the specific work that the turbine extracts from the flow. Because the magnitude of the objective function had changed compared to the efficiency values[1], a value of $\lambda = 0.01$ was used. As can be seen in Fig. 4 a), the absolute value of the specific work could be maximized, starting from approximately $-3.05\,\mathrm{kJ\,kg^{-1}}$ to almost $-3.6\,\mathrm{kJ\,kg^{-1}}$, resulting in an increase of about 15%. A similar clustering of firing instants, as in the previous experiment, can be seen in Fig. 4c). This can be explained because the numerator in Eq. (2) is linearly correlated to the specific work, as calculated in Eq. (1), and the change is mainly driven by the development of T_5, as can be seen in Fig. 4d). The denominator in Eq. (2) saw an almost negligible change throughout this experiment, as can be seen from the corresponding signals in Fig. 4b), resulting in a similar dependency of the objective function to the control input, as in the previously shown experiment.

5.2 Start-Up Process

Finally, an experiment using the proposed start-up strategy for PDC-based gas turbines was performed. The pressure variance in the plenum in front of the turbine stage was used as the characteristic state of the system that should be optimized through a start-up-σ-ESC. As in the previous experiments, the ESC

[1] The range of the calculated efficiency was [0, 1].

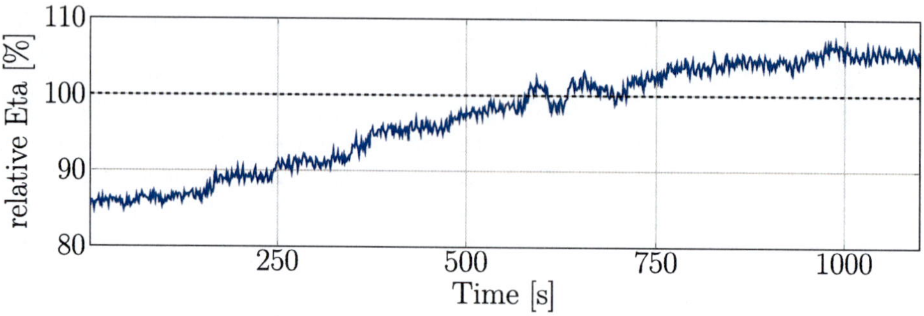

a) Efficiency development using η-ESC.

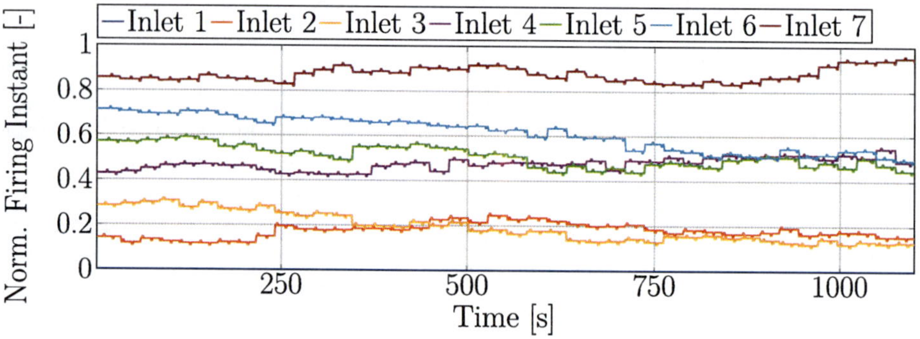

b) Firing pattern using η-ESC.

Fig. 3. Development of efficiency and corresponding firing pattern over time in an experiment using η-ESC.

could only change the firing pattern, but the firing tubes were added sequentially to the operation. Unlike in the previously shown experiments, the air supply through the HAT had to be changed manually during the experiment to ensure a desired flow of $0.27\,\mathrm{kg\,s^{-1}}$ to the turbine. A less aggressive parameterization of λ was used in the ESC, compared to the one chosen in [6], which was based on the same objective function. This was done to ensure a smooth evolution even under pronounced varying inflow conditions through the changed mass flow. Here, $\lambda = 0.5$ was set. The first phase of the start-up was realized through a gradual increase in the supply mass flow by the HAT with all SVs opened and is not shown here. Therefore, the experiment starts with the second phase of the start-up. Figure 5 shows the variance as calculated by Eq. (3), which was minimized by the start-up-σ-ESC, as defined in Sect. 4.2. The black dashed line indicates when a firing tube was added to the operation, while the colored dashed lines indicate the starting pattern for the non-pulsating tubes until they are added to the unsteady type of operation. The experiment began bypass mass flow only. After each added firing inlet, the inflow mass flow was manually

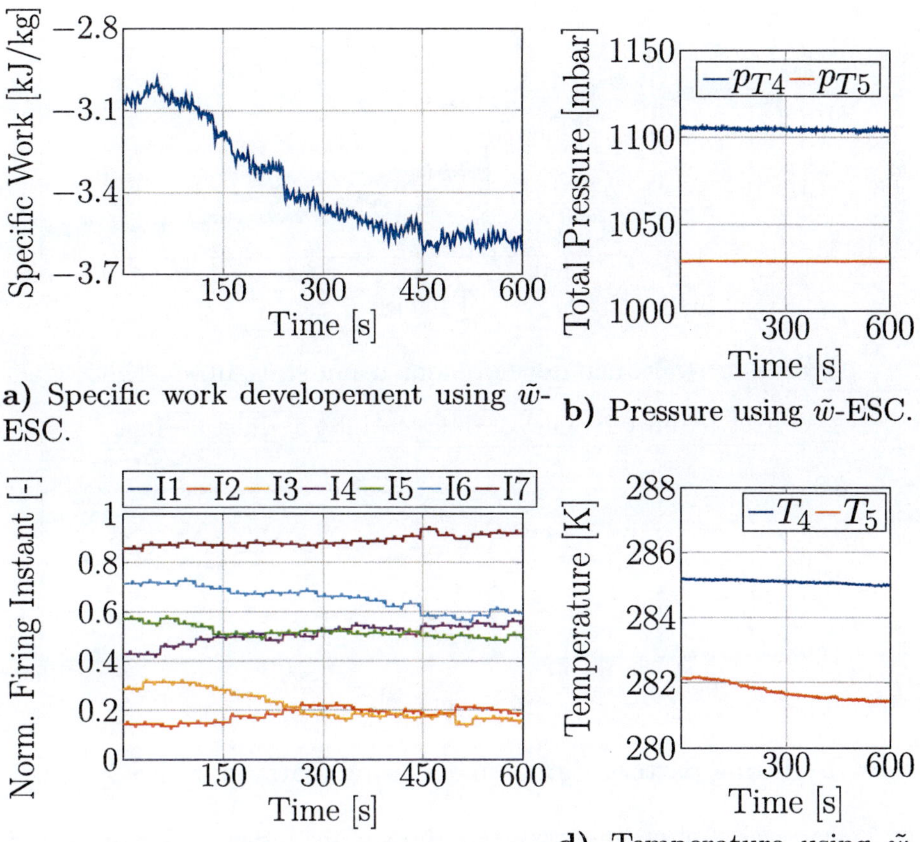

a) Specific work developement using $\tilde{w}$-ESC.

b) Pressure using $\tilde{w}$-ESC.

c) Firing pattern using $\tilde{w}$-ESC.

d) Temperature using $\tilde{w}$-ESC.

Fig. 4. Development of specific work and corresponding firing pattern, temperature, and pressure over time in an experiment using $\tilde{w}$-ESC.

adjusted by the machine operator in such a way that the supplied mass flow to the turbine was brought back to $0.27\,\mathrm{kg\,s^{-1}}$, as noted above. The instants at which an additional tube should be added were also chosen manually. This was done when a minimization of the variance through the ESC given the current number of firing tubes was regarded as mostly completed. In future work, this step could be automated as well.

The first inlet was added to the operation at 11 s. The drop in variance that can be seen thereafter arose because of change in the flow resistance. The rise in variance afterwards is caused by the manual change of the mass flow. Since the first inlet's firing instant is fixed in the actuation period, this ESC does not yet interfere. Starting at 40 s, the second inlet is added to the firing pattern and the mass flow is adjusted. The ESC immediately tries to adapt the firing instant

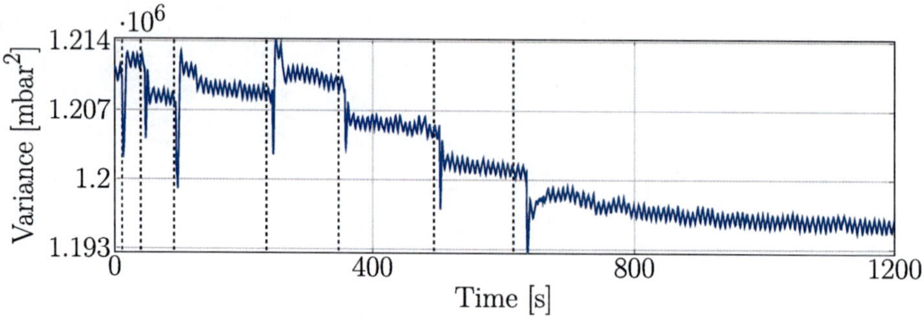

a) Pressure variance development using start-up-σ^2-ESC.

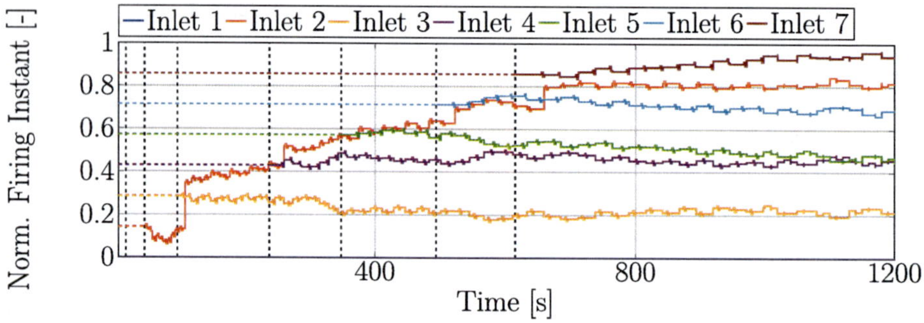

b) Firing pattern development using start-up-σ^2-ESC.

Fig. 5. Development of pressure variance in front of the turbine and corresponding firing pattern in an experiment using start-up-σ^2-ESC.

of this tube, but since no significant improvement is seen, the next firing tube is added to the operation at 91 s. The bypass mass flow had to be readjusted twice afterwards because of overshoots in the supplied mass flow, which explains the hard drop in variance shortly after the rise at around 100 s. The ESC then minimized the variance until the next tube was added at 235 s. Here, two mass flow adjustments had also to be done to ensure the desired flow. The next firing tubes were added to the operation at 347 s, 494 s, and, finally, at 615 s, with just one adaptation of the mass flow after the addition of each firing inlet.

The ESC successfully minimized the variance of pressure sensor signals in front of the turbine stage between each added firing event. This is especially true after all firing tubes had been added at the end of the experiment.

Looking at the firing pattern, it can be seen that the firing instant of the second firing inlet changed significantly over time. This is assumed to be due to the effect of the manually adapted mass flow that was not considered in the evaluation of this ESC and because an adaptation of the supply mass flow was always performed while the firing instant of the second inlet was perturbed.

Hence, this affected the gradient estimation of the second inlet. Further, an increase can be seen in the time interval between perturbations. This is due to the fact that each added firing tube increased the number of parameters that had to be changed for the gradient estimation of the ESC. After the start-up, the η- or $\tilde{w}$-ESC could be activated to reach another goal.

6 Conclusion

In this contribution, different applications of ESC to the operation of a PDC-based gas turbine were introduced and applied to a cold mock-up test rig. It was designed specifically for the analysis of turbine behavior under pulsating inflow conditions. Using an ESC, which can only adapt the synchronization times between different firing tubes, an efficiency increase of the turbine of about 20% points was reached within the experiment and compared to a non-pulsating operation using the same inflow characteristics an efficiency increase of 5% points was obtained. Further, a maximization of the specific turbine work in operation was presented using the same control approach, resulting in a similar synchronization at the end. Additionally, a possible ESC-supported, sequential start-up strategy for PDC-based gas turbines was introduced. This also was applied to the test rig, where it could successfully minimize the variance of pressure fluctuation in front of the turbine to protect the turbine from too-harsh changes in its inflow conditions.

Acknowledgments. The authors acknowledge the support for this research by the Deutsche Forschungsgemeinschaft (DFG) in the context of the Collaborative Research Center CRC 1029 'Substantial Efficiency Increase in Gas Turbines through Direct Use of Coupled Unsteady Combustion and Flow' through sub-project A05 and B05.

References

1. Gülen, S.: Étude on gas turbine combined cycle power plant - next 20 years. J. Eng. Gas Turb. Power **138**(5), 051701-1–051701-10 (2015)
2. Bussing, T., Pappas, G.: An introduction to pulse detonation engines. In: 32nd Aerospace Sciences Meeting & Exhibit, p. 263 (1994)
3. Li, J.-M., Teo, C.J., Khoo, B.C., Wang, J.-P., Wang, C.: Detonation Control for Propulsion: Pulse Detonation and Rotating Detonation Engines. Springer (2018)
4. Bobusch, B.C., Berndt, P., Paschereit, C.O., Klein, R.: Shockless explosion combustion: an innovative way of efficient constant volume combustion in gas turbine. Combustion science and technology, vol. 186(10–11): The 24th International Colloquium on the Dynamics of Explosions and Reactive Systems, pp. 1680–1689 (2014)
5. Nicholls, J.A., Wilkinson, H.R., Morrison, R.B.: Intermittent detonation as a thrust-producing mechanism. J. Jet Propul. **27**(5), 534–541 (1957)
6. Topalovic, D., Wolff, S., King, R., Heinrich, A., Peitsch, D.: Minimization of Pressure Fluctuations in an Axial Turbine Stage Under Periodic Inflow Conditions. In: AIAA, pp. 2019–4213 (2019)

7. Heinrich, A., Herbig, M., Peitsch, D., Topalovic, D., King, R.: A testrig to evaluate turbine performance and operational strategies under pulsating inflow conditions. In: AIAA, pp. 2019-4039 (2019)
8. Knobloch, K., Lahiri, C., Enghardt, L., Bake, F., Peitsch, D.: Hot-acoustic-testrig (HAT): a unique facility for thermoacoustic research'. In: ASME 2011 Turbo Expo: Turbine Technical Conference and Exposition, vol. 7, pp. 1023–1032 (2011)
9. Suresh, A., Hofer, D.C., Tangirala, V.E.: Turbine efficiency for unsteady, periodic flows. ASME J. Turbomach. **134**(3), 034501-1–034501-6 (2012)
10. George, A.S., Driscoll, R., Munday, D., Gutmark, E.J.: Experimental comparison of axial turbine performance under pulsed air and pulsed detonation flows. In: 49th AIAA/ASME/SAE/ASEE Joint Propulsion Conference (2013)
11. Khong, S.Z., Nešic, D., Krstic, M.: Iterative learning control based on extremum seeking. Automatica **66**, 238–245 (2016)
12. Neuhäuser, K., King, R.: About the influence of periodic disturbance patterns on the efficiency of active flow control in a linear stator cascade. Act. Flow Combust. Control (submitted, 2021)

Part VI: Active Flow Control Methods

Experimental Investigations of Active Flow Control Using a Piezo Adaptive Blade in a Compressor Cascade Under Periodic Boundary Conditions with High STROUHAL-Number

Tobias Werder(✉), Lukas Kletschke, and Robert Liebich

Technische Universität Berlin, Straße des 17. Juni 135, 10623 Berlin, Germany
tobias.werder@tu-berlin.de

Abstract. As a consequence of constant volume combustion in gas turbines, pressure waves are generated that propagate upstream the main flow into the compressor system are generated leading to incidence variations. Numerical and experimental investigations of stator vanes have shown that Active Flow Control by means of adaptive blade geometries is beneficial when such periodic incidence variations occur. A less susceptible to stall and choking for compressors dealing with periodic disturbances can be achieved. Experimental investigations with high STROUHAL numbers using such a method have not yet been done in order to demonstrate the effects. Therefore, this work investigates a linear compressor cascade that is equipped with a piezo-adaptive blade structure utilizing macro-fiber-composite actuators. A throttling device is positioned downstream of the trailing edge to emulate an unsteady combustion process. Periodic transient throttling events with STROUHAL numbers up to 0.144 were being investigated due to incidence changes. Consequently, pressure fluctuations on the blade's surface occur, having a significant impact on the pressure recovery downstream of the stator cascade. Experimental results of harmonically actuating the piezo-adaptive blade with STROUHAL numbers of up to 0.72 show that the impact of disturbances at resonance can be nearly reduced to zero. Therefore, the blade design must be matched to the type of disturbances to achieve further improvements in the technology.

Keywords: Piezo · Blade · Flow control · Active · AFC · CVC · Compressor

1 Introduction

Modern gas turbines are highly efficient and powerful machines, which are used in a variety of applications, e.g. airplane propulsion and power plants to generate electrical energy and heat. Efforts raising the efficiency of gas turbines

R. King and D. Peitsch (Eds.): AFCC 2021, NNFM 152, pp. 305–320, 2022.
https://doi.org/10.1007/978-3-030-90727-3_19

for reasons of environmental and economical issues are continuously increasing while nowadays small improvements usually demand substantial investments in component research and development. A major breakthrough for the overall gas turbine efficiency cannot be expected without disruptive technologies.

One of those technologies that aims at a significant increase in efficiency is constant volume combustion (CVC). By applying a pressure gain combustion process, an inherently higher thermodynamic efficiency can achieved. Nevertheless, such improvement comes at the cost of unsteady effects, such as pressure fluctuations originating from the combustor. This leads to implications for compressor and turbine, such as incidence variations with an increased risk of stall or choke. An effective way of dealing with any kind of disturbances in turbomachinery is Active Flow Control (AFC). In the case of a CVC, AFC can reduce the impact of disturbance and ensure robust and stable operation.

The application of AFC may present one means to gain higher efficiency. Many of them use pressurized air to manipulate the passage airflow or conditions on the blade's surface. Staats and Nitsche [1] and Steinberg et al. [2] investigated a compressor cascade with periodic unsteady outflow conditions. Fluidic actuators on the blade's suction side and the adjacent sidewall near the leading edge were used to control static pressure distribution on the blade's surface and the pressure rise as well as the total pressure loss coefficients in the wake. A separation suppression of the flow and less sidewall effects lead to effective disturbance rejection.

Another AFC method comprises an adaptive blade geometry. Hammer et al. [3] showed that adjustable shapes can increase the operating range of a compressor stage and thus enable a more stable operation of the stage. They investigated a mechanically adjustable front part of piezo adaptive blades and showed that this method can enlarge the operating range of a compressor stage. Phan et al. [4] had shown in a numerical study that an oscillating leading edge of a piezo adaptive blade can reduce the negative effects of oscillating incidences, e.g. a suction peak, and thus the associated higher susceptible of stall. Furthermore, they designed and tested a similar adaptive blade design with Macro Fiber Composite Actuators (MFC) for cascade experiments [5]. Krone et al. [6] investigates the benefits of MFC in a linear cascade to extend the workspace as well. An other moving leading edge investigation from Ko et al. [7] is done to control the reverse flow of a blade.

An adaptive blade with oscillating leading edge in an open wind tunnel with throttle-induced pressure fluctuations was already investigates before by [8]. But the case of high STROUHAL-numbers in actuation compared with throttling has not yet been examined experimentally. The present work experimentally demonstrates the benefits of this system under conditions similar to those of CVC-capable gas turbines.

2 Theoretical Considerations

The theoretical background for the idea of harmonically varying the shape of the front part of a compressor vane will be shown in this section. A compressor vane is normally designed for one angle of attack α with minimal angle of incidences

i. Under these normal circumstances, minor losses are to be expected for a little range of incidences. However, for operating a gas turbine with CVC a fluctuation in angle of incidence of up to 3 degrees occurs [1].

This incidence called i_{CVC} leads to much greater losses, which can be compensated by generating a dynamic blade angle $\beta = \omega t + \varphi$, which represents the stationary incidence angle i_{stat}. The ω stands for the angular frequency of excitation ($\omega = 2\pi f_a$), t is time and φ is a phase shift, resulting in an so called blade incidence i_{blade} described in Eq. 1. If a blade with i_{blade} is designed in such a way that i_{res} disappears, a disturbance-free flow around the blade is obtained. For a slowly moving blade in a undisturbed flow, the angle of incidence changes in a more stationary manner as described in Eq. 3, comparable to an airplane wing or moving blades in stationary gas turbines for start-up operations, turbines for ramp-up operations or load changes. If the blade – or in this case the leading edge – moves fast, however, we can not neglect the induced velocity perpendicular to the chord. This velocity leads to an angle of incidence i_{dyn} in addition to the stationary i_{stat}. It will be called dynamic incidence henceforth and is described in Eq. 4. The negative sign in Eq. 1 designates from the definition of the direction of angles. In this equation $A(\omega)$ is the dynamic displacement amplitude of the leading edge at an applied harmonic voltage to a MFC Actuator of $V = \hat{V} sin(\omega t)$ with excitation frequency ω. This amplitude changes like a forced harmonic oscillator. The oscillation is described in Eq. 5 with natural frequency ω_0. The function $f(\hat{V})$ in Units of mm describe a magnification of $A(\omega)$ for different excitation amplitudes $\hat{V}$ in Units of volts. The parameter D stands for structural damping and A_0 is the offset of amplitude considering a different stiffness in up and down movement, if necessary. Measurements of i_{CVC} result for an applied case in a blade design and blade material that gives the parameters for the moving function of the blade $A(\omega)$ and a driving pattern to adjust the phase shift in an closed loop control system. In a best case scenario with perfectly matched parameters i_{res} vanishes at CVC.

$$i_{blade} = i_{stat} - i_{dyn}, \tag{1}$$

$$i_{res} = i_{CVC} - i_{blade}, \tag{2}$$

$$i_{stat} = \arcsin\left(\frac{A(\omega)\sin(\omega t + \varphi) + A_0}{c}\right), \tag{3}$$

$$i_{dyn} = \arctan\left(\frac{A(\omega)\cdot\omega\cdot\cos(\omega t + \varphi)}{v_\infty}\right), \tag{4}$$

$$A(\omega) = \frac{f(\hat{V})}{\sqrt{(1-\eta^2)^2 + (2D\eta)^2}}, \text{ with } \eta = \frac{\omega}{\omega_0}. \tag{5}$$

Figure 1 shows changes in relative velocity and incidence with their relation to an induced velocity w that occurs because of the moving tip.

Assuming a simple linear oscillator and values given in Table 1 referring to our system described later, the calculated frequency response of the incidence angles and the spectrum of movement of the blade's leading edge are shown in Fig. 2. What can be seen in Fig. 2 is an increasing magnitude at resonance causing the

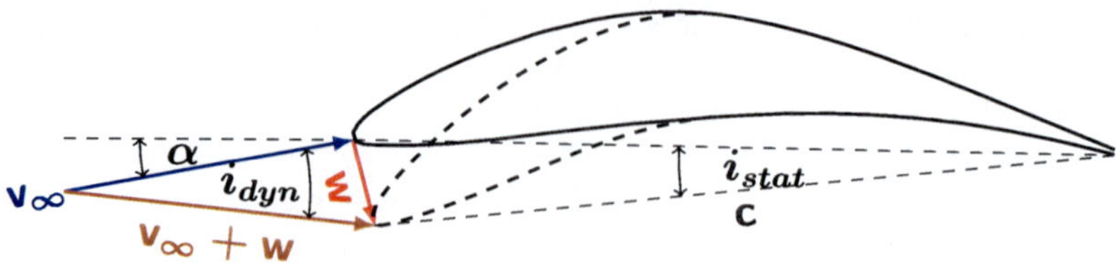

Fig. 1. Schematic velocity triangle of dynamically deformed blade

Table 1. Data of calculation

Parameter	Symbol	Value	Unit
Chord length	c	150	mm
Inflow velocity	v_∞	25	m/s
Angular frequency of blade	ω_0	122π	rad/s
Expected amplitude of excitation driven by MFC	$f(\hat{V})$	0.03	mm
Structural damping ratio of the blade	D	0.015	–

oscillations of the blade. The amplitudes of i_{res} dominate over all frequencies. The reason is the phase shift between i_{stat} and i_{dyn}. For low frequencies the phase angle φ is zero neglecting the influence of the dynamic incidence. At high frequencies the dynamic incidence prevails due to the very low amplitude but high momentum of the blade's leading edge. Looking at the angle of maximum amplitudes given by

$$\frac{\partial i_{res}}{\partial t} = 0, \tag{6}$$

$$\Rightarrow \varphi_{max} = \arctan(\frac{-2v_\infty}{c\omega}). \tag{7}$$

at which these incidences occur, there is a phase shift from $\frac{\pi}{2}$ to π pictured in Fig. 2 below. This is due to the zero phase shift of sinus for low and a phase shift of $\frac{\pi}{2}$ for cosinus at high frequencies. Near resonance, it considering both phase angles is important. This considerations are necessary to understand the fact that a pressure signal from a measurement can be separated into different transfer functions. For an open loop control system the only variables here are the voltage and frequency of the excitation. Figure 3 shows the signal path of the measurement system in a block diagram. The figure shows a schematic of signal processing by using transfer functions. All of these subsystems have to be identified to obtain the actual behavior of the blade and the correct interpretation of the measured signals. The four subsystems are represented as

- $\mathbf{G}_b$ as the transfer function for blade motion,
- $\mathbf{G}_i$ as the transfer function for the angles of incidence,

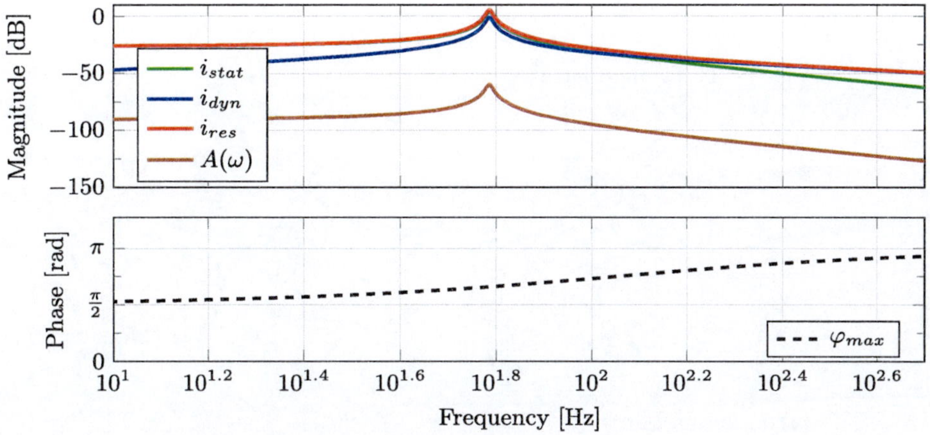

Fig. 2. Frequency response of the different incidence angles and phase shift of i_{res} for an blade design made of aluminum and parameters from Table 1

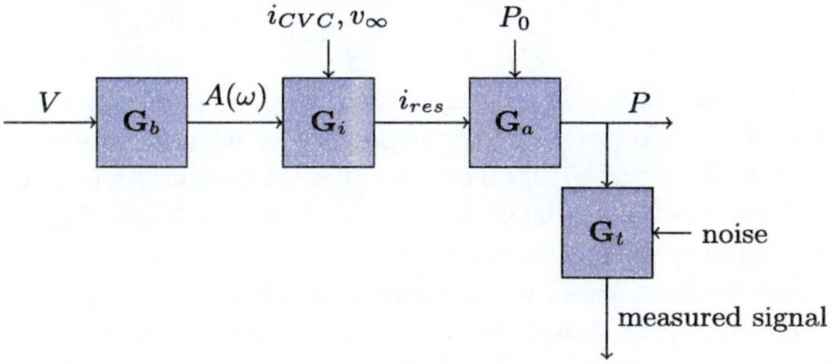

Fig. 3. Signal path in block diagram with different transfer functions of the whole measurement

- $\mathbf{G}_a$ for the ambient air and
- $\mathbf{G}_t$ as the transfer function of a tube from the blade surface to the sensor.

3 Experimental Setup

All experimental results in this paper are based on a similar setup to [8]. They were obtained from a low-speed compressor cascade, which consists of nine linearly arranged compressor stator blades (see Fig. 4a). This design guarantees symmetric conditions at the measurement passage, while minimizing the influence on the top and bottom walls. Measurements and actuation were carried out on blade number 5 and its wake. A boundary layer suction device was installed on the top and the bottom side walls of the cascade and adjusted such that the static inflow pressure is uniform. Static pressure taps at half the chord length c

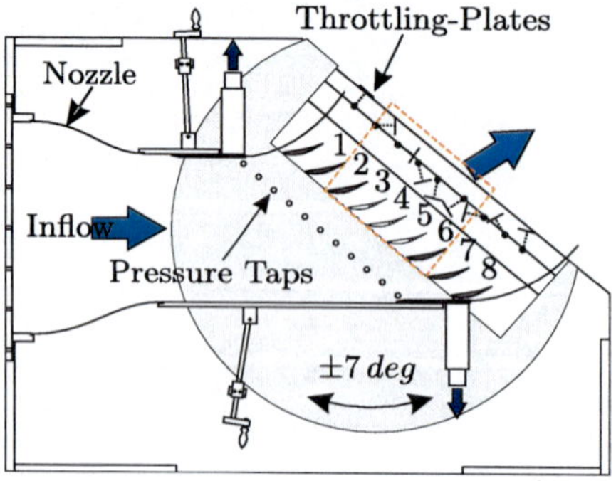

(a) Linear cascade

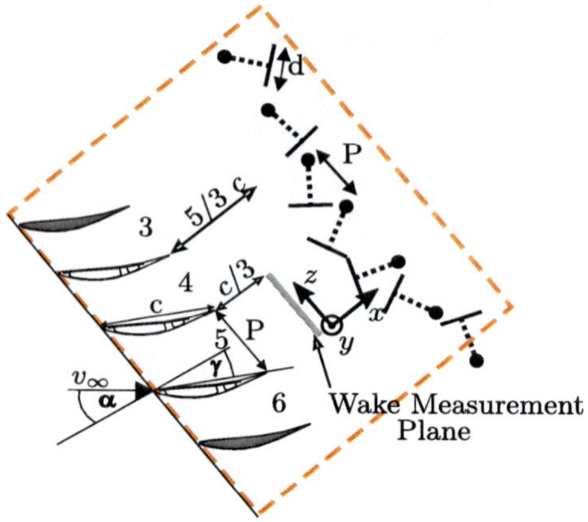

(b) Detail of measurement domain and throttling mechanism

Fig. 4. Experimental setup

upstream of the blade's leading edge measured the pressure distributed over the entire passage width. For preliminary experiments with this blade, the inflow angle can be varied up to ±7°. The Controlled Diffusion Airfoil (CDA) has a deflection angle of 26.4°. It is designed as a laminar profile shifting the laminar-to-turbulent boundary layer transition further downstream and thus reducing the amount of secondary flow in the cascade as described in [9]. Three of the nine blades were designed adaptable for nearly symmetrical conditions in passage 4 and 5, but only one is used in this investigation to depict the potential of the design. The drive of all three blades caused a resonance failure and has to be investigated by aeroelastic specialists before an other drive with more than one adaptive blade will be done.

Table 2. Data of cascade

Parameter	Symbol	Value	Unit
Pitch	P	102	mm
Plate depth	d	55	mm
Plate height	L_p	300	mm
Stagger angle	γ	30	deg
Inlet flow angle	α	47	deg
Reynolds number	Re	$2.5 \cdot 10^5$	–
Mach number	Ma	0.07	–
Blade height	L	300	mm

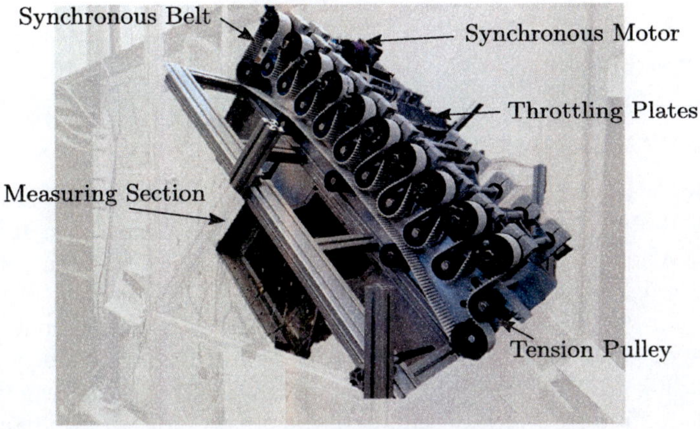

Fig. 5. Throttling device

Figure 4 shows the installed throttling device. It was located at a distance of 5/3 c in the wake of the cascade. The device consists of nine throttling blades eccentrically mounted on nine rotor shafts driven by a synchronous motor. A synchronous belt with tension pulley ensures the exact positioning of the throttling blades and an exact blade-to-blade sequence without collisions, see Fig. 5. The throttling blades block the flow through the passages being arranged in an angle of 45 degrees to each other. The blockage area covers 95% of the passage area. The permanent blockage of the entire device without driving amounts to 25 % of the cascade's wake. All plates rotating in the same direction so the throttling pattern is 1-2-3-4-5-6-7-8-1. Thus it blocks one passage after the other and again from the start. Although the device is designed for frequencies up to $f_c = 100$ Hz, the highest driving frequency here is unfortunately about 25 Hz because of not existing safety system for "plate of" events.

All measurements were taken at a REYNOLDS number of $Re = 2.5 \cdot 10^5$ and an averaged inflow velocity of $v_\infty = 25$ m/s. This can be expressed in the STROUHAL number defined by

$$Sr = \frac{f\,c}{v_\infty} \tag{8}$$

of $Sr = 0.6$ with the throttling frequency being $f = f_c$, the chord length of the blade c and the inflow velocity v_∞. At machine scale this would correspond to throttling frequencies of approx. $f_c = 1575$ Hz at Mach number of $Ma = 0.4$ and a chord of $c = 0.08$ m, which will represent a drive of a Pulse Detonation Combustion (PDC) with a tube length of about 0.5 m. These values represent an industrial high pressure compressor stage [10] and a PDC operation frequency has been shown by Asahare et al. [11]. In the presented work, measurements with STROUHAL numbers from 0.012 to 0.12 for throttling with $f_c = 2...24$ Hz, and $Sr = 0.006 \ldots 0.72$ for actuation frequencies of $f_a = 1...120$ Hz were performed. These high frequencies could occur at Rotating Detonation Combustion (RDC) as shown in the work of Anand et al. [12]. All relevant parameters of the passage are summarized in Table 2.

3.1 The Piezo-Adaptive Compressor Stator Vane

An adaptive blade structure with a hollow design utilizing MFC actuators (see Fig. 6 and Table 3) is used to generate leading edge deflections. It is made of aluminum. When actuated harmonically in a sinusoidal form with a supply voltage of up to $V_{max} = \pm 500$ V, the MFCs cause a deformation of the leading edge and thus can react to incidence variation. The actuated front part of the blade changes the blade's designed inflow blade angle by up to 5 deg. The design ensures leading edge pitching like a harmonic oscillator in the region below the second natural frequency. In consequence, the pressure distribution and flow conditions are changed because of incidence variations of values up to 3 deg expected according to Eq. 1. The actuated blade itself consists of three separately manufactured parts: the suction side part including the leading and trailing edge as well as the mounting holes, the pressure side part to generate a structural gap to allow high deflections at the leading edge, and a cover on the pressure side for a supply cable channel. Deformation of the blade's front part is mechanically achieved by stretching and compressing the adhesive-bonded MFCs. This results in a structural bending moment causing a deflection of the leading edge and the whole front part. The structural gap leads to a step on the pressure side and thus in an transition of the flow in it's region an a bit more wide wake depression [5], but the pressure side flow is very stable so that the flow does not detach before the second eigenmode occurs. To measure the angle of incidence caused by the moving tip during wind tunnel measurements we added two strain gauges on the inner side of the suction surface near the mount. These sensors are only measuring strain at the point of application. To obtain the tip deflection, calibration of these sensors is first required. For the calibration, a test rig has been designed to validate the blade's structural properties before using in the cascade. This rig allows clean measurements of the frequency response, structural life time, natural frequencies and shapes without any influence from flow. The rig was mounted on a massive damped base for vibration isolation. The blade was mounted using the holes on the suction side part comparing to the

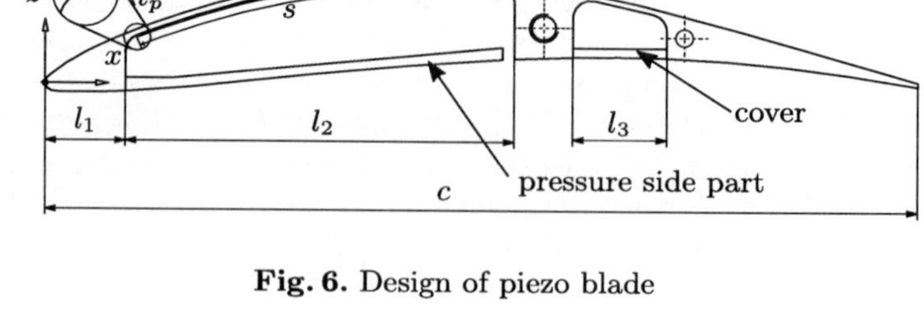

Fig. 6. Design of piezo blade

mounting in the cascade. In addition, the rear part was clamped, so that there is no additional influence of its vibration. The clamping has no effect on the firs natural mode and frequency so it is comparable with the mount in the cascade. This test rig enables oscillation measurements of the blade's front part only. A laser triangulation sensor as well as an eddy current sensor were used for leading edge displacement measurements under MFC actuation to calibrate the strain gauges. The frequency responses at various supply voltages were measured.

Table 3. Data for the Piezo Blade

Parameter	Symbol	Value	Unit
Nose	l_1	14	mm
Hollow part length	l_2	67	mm
Supply cable channel	l_3	16	mm
Patch length	s	67	mm
Patch thickness	t_p	0.3	mm
Blade active part thickness	t_a	1	mm

4 Measurement Methods

The value of interest for a vane is the static pressure distribution on the surface given by the C_P value

$$C_P = \frac{p - p_\infty}{q}. \tag{9}$$

Where p is the local static pressure on the blade's surface, p_∞ the static inflow pressure measured at the wall, and q the dynamic inflow pressure given by

$$q = \frac{\rho}{2} v_\infty^2, \qquad \text{with } \rho = \frac{m\, p_{amb}}{RT}, \tag{10}$$

measured at an undisturbed position at the end of the nozzle outlet of the channel, see Fig. 4a. Here is m the molar mass of air, R is the general gas constant and T is temperature. The value v_∞ is the undisturbed inflow velocity and ρ stands for the density of the medium, which in this case is air. The values for p_{amb} and T were also measured. For an inflow velocity of $v_\infty = 25$ m/s, calculated from Eq. 10, and a density of $\rho = 1.25$ kg/m^3, a Mach number of $Ma < 0.3$ applies throughout the range. Figure 7 shows the pressure distribution of measurement data from the adaptive blade and numerical results at Reynolds number of $Re = 2.5 \cdot 10^5$. What can be seen is a good match of the measurement with the numerically calculated ones. A slight deformation of the blade at leading edge occurred by having a look on tap no. 17. No stagnation was not measured, but a negative static pressure. This means that the nose was moving towards the suction side. Also, the separation is shifted downstream, but the pressure gain is almost the same. Tap No. 15 is marked because all following results are shown as an example for this tap.

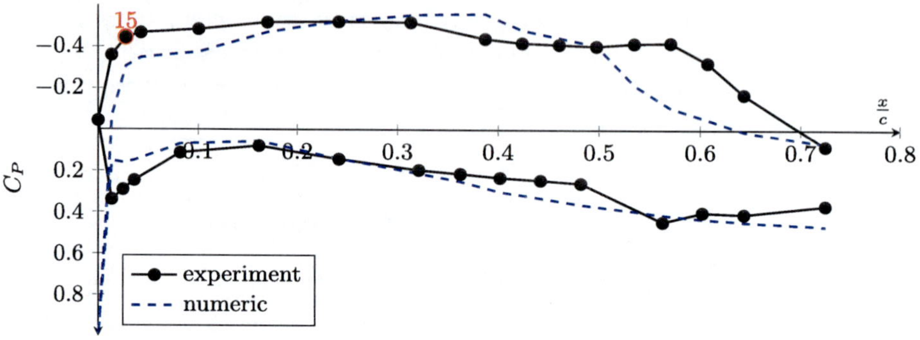

Fig. 7. Pressure distribution of the undeformed blade at $Re = 2.5 \cdot 10^5$, subsequent results shown for tap No. 15

Characterization of the static pressure distribution of the piezo adaptive blade was performed using 32 amplified differential low pressure sensors from First Sensor's HCLA series, specifically model designation HCLA0025DB. Each of the sensors has a measurement range of ±25 mbar at a supply voltage of 5 V. The uncertainty of non-linearity and hysteresis is max. ±0.25% at full scale span. These sensors were placed outside the flow of the measuring section and connected to a long elastic tube and a short metallic tube inside the blade with orthogonal access to the point of interest perpendicular to the surface. The small metallic tubes have an inner diameter of 0.3 mm and a length of approx. 10 mm. The elastic tubes have an inner diameter of 1 mm and a length of approx. 3 m. The traveling pressure signals from the surface to the sensor within the elastic tube does not show much damping in a frequency range 120 Hz. The small inner diameter of the metallic tube but represents a challenge in this context because the whole tube cavity system has a high damping effect causing by wall friction. Consequently, a test bench was designed to calibrate the behavior of all of the tubes. With reference to [13] it is given with a linear approach by the transfer function

$$\mathbf{G}_t = \frac{1}{T_2 s^2 + T_1 s + 1}, \text{ with } T_2 = \frac{l \rho C}{A}, \, T_1 = \frac{k_1 C}{A}, \, k_1 = \frac{32 \mu l}{d^2}, \, C = \frac{V}{\kappa P_0}, \quad (11)$$

with parameters given in table 4. By applying a binary signal to a magnetic valve with increasing frequency, the transfer function can be identified. As example of the measurements of all 32 tubes, an extract of the signal for the tube 15 is shown in Fig. 8. The measured input pressure signal is an oscillating signal with an overshoot. The output is damped because of the long traveling distance through the tube. At higher frequencies, the output is highly damped but the signal to noise ratio is high enough and a dynamic is still recognizable. Equation 11

Table 4. Data of tube system

Parameter	Symbol	Value	Unit
Average of the two tube diameters	d	0.6	mm
Average area	A	$2\pi(\frac{d}{2})^2$	mm^2
Volume of cavity taken from [13]	V	310	mm^3
Dynamic viscosity of air at 20 °C	μ	$18.2 \cdot 10^{-6}$	Pa s
Atmospheric pressure of air at 20 °C	P_0	1013	hPa
Heat capacity ratio	κ	1.4	–

is used for calibration with unknown tube length l. A simple gradient decent algorithm is used to fit the model to the data. As an example, Fig. 9 shows the fit for one tube. The values of the linear model deviate slightly from the measured data. After the calibration, the drop of amplitudes at high frequencies are out of the interesting range and thus they can be neglected. This backs the importance of frequency calibration. To suppress the amplifying effect of noise, the measured signal was first filtered with a butterworth low-pass filter of order 3. This is even more important because of the calibration mentioned above. Without filtering, the amplitudes of noise in particular would increase at high frequencies. To analyze the effect of throttling, the angle of incidence i_{CVC} was not be measured directly. The value of peak-to-peak pressure fluctuation $\hat{P}$ of all pressure taps were taken as comparable value for the throttling effect and the effect of the actuated blade.

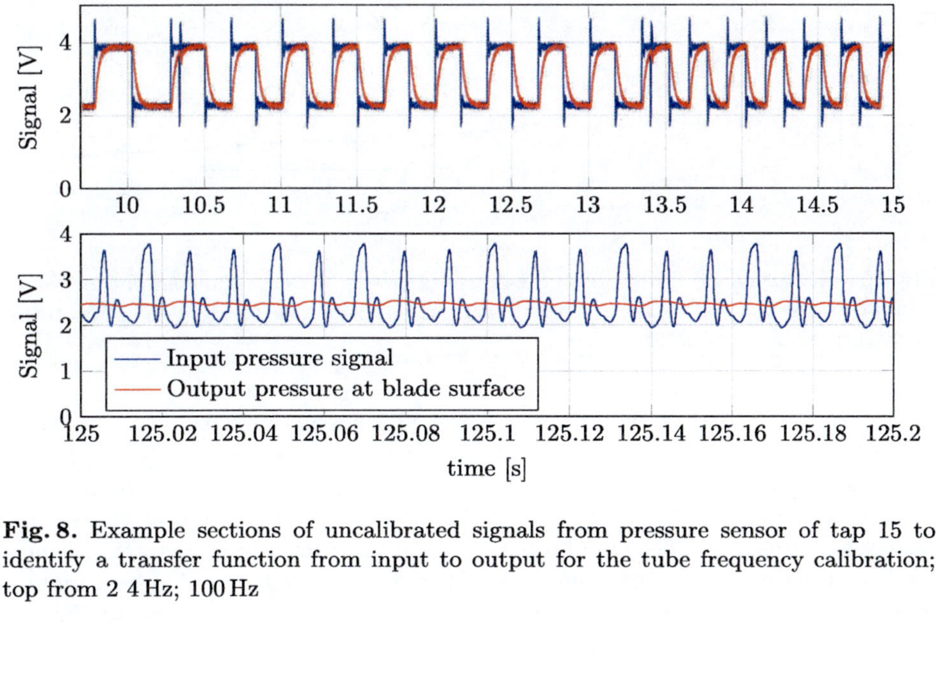

Fig. 8. Example sections of uncalibrated signals from pressure sensor of tap 15 to identify a transfer function from input to output for the tube frequency calibration; top from 2 4 Hz; 100 Hz

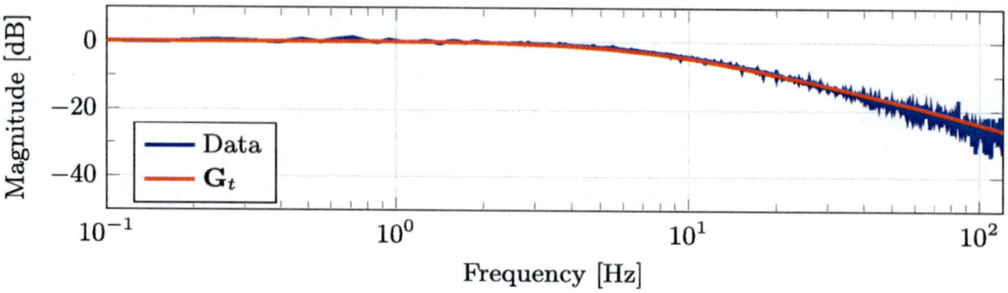

Fig. 9. Frequency response of the signal and fitted model for tube 15

5 Results

The peak-to-peak pressure fluctuation $\hat{P}$ at different throttling frequencies at one pressure tap (see Fig. 7) can be seen in Fig. 10. The behavior is comparable to wave propagation in a tube with fixed length and different boundary conditions as described in Eq. 11. The peak-to-peak pressure fluctuation drop at higher frequencies with proportion to the inverse of the frequency with all other parameters unchanged. This is the same behavior as the 20dB per decade pressure drop pictured in Fig. 9 starting at cutoff frequency because of the distance between throttling and measurement. It means that pressure fluctuations with high frequencies ($Sr > 0.144$) will not be recognizable at compressor stator stages with a distance to combustor of 5/3c.

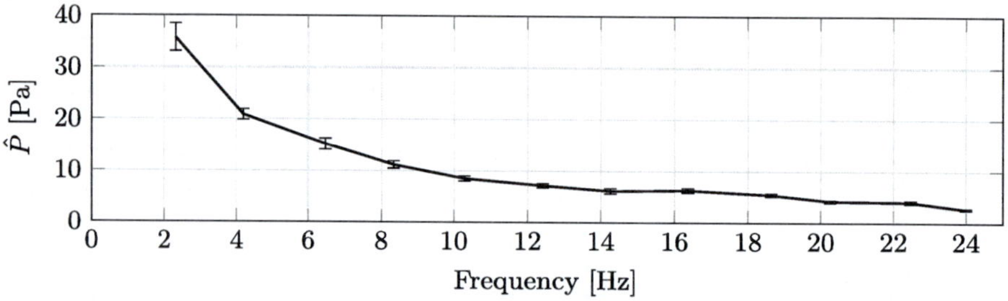

Fig. 10. Measurement results of the calibrated pressure fluctuation amplitudes and standard deviation for different throttling frequencies at tap 15 for a set of 50 measurements without actuation

Figure 11 shows an example of the averaged frequency spectrum of calibrated pressure measurements of tube 15 at an throttling frequency of $f_c = 24$ Hz ($Sr = 0.144$) from which the data for Fig. 10 was obtained. Although the fluctuations are in the hysteresis an non-linearity range, the fluctuation can be clearly measured. The second throttling frequency comes from averaging, since the throttling fluctuates, but this does not exceed two percent. The high amplitudes at twice and three times the throttling frequency are an indicator for an additional disturbance of the rotating plates when they block the flow downstream again because of their rotating design. They are for this case higher than the throttling itself. However, there is no effect for four times the throttling frequency. It could also be an effect of the reflection of a pressure wave at the cascade and amplification after reflection.

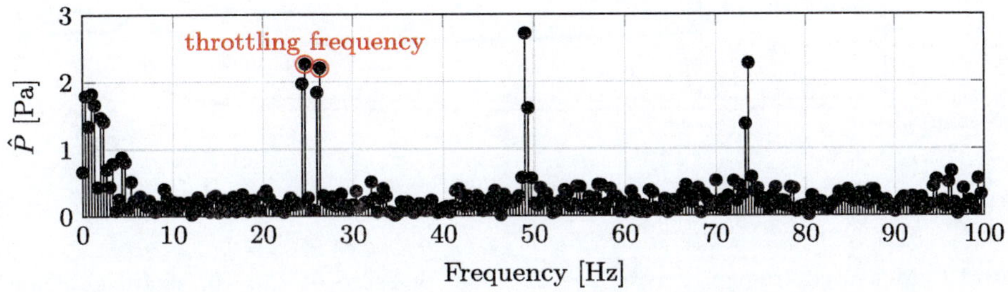

Fig. 11. Mean FOURIER plot of pressure fluctuations measurements for throttling with $f = 24$ Hz and no actuation measured at tap 15 for a set of 50 measurements

In Fig. 12 depict the frequency responses of the strain gauge and the pressure sensor of tap 15. Resonance occurs at about $f_a = 61$ Hz. Furthermore, it can be seen that we can generate pressure amplitudes of up to 25 Pa/dB or about 18 Pa at resonance frequency with the piezo adaptive blade at this location. In addition the calculated frequency response of incidence angle i_{stat} after calibration at the structural test rig is shown. The result is similar to the calculated one also depict here. The frequency and magnitude of the first natural mode is driven by the choice of the material. Consequently lower natural frequencies and higher actuation magnitudes can be achieved with other materials. The highest impact on the flow can be achieved near by the first natural frequency due to a high deflection angle there. The second natural frequency is about twice as high and thus has no significant influence. It can be seen that the peak-to-peak pressure fluctuations reach a plateau between $f_a = 70$ Hz and the second natural frequency at around $f_a = 120$ Hz with values above the expected level. This effect results from the dynamic incidences described above and also seen in the calculated incidence angles in Fig. 2. It further shows, that pressure fluctuation of this high frequency could be extinguished.

Figures 13a to 13c shows the non actuated blade pressure distribution for min. and max. incidence at different throttling frequencies without any excitation of the blade to show the influence of throttling. Comparing the pressure distribution measurements at different angles of attack shown in [5,9], the angle of incidence at low frequency behaves like a change in the angle of attack. The differences between min and max almost disappear for high STROUHAL numbers in throttling because of the reduced maximum effective incidence changes showed in Fig. 10. This means less instabilities at higher CVC-frequencies because of reduced i_{CVC}. The effect of only actuating the blade in resonance ($f_a = 61$ Hz,

$Sr = 0.366$) without any throttling but same flow conditions is shown in Fig. 13d. The min. and max. effect to the pressure distribution is comparable to throttling without actuation at $f_c = 10$ Hz ($Sr = 0.06$).

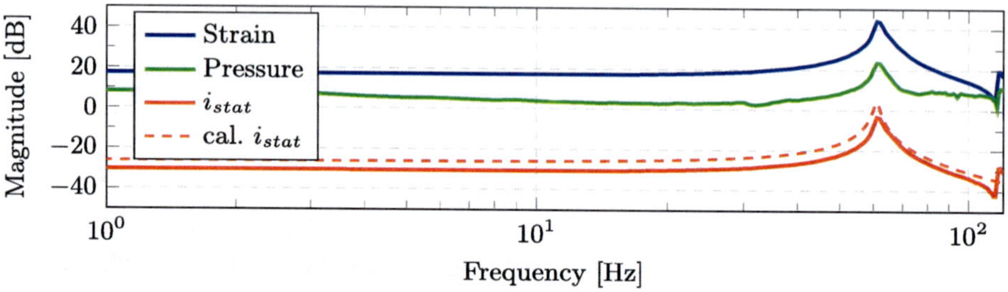

Fig. 12. Measured frequency response of static pressure at tap 15, strain gain and static incidence for the actuated blade

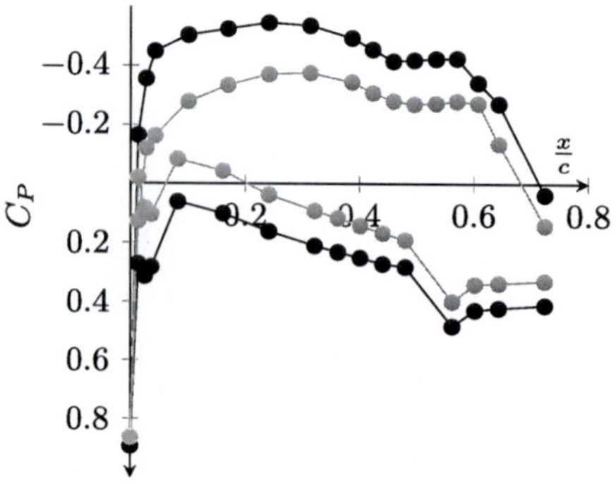

(a) Throttling with $f_c = 2$ Hz ($Sr = 0.012$)

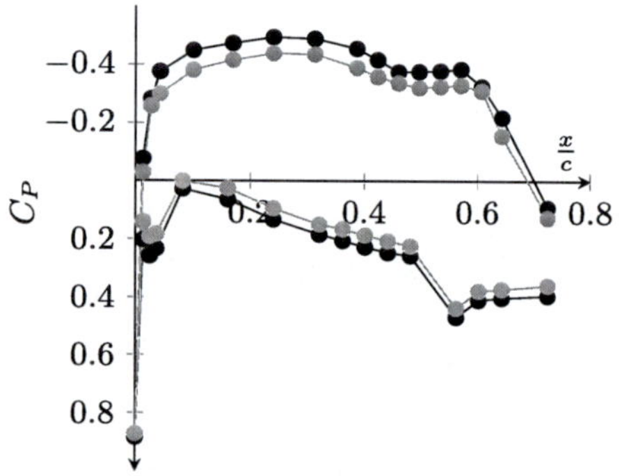

(b) Throttling with $f_c = 10$ Hz ($Sr = 0.06$)

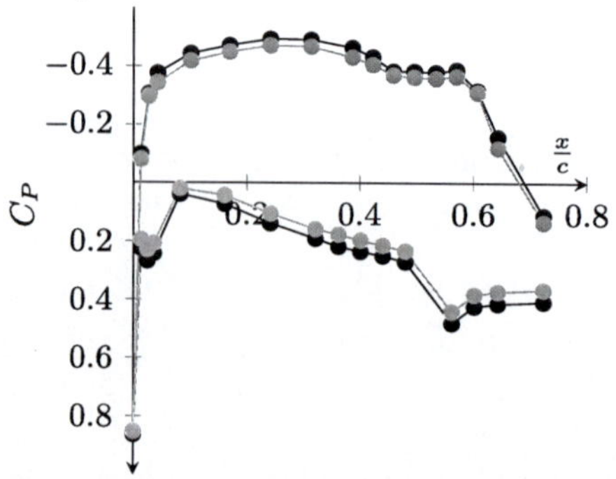

(c) Throttling with $f_c = 20$ Hz ($Sr = 0.12$)

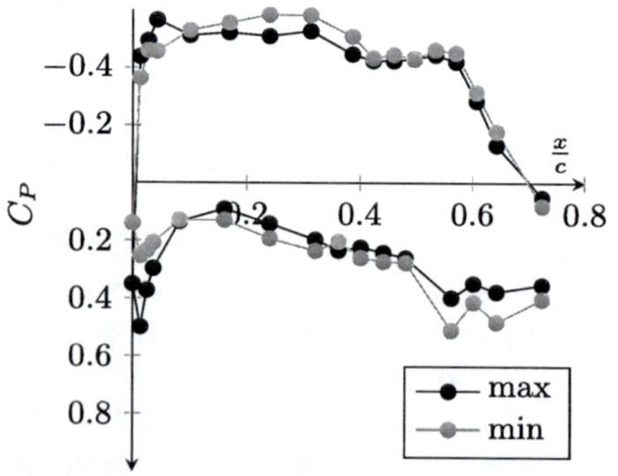

(d) Actuated with $f_c = 61$ Hz ($Sr = 0.366$)

Fig. 13. Different pressure distribution of static blade with throttling and actuated blade without any throttling at $Re = 2.5 \cdot 10^5$ and different STROUHAL numbers

6 Conclusion

In this contribution, a piezo-adaptive blade in a periodically throttled stator cascade at high STROUHAL-numbers was demonstrated. MFC actuators on the inside of a hollow blade structure were used to bend the blade's front part covering in a high frequency range. A throttling device was used. By subsequent passage blocking, it generated pressure fluctuations mapping expected disturbances arising from a CVC process. These are causing incidence variations and thus an increased risk of choking or stalling. The effects of throttling and excitation were been demonstrated separately. The calculations in Sect. 2 and numerical results with other boundary conditions in [4] prove that both together leads to a resulting incidence i_{res} of nearly zero. Furthermore, this work shows that the effect of an excited aluminum piezo adaptive blade at the natural frequency is much higher than the effect of throttling at higher frequencies, seen in Fig. 13. A blade system with resonance near the CVC frequency could minimize the effect of throttling but the system properties have to be tailored very precisely and a control system could be necessary.

Looking at the pressure signal of the throttled case it is nearly harmonic. This means that a control system has only to excite the blade harmoniously with right voltage and phase shift. So it should be possible to design an easier open loop control system as an alternative to the previous work [8].

The experimentally obtained result shows that it is possible to minimize the effects of pressure gain combustion or other flow disturbances by adaptive blade morphing. An engine that uses this type of combustion to operate is most likely more effective. In summary, the actuation of the front part of a vane can reduce fluctuations in the pressure distribution without any compressed air. Further improvements in reducing fluctuations as well as raising static pressure behind the stage could be achieved with various blade designs and control systems.

Acknowledgement. Financial support from the German Research Foundation through the collaborative research center "Substantial efficiency increase in gas turbines through direct use of coupled unsteady combustion and flow dynamics" (SFB 1029) is gratefully acknowledged.

References

1. Staats, M., Nitsche, W.: Active control of the corner separation on a highly loaded compressor cascade with periodic nonsteady boundary conditions by means of fluidic actuators. J. Turbomachinery **138**(3), 12, 031004 (2015)
2. Steinberg, S., King, R., Staats, M., Nitsche, W.: Iterative learning active flow control applied to a compressor stator cascade with periodic disturbances. J. Turbomachinery **137**(11), 111003 (2015)
3. Hammer, S., et al.: Active flow control by adaptive blade systems in periodic unsteady flow conditions. In: Active Flow and Combustion Control. Springer (2014)
4. Phan, T.D., Springer, P., Liebich, R.: Numerical investigation of an elastomer-piezo-adaptive blade for active flow control of a nonsteady flow field using fluid-structure interaction simulations. J. Turbomach. **139**(9), 04, 091004 (2017)

5. Phan, T.D.: Experimentelle und numerische Untersuchung des strukturdynamischen und strömungsmechanischen Verhaltens einer piezoadaptiven Verdichterschaufel zur Strömungsbeeinflussung. Doctoral thesis, Technische Universität Berlin (2019)
6. Krone, J.H., et al.: Experimental investigation and design of a shape-variable compressor cascade. CEAS Aeronaut. J. **8**(1), 105–127 (2017)
7. Ko, D., Guha, T.K., Amitay, M.: Control of reverse flow over a cantilevered blade using passive camber morphing. AIAA J. 1–22 (2021)
8. Werder, T., Liebich, R., Neuhäuser, K., Behnsen, C., King, R.: Active flow control utilizing an adaptive blade geometry and an extremum seeking algorithm at periodically transient boundary conditions. J. Turbomach. ***143***(2), 021008 (2021)
9. Steinert, W., Starken, H.: Off-design transition and separation behavior of a CDA cascade. J. Turbomach. **118**(2), 04, 204–210 (1996)
10. Bräunling, W.: Flugzeugtriebwerke. Springer, Heidelberg (2004). https://doi.org/10.1007/978-3-540-76370-3
11. Asahara, M., Kasahara, J., Matsuoka, K., Kawasaki, A., Matsuo, A., Funaki, I.: Pressure and visualization measurements on pulsed combustion thrustor. In: AIAA SciTech 2020 Forum, American Institute of Aeronautics and Astronautics (2020)
12. Anand, V., Gutmark, E.: Types of low frequency instabilities in rotating detonation combustors. In: King, R. (ed.) Active Flow and Combustion Control 2018, pp. 197–213. Springer, Cham (2019). https://doi.org/10.1007/978-3-319-98177-2_13
13. Kiesner, M.: Geregelte Strömungskontrolle einer Statorkaskade mit Seitenwand- und Hinterkantenaktuation. Doctoral thesis, Technische Universität Berlin, Berlin (2018)

A Comparison of Optimal, Binary Closed-Loop Active Flow Control Applied to an Annular Compressor Stator Cascade with Periodic Disturbances

Benjamin Fietzke[1(✉)], Jan Mihalyovics[2], Rudibert King[1], and Dieter Peitsch[2]

[1] Department of Process Engineering, Chair of Measurement and Control, Technische Universität Berlin, 10623 Berlin, Germany
B.Fietzke@tu-berlin.de

[2] Institute of Aeronautics and Astronautics, Chair for Aero Engines, Technische Universität Berlin, 10587 Berlin, Germany
Jan.Mihalyovics@tu-berlin.de

Abstract. Pressure gain combustion has been proposed to exploit superior thermodynamic cycles in gas turbines. However, further research on their integration is needed to reduce the induced negative effects on the last stages of a compressor. In this contribution, mitigation results on the effects of periodic disturbances on an annular compressor stator rig are presented and compared for different closed-loop controllers. Instead of a real, unsteady combustion setup, a rotating disc was installed to create periodic disturbances downstream of each passage. Pneumatic active flow control served to influence the suction side of each stator blade.

With steady blowing actuation, the effects of periodically induced disturbances could not be explicitly addressed and led to worse results compared to the closed-loop versions. For closed-loop control, a clear recommendation for a class of learning approaches can be given. Finally, an evaluation of the efficiency of flow control is presented with a refined characterization of the actuation effort.

Keywords: Closed-loop control · Active flow control · MPC · ILC · RMPC · Annular compressor stator cascade · Pulsed jets

1 Introduction

The performance of modern gas turbine compressors is limited by flow separation from the suction side of the stator and rotor blades. Therefore, in general, a compressor blade is designed to avoid boundary layer separation, limiting the amount of achievable flow turning and pressure rise from a single blade row and hence an axial compressor stage. However, the loading of such blades and the risk of separation increases significantly when pressure gain combustion (PGC) is integrated due to its unsteady effects on the upstream flow field. To mitigate these effects and still benefit from PGC, the application of active flow control (AFC) is proposed.

R. King and D. Peitsch (Eds.): AFCC 2021, NNFM 152, pp. 321–335, 2022.
https://doi.org/10.1007/978-3-030-90727-3_20

Previous studies have shown that actively controlling the secondary flow features within a passage can lead to higher pressure recoveries of a stator vane [1]. A comprehensive review of a wide variety of possible actuator concepts is presented by Cattafesta et al. [2]. Of those concepts, pneumatic actuation by air injection is particularly interesting for axial compressors, as bleed air can be used.

However, recent studies [11] of a linear stator cascade have shown an overall efficiency drop with increasing actuation amplitudes for blowing. Therefore, the actuation effort should be as low as possible while damping the induced disturbances of PGC.

In this paper, the results of an annular compressor stator cascade with AFC are discussed. An annular test rig was equipped with highly loaded stator vanes and an end-wall actuator at the hub side of each passage. In the experiments, air blowing out of rectangular actuation slots was modulated using solenoid valves, thus requiring a binary control signal. In addition to steady blowing, different closed-loop concepts were applied and compared—two variants of model predictive controllers (MPCs), a quadratic iterative learning controller (QILC), and a repetitive model predictive controller (RMPC). Both the QILC and RMPC exploit the periodic character of the induced disturbances and have already been successfully adapted with a real-valued control signal and proportional valves in a linear stator cascade [10]. For the MPCs, a simple disturbance model and an extended disturbance model are proposed here to estimate and predict the disturbance.

Within all four closed-loop algorithms, an optimal binary control signal had to be calculated with respect to a chosen cost function. This led to a binary quadratic program (BQP) and was solved with a classical branch-and-bound (B&B) [6] algorithm combined with an underlying quadratic program solver [5] to obtain the optimal solution of the BQP in real time.

2 Experimental Setup

In this section, the setup for the annular stator cascade experiments is briefly explained. All experimental investigations were conducted using a low-speed, open circuit wind tunnel at the Chair for Aero Engines. The annular design was chosen to create enhanced three-dimensional flow characteristics and thus enable investigation of the effects of unsteady disturbances at a high spatial resolution.

The annular setup consisted of a highly loaded compressor stator cascade equipped with controlled diffusion airfoils. A schematic of the cascade and the airfoil geometry is presented in Fig. 1a. The blades were designed to produce an axial outflow with a chord-based Reynolds number of $Re = 6 \times 10^5$ without any gap at the hub or tip and with a total mass flow of $\dot{m}(k) \approx 9.3\,\text{kg/s}$ at every time step k. Downstream of the cascade, a rotating throttling disc was installed (Fig. 1b) to mimic periodic disturbances that could occur using downstream PGC. During the experiments, the disc ran at a frequency of 3.7 Hz, which led to a disturbance frequency of $f_d = 7.4\,\text{Hz}$ and a resulting Strouhal number of

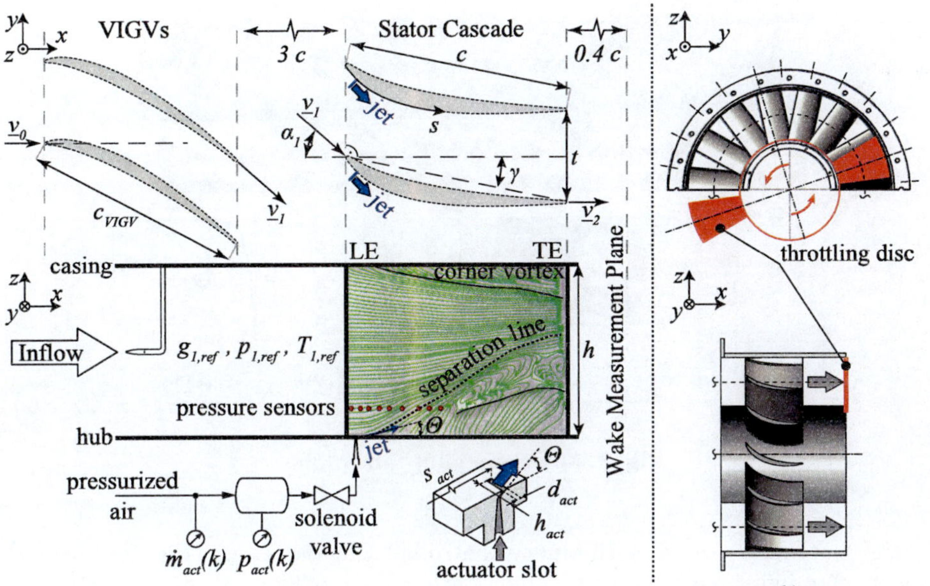

Fig. 1. (a) Actuation concept and geometry; (b) location of the throttling disc

$Sr_d \approx 0.03$. More geometric details and the parameters of the wind tunnel were presented in a previous publication [4].

2.1 Instrumentation and Data Acquisition

All measurements were conducted with instrumentation amplifiers [DEWETRON DAQP-STG] and 24-bit digitization [DEWETRON ORION-1624-200]. A control blade with ten suction-side embedded pressure transducers [KULITE XCS-062] was used for the closed-loop experiments. The corresponding transducer positions are marked in red in Fig. 1a, with a span-wise position at $z/h = 0.05$. For the closed-loop experiments, the control algorithms were implemented with Matlab/Simulink 2017a on a real-time system featuring a multi-function I/O device [NATIONAL INSTRUMENTS PCIe 6259]. The sampling rate of the real-time system was set to $n_s \cdot f_d = 111\,\text{Hz}$, with $n_s = 15$ as the number of stator blades. This choice implies that the blockage produced by the throttling disk moves one passage per sampling instant. As a result, the control signal applied to the control blade's passage could be used to actuate the next passage for the next time instant. By repeating this for every consecutive passage, the information from the control blade could be used to control the whole annular test rig.[1]

[1] This, of course, assumes that the cascade is rotational symmetric, all passages behave exactly the same, and other local disturbances do not occur.

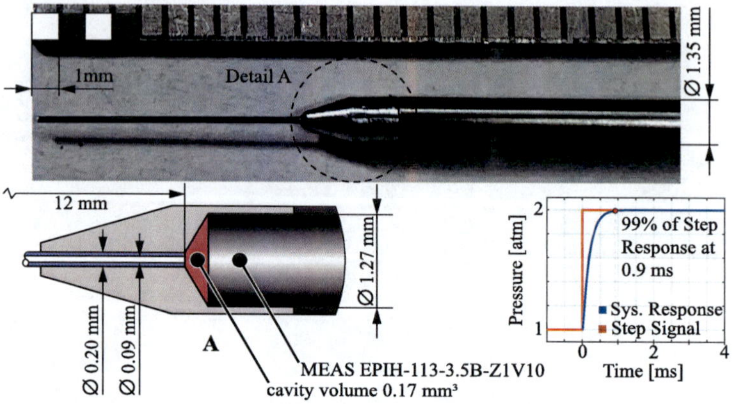

Fig. 2. Miniature pitot tube probe

A jet actuator system with binary controlled solenoid valves (Fig. 1a; bottom) was used to influence the flow via pneumatic blowing. The actuation system was located on the hub wall side and consisted of a rectangular outlet orifice, with slot height $h_{act}/c = 0.053$ and slot width $d_{act}/c = 0.001$ in relation to the blade chord length c (Fig. 1a). The outlet orifices had a blowing angle of $\Theta = 15°$ relative to the passage end wall and were oriented perpendicular to the blade's surface. Further information on the jet actuator system can be found in [4].

To assess the outflow conditions of the additively manufactured actuators, a miniature pitot tube (MPT) was used during additional actuator slot experiments. The MPT featured a sub-miniature pressure sensor [MEAS EPIH-113-3.5B-Z1V10] with an outer diameter of only 1.27 mm specifically designed for dynamic and high-frequency measurements. The MPT's slender inlet tube had an outer diameter of 0.2 mm, an inner diameter of 0.09 mm, and a length of 12.00 mm. Thus, the MPT had a blockage of less than 1% with respect to actuator slot height of 0.4 mm and length of 10.00 mm. The pressure sensor did not have a screen or any protective coatings on the diaphragm, and the cavity volume in the MPT was kept to a minimum of only 0.17 mm^3 (see Fig. 2). Using the analysis of line-cavity systems presented in [3] to model the linearized dynamic of a pneumatic measuring system, one can state that the geometric parameters of the MPT yield a bandwidth of approximate 1070 Hz, thus being sufficiently fast for the presented measurements. The MPT was traversed in a cartesian grid covering 30 points in the exit plane of the actuator slot. The grid had three points in d- and ten points in the h-direction.

The amplitude of actuation is represented here by the momentum coefficient $c_\mu(k)$. It describes the momentum of an actuated passage with actuation mass flow $\dot{m}_{jet}(k)$ and properly defined velocity $v_{jet}(k)$; (see below) in relation to the momentum of a passage flow in the cascade:

$$c_\mu(k) = \frac{\dot{m}_{jet}(k) \cdot v_{jet}(k)}{A_{psg} \cdot q_{1,ref}(k)} = \frac{\dot{m}_{jet}^2(k)}{A_{psg} \cdot q_{1,ref}(k) \cdot \rho \cdot d_{act} \cdot h_{act}}. \tag{1}$$

with:

$$\dot{m}_{jet}(k) = \begin{cases} \frac{\dot{m}_{act}(k)}{n_{act}(k)} & n_{act}(k) > 0 \\ 0 & n_{act}(k) = 0 \end{cases}. \tag{2}$$

Hereby, $n_{act}(k)$ represents the number of actuated passages, A_{psg} the cross sectional area of one passage, $q_{1,ref}$ the dynamic pressure measured upstream of the passage on the reference location, and ρ represents the density of air. During the closed-loop and open-loop wake measurements, it was not possible to measure v_{jet} and verify whether it was homogenous over the actuator slot. Therefore, a mean v_{jet} was estimated via linear interpolation, using the results of the actuation slot measurements obtained from a separate experiment (see Fig. 5). Finally, the resulting values of c_μ could be calculated.

3 Closed-Loop Control

In [4], without the refined actuation slot measurements, we already compared open-loop actuation against RMPC. Additionally, more details with respect to RMPC and the chosen surrogate control variable for the test rig were given. In the following, we repeat only some of the details and introduce other controllers used in the comparison.

3.1 Model Identification

With a surrogate control variable, a scalar was introduced that best reflected the results of the flow disturbances. For time step k, it describes the disturbance impact on the static pressure coefficients of the control blade and reads as

$$y_k = \underline{p}_1^T \cdot \underline{c}_p(k) - y_s \quad . \tag{3}$$

Here, $\underline{p}_1^T$ is the first principle component of the control blade's disturbed static pressure coefficients. The vector $\underline{c}_p(k)$ comprises measurements from ten sensor locations (see Sect. 2.1), and y_s describes a constant operating point. As a consequence, when $y_k = 0$, the variance of the disturbance's impact along the direction given by $\underline{p}_1$ is zero, which is the overall goal of the closed-loop control. Therefore, the reference for closed-loop control is set to $r_k = 0$.

In Eq. (3), $\underline{p}_1$ can be understood as a weighting vector for the influence of the control blade's static pressure sensors on y_k. Figure 3a presents a plot of $\underline{p}_1$ over the sensor positions of the control blade. Because $\underline{p}_1$ decreases over the suction side coordinate s, the last sensors have a significantly less effect on y_k than the first ones. In Fig. 3b, y_k is shown over the cycle of one disturbance period.

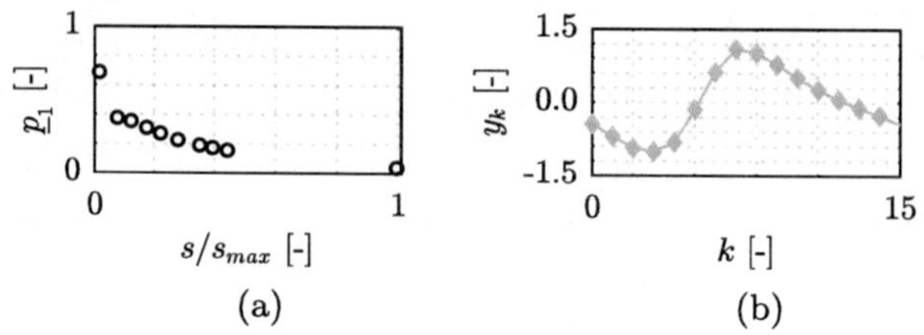

Fig. 3. First principle component of disturbance influence on $\underline{c}_p(k)$ (a) and its time-dependent projection on $\underline{p}_1$ (b).

For modeling, a system excitation was performed for different actuation amplitudes and blowing distributions with respect to the actuated passages. Pseudorandom binary signals were used as valve control signals to identify a linear, time-invariant, discrete-time single-input single-output state-space model for the actuation valves' physically possible frequency range.

All control algorithms applied in this study will be based on a prediction of the future evolution of the systems' state and control error. With the identified model, these predictions will only depend on the actual state at time k and the future undetermined evolution of the control input. The predictions can be given in a very compact form introducing appropriately built matrices $\mathbf{F}$ and $\mathbf{G}$ to describe the influence of the actual state and predicted control input on the future evolution of the control error; see [4]. In all control algorithms, an optimization problem will be solved based on optimization criteria I evaluating the outcomes of the predictions.

3.2 Controller Formulation

In the following, the formulation of all used control algorithms is shown. We start with the RMPC formulation, the derivation of which is shown in more detail in [4]. With relatively minor adjustments, the formulations of the MPC and the QILC can finally be derived by modifying the optimization criteria.

RMPC. In an RMPC, the prediction of the system's output error $e_k = r_k - y_k$ makes use of the knowledge of the periodic disturbance $\underline{\mathbf{d}}_{k-n}$ of the last cycle and assumes that this disturbance will occur again. The matrix $\mathbf{G}$ maps the supposed control input trajectory $\underline{\mathbf{u}}_{k|k}$ and $\mathbf{F}$ maps the initial state error $\underline{\varepsilon}_k$ to the predicted error trajectory $\underline{\mathbf{e}}_{k+1|k}$. In the following, a vector with the notation $(\cdot)_{k|k}$ and $(\cdot)_{k+1|k}$ indicates a prediction at current time step k for future values starting at k and $k+1$, respectively. With the special notation $(\cdot)_{k+n|k}$, the last entry of a prediction vector is addressed. $(\cdot)_{k-n}$ refers to the corresponding trajectory of the last cycle, $\Delta(\cdot)$ refers to the change in comparison to the last cycle, and $\breve{(\cdot)}$ refers to the last row of the corresponding matrix.

The cost function $I(\underline{\mathbf{u}}_{k|k})$ of the optimization problem, depending on $\underline{\mathbf{u}}_{k|k}$, is defined by the summation of the prediction's running costs and end costs. As our identified system is asymptotically stable, we can ensure stability for the closed-loop system with an adequate set of weight matrices. These are for the future error $\mathbf{W_e}$, future control input $\mathbf{W_u}$, change in control input with respect to the last cycle $\mathbf{W}_\Delta$, and the end costs $\mathbf{W}_e$. $I(\underline{\mathbf{u}}_{k|k})$ can finally be written as a quadratic function[2] with the Hessian matrix $\mathbf{H}$ and a linear term $\underline{f}_k^T$. Thus, the binary optimization problem to calculate the optimal control trajectory $\underline{\mathbf{u}}_{k|k}^*$ is

$$\underline{\mathbf{u}}_{k|k}^* = \underset{\underline{\mathbf{u}}_{k|k}}{\operatorname{argmin}} \overbrace{\{\underline{\mathbf{u}}_{k|k}^T \mathbf{H}\, \underline{\mathbf{u}}_{k|k} + \underline{f}_k^T \underline{\mathbf{u}}_{k|k}\}}^{I(\underline{\mathbf{u}}_{k|k})} \quad , \tag{4a}$$

subject to

$$\underline{\mathbf{u}}_{k|k} \in \mathbb{B}^n \quad , \tag{4b}$$

with

$$\mathbf{H} = \mathbf{G}^T \mathbf{W_e}{}^T \mathbf{G} + \mathbf{W_u} + \mathbf{W}_\Delta + \frac{1}{2} \check{\mathbf{G}}^T \mathbf{W}_e \check{\mathbf{G}} \quad , \tag{5a}$$

$$\underline{f}_k^T = -2 \left(\underline{\mathbf{s}}_{k+1|k}^T \mathbf{W_e} \mathbf{G} + \underline{\mathbf{u}}_{k-n}^T \mathbf{W}_\Delta + \frac{1}{2} t_{k+n|k}^T \mathbf{W}_e \check{\mathbf{G}} \right) \quad , \tag{5b}$$

$$\underline{\mathbf{s}}_{k+1|k} = \underline{\mathbf{e}}_{k-n} + \mathbf{F} \Delta \underline{\varepsilon}_k + \mathbf{G} \underline{\mathbf{u}}_{k-n} \quad , \quad t_{k+n|k} = e_k + \check{\mathbf{F}} \Delta \underline{\varepsilon}_k + \check{\mathbf{G}} \underline{\mathbf{u}}_{k-n} \quad . \tag{5c}$$

As solenoid valves were used for actuation, the design variables of this and the following optimization problem are binary (Eq. (4b)). The first input of the optimal control trajectory $\underline{\mathbf{u}}_{k|k}^*(1)$ is used as control input u_k for the control blade's passage valve, while at the next time step $k+1$ the optimization problem has to be solved again. This "principle of receding horizon" is part of every model predictive control used in this study.

MPC. In contrast to the RMPC, the past cycle is not taken into account in an MPC, so that the cyclic character of the disturbance induced by the throttling disc is not exploited. Instead, a prediction of the disturbance is used, see below. Moreover, all terms relating to the last cycle are dropped. Thus, the structure of the BQP is the same (Eq. (4)), but with slightly different matrices

$$\mathbf{H} = \mathbf{G}^T \mathbf{W_e}{}^T \mathbf{G} + \mathbf{W_u} + \frac{1}{2} \check{\mathbf{G}}^T \mathbf{W}_e \check{\mathbf{G}} \quad , \tag{6a}$$

$$\underline{f}^T = -2 \left(\underline{\mathbf{s}}_{k+1|k}^T \mathbf{W_e} \mathbf{G} + \frac{1}{2} t_{k+n|k}^T \mathbf{W}_e \check{\mathbf{G}} \right) \quad , \tag{6b}$$

$$\underline{\mathbf{s}}_{k+1|k} = \mathbf{F} \underline{\varepsilon}_k - \underline{\mathbf{d}}_{k+1|k} \quad , \quad t_{k+n|k} = \check{\mathbf{F}} \underline{\varepsilon}_k - d_{k+n|k} \quad . \tag{6c}$$

QILC. In contrast to the two other controllers, QILC is not based on the receding horizon principle, but the control trajectory of the upcoming cycle is

[2] The offset of the cost function is ignored here because it is not essential for the optimization program.

calculated from the control trajectory and the remaining control error of the last cycle by an optimization program [10].

$$I(\underline{\mathbf{u}}_j) = \underline{\mathbf{u}}_j^T \mathbf{H}\, \underline{\mathbf{u}}_j + \underline{f}_j^T \underline{\mathbf{u}}_j \tag{7a}$$

with

$$\mathbf{H} = \mathbf{G}^T \mathbf{W_e}^T \mathbf{G} + \mathbf{W_u} + \mathbf{W}_\Delta \quad , \tag{8a}$$

$$\underline{f}_j^T = -2\left(\underline{\mathbf{s}}_j^T \mathbf{W_e}\, \mathbf{G} + \underline{\mathbf{u}}_{j-1}^T \mathbf{W}_\Delta\right) \quad , \tag{8b}$$

$$\underline{\mathbf{s}}_j = \underline{\mathbf{e}}_{j-1} + \mathbf{F}\Delta\underline{\varepsilon}_{0,j} + \mathbf{G}\underline{\mathbf{u}}_{j-1} \quad . \tag{8c}$$

This BQP is solved only once at the start of each cycle, and the entire control trajectory is applied. In the case of QILC, the notation $(\cdot)_j$ indicates a trajectory for the current cycle and $(\cdot)_{j-1}$ indicates a trajectory for the last cycle. Additionally, within the QILC formulation there is no term regarding the end costs because when using a QILC it is usually assumed that the initial conditions are always constant, which was approximately the case during the experiments.

Finally, the setup for the learning weight $\mathbf{W}_\Delta$ for the RMPC and QILC, with equal diagonal entries of $w_\Delta \in \mathbb{R}_{>0}$ will be discussed here. As suggested in [7], w_Δ is implemented as an adaptive parameter to account for peculiarities of the binary domain. In contrast to the case of real-valued control signals, for a fixed $w_\Delta >> 0$, the binary solution for $\underline{\mathbf{u}}_{k|k}^*$ is likely to get stuck in a sub-optimal solution. Therefore, it is checked whether the control signal trajectory has changed at every time step compared to the previous cycle. If this is not the case, w_Δ is reduced by 5%, to allow for a possible higher change for the next optimal control trajectory.

Since in case of the QILC the corresponding BQP only has to be solved once per cycle, in every remaining time step[3] w_Δ can be adapted to solve the BQP of the last cycle start, with lowered w_Δ until a change in the optimal control signal trajectory occurs. The changed w_Δ then influences the trajectory in the next cycle. The weight adaptation is done until w_Δ falls below a small threshold of 0.01, where both the RMPC and QILC are likely to be converged. After that, w_Δ is set back to the initial value of $w_{\Delta,0} = 100$.

3.3 State Estimation

Since the state vector is not directly measurable, a Kalman filter is used to calculate the estimated state $\hat{\underline{x}}_k$ and, by this, the estimated state error $\hat{\underline{\varepsilon}}_k$ needed for the MPC and RMPC. To improve the quality of the estimation, we introduce disturbance models. In the case of the MPC, a disturbance model is even mandatory to estimate the disturbance d_k and predict the future disturbance trajectory $\underline{\mathbf{d}}_{k+1|k}$, Eq. (6).

Two approaches for the disturbance model were investigated. The first one assumes the disturbance as a constant offset δ_k. Therefore, within the MPC, the

[3] Meaning the $n-1$ time steps that are not used for solving the original BQP.

vector $\mathbf{d}_{k+1|k}$ is held constant at δ_k during a prediction horizon. With $\mathbf{A}$ being the matrix of the identified state-space model, $\underline{b}$ and $\underline{c}$ being the input and output vectors, and $\underline{x}_k$ being the corresponding state-space vector, the augmented state-space model, including the constant disturbance model, reads

$$\begin{pmatrix} \underline{x}_{k+1} \\ \delta_{k+1} \end{pmatrix} = \begin{bmatrix} \mathbf{A} & \mathbf{O} \\ \mathbf{O} & 1 \end{bmatrix} \begin{pmatrix} \underline{x}_k \\ \delta_k \end{pmatrix} + \begin{bmatrix} \underline{b} \\ 0 \end{bmatrix} u_k \quad , \tag{9a}$$

$$y_k = \begin{bmatrix} \underline{c}^T & 1 \end{bmatrix} \begin{pmatrix} \underline{x}_k \\ \delta_k \end{pmatrix} \quad , \tag{9b}$$

with $\mathbf{O}$ as the zero matrix/vector of the corresponding dimension.

Inspired by the similarity of the induced disturbances with a harmonic oscillation (compare Fig. 3b), an oscillating second-order system with zero damping and a state vector $\underline{z}_k \in \mathbb{R}^2$ is included in Eq. (9) to define a second dynamic disturbance model within the augmented state-space model:

$$\begin{pmatrix} \underline{x}_{k+1} \\ \underline{z}_{k+1} \\ \delta_{k+1} \end{pmatrix} = \begin{bmatrix} \mathbf{A} & \mathbf{O} & \mathbf{O} \\ \mathbf{O} & \acute{\mathbf{A}} & \mathbf{O} \\ \mathbf{O} & \mathbf{O} & 1 \end{bmatrix} \begin{pmatrix} \underline{x}_k \\ \underline{z}_k \\ \delta_k \end{pmatrix} + \begin{bmatrix} \underline{b} \\ 0 \\ 0 \\ 0 \end{bmatrix} u_k \quad , \tag{10a}$$

$$y_k = \begin{bmatrix} \underline{c}^T & 1 & 0 & 1 \end{bmatrix} \begin{pmatrix} \underline{x}_k \\ \underline{z}_k \\ \delta_k \end{pmatrix} \quad . \tag{10b}$$

Here, $\acute{\mathbf{A}}$ is chosen so that the eigenfrequency of the second-order system is equal to f_d. Within the MPC, the matrix $\acute{\mathbf{A}}$ and the estimated disturbance state vector $\underline{z}_k$ can be used to calculate a dynamic prediction $\mathbf{d}_{k+1|k}$, which is more accurate than the first approach above.

Finally, each augmented state-space model is used in a corresponding Kalman filter to calculate the augmented state vector. The equations and settings for a Kalman filter are well documented in the literature (e.g., see [8]).

4 Results

The outcome of the actuation slot measurements are presented first, followed by the results of the closed-loop experiments. Finally, a series of open-loop wake measurements are presented, comparing the impact of steady blowing and the RMPC control trajectory on important characteristics of the considered passage.

4.1 Actuation Slot Experiments

Figure 4 shows an example of the time-averaged velocity field measured at the actuator slot outlet using the MPT for the lowest ($p_{act} = 2\,\text{bar}$) and the highest set pressure in the actuator pressure tank ($p_{act} = 4\,\text{bar}$). It can be seen that the velocity field is relatively uniformly distributed, especially at a higher pressure.

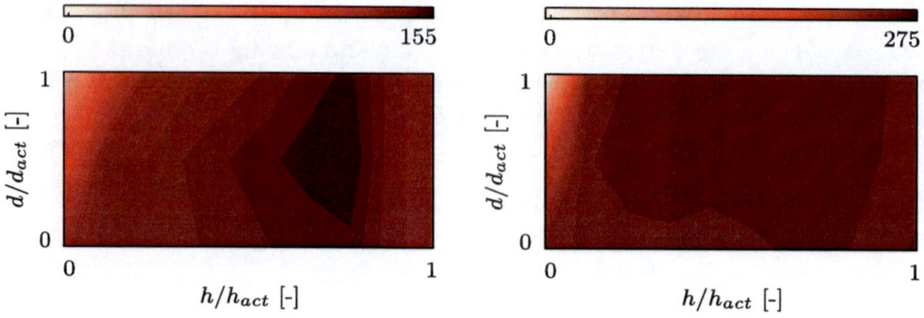

(a) $\overline{v}_{jet}(h, d)$ [m/s] (b) $\overline{v}_{jet}(h, d)$ [m/s]

Fig. 4. Time-averaged velocity field in the actuator slot for (a) $p_{act} = 2\,$bar and (b) $p_{act} = 4\,$bar.

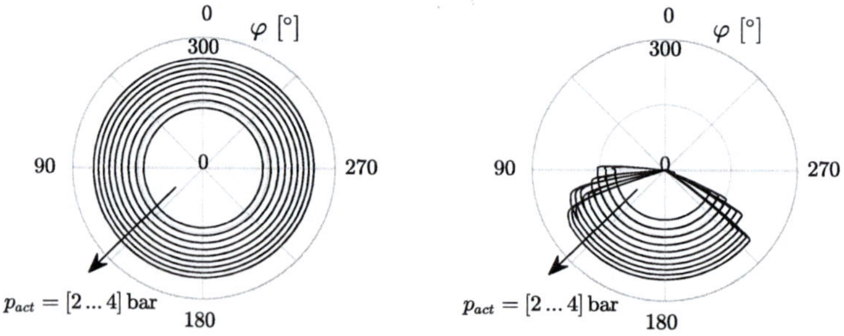

(a) $\overline{v}_{jet}(\varphi)$ [m/s] (b) $\overline{v}_{jet}(\varphi)$ [m/s]

Fig. 5. Area-averaged actuator slot velocity over a mean period φ over $p_{act} = [2...4]\,$bar for (a) steady actuation and (b) applied RMPC trajectories.

The velocity drops to nearly zero only at a measurement point in the upper left corner, as this actuator slot was blocked due to manufacturing tolerances decreasing the effective outlet area.

The area-averaged jet velocity for a mean period[4] $\varphi = [0...360]°$ for a range of different p_{act} is displayed in Fig. 5. Figure 5a shows the results for steady blowing, and Fig. 5b shows the results for several RMPC trajectories and pressure levels. The outlet velocities can be estimated using linear interpolation, and the corresponding c_μ values can be calculated retrospectively for the open-loop measurements previously presented in [4].

[4] Note: Because the disturbance induced by the throttling disc is periodic, a period angle φ can be defined.

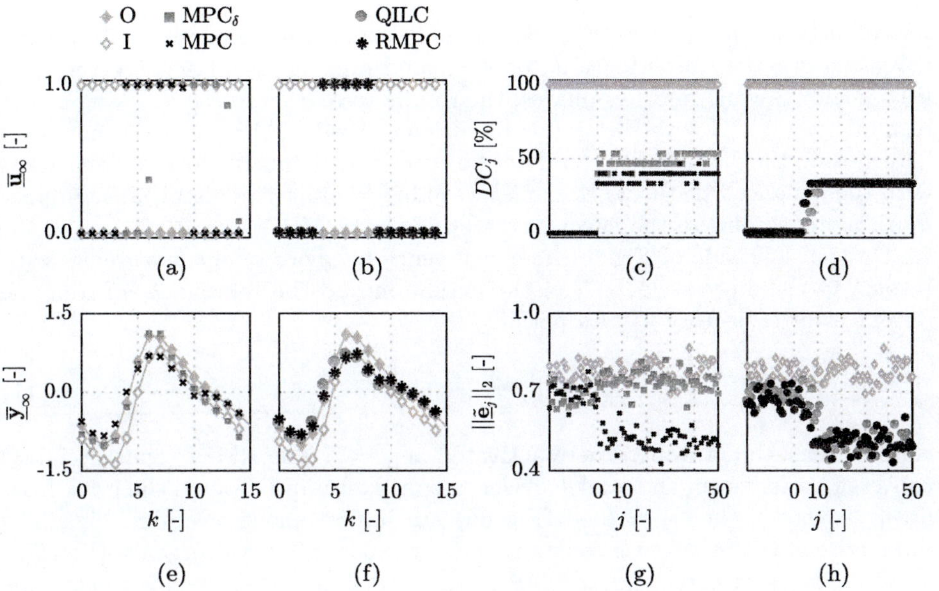

Fig. 6. Comparison of different closed-loop actuations. On the left, the two MPCs (a, e) and learning controllers (b, f) are shown and compared with the non-actuated case (O) and the permanently actuated case (I) over an averaged cycle. On the right, the duty cycle and normalized control error norm are shown over several cycles for the two MPCs (c, g) and the learning controllers (d, h).

4.2 Closed-Loop Experiments

Figure 6 shows the results of the closed-loop experiments with different controller configurations compared with the base case (O - *grey diamonds*) and the case with steady actuation (I - *grey diamonds*). The *squares* show the data for the MPC with the Kalman filter using the simple disturbance model (Eq. (9)). The *crosses*, *circles*, and *stars* show the data for MPC, QILC, RMPC with the Kalman filter using the extended disturbance model (Eq. (10)). On the left side, the averaged, converged control trajectories $\overline{\mathbf{u}}_\infty$ and output trajectories $\underline{\overline{\mathbf{y}}}_\infty$ over the time steps k of a disturbance cycle j are presented. The notation $(\cdot)_\infty$ indicates that the considered number of cycles before averaging was high enough so that the learning controllers could converge. In terms of the two MPCs, the control trajectories were still slightly fluctuating because, in contrast to the learning controllers, the change of control trajectory is not penalized. This can be observed from the non-binary[5] values of $\overline{\mathbf{u}}_\infty$ of the MPCs (Fig. 6a), which is not the case for QILC and RMPC that reached a converged state (Fig. 6b).

As discussed above, $\underline{\overline{\mathbf{y}}}_\infty = \underline{0}$ means that the influence of the throttling disc on the first principle component of the pressure readings on the suction side disappears. Without control (O), $\underline{\overline{\mathbf{y}}}_\infty$ clearly deviates from zero, which indicates

[5] This results when fluctuating binary trajectories are averaged over several cycles.

the disturbance's influence (Figs. 6e and 6f). For open-loop control with steady actuation (I), the absolute value of $\underline{\overline{\mathbf{y}}}_{\infty}$ could only be decreased for very few time instants, while in other phases, the results were even worse. Looking at the results of the MPC with the simple disturbance model (MPC$_\delta$), the same holds here as well: the disturbance could not be effectively damped. However, the MPC with the enhanced disturbance model was able to lower the disturbance impact much more, similar to the effect of the QILC and RMPC (Fig. 6f).

On the right side of Fig. 6, the convergence behavior of the controllers with respect to the duty cycle DC_j of the actuation and the relative 2-norm of the output error trajectory $\underline{\mathbf{e}}_j$ of a period j

$$||\underline{\tilde{\mathbf{e}}}_j||_2 = ||\underline{\mathbf{e}}_j||_2 \Big/ \sqrt{n} \tag{11}$$

can be seen. All controllers were activated at $j = 0$. The RMPC and the QILC converged after seven and twelve cycles, respectively, and lowered the error from $||\underline{\tilde{\mathbf{e}}}_j||_2 \approx 0.68$ to $||\underline{\tilde{\mathbf{e}}}_j||_2 \approx 0.50$ (Fig. 6h), an improvement of about 25%. The duty cycle of the valves was constant at $DC_j = 33\%$ after convergence (Fig. 6d). With steady actuation and the MPC with the simple disturbance model, the error even increased (Fig. 6g), while DC_j, and thus the actuation effort, was higher (Fig. 6c). The MPC with enhanced disturbance model lowered the error similar to the learning controllers but was faster due to the missing costs of control trajectory change. In return, DC_j was on a slightly higher level and fluctuated more.

The closed-loop experiment with the RMPC, presented in Fig. 6, was conducted five times for every considered actuation amplitude. The converged RMPC trajectories were then averaged and used for the following wake measurements.

Open-Loop Wake Experiments

In this section, the effect of the RMPC trajectory on the wake of a passage and particularly three important characteristics of the passage will be shown. The c_μ-values were calculated with the data of Fig. 5 and are more accurate than the c_μ-values in our previous publication [4].

Keeping in mind the overall purpose of a stator vane—the conversion of dynamic pressure into static pressure—the static pressure rise coefficient $C_p(z, y, k)$ of a passage is an important parameter. It is defined as the difference between the local static pressure downstream $p_2(z, y, k)$ and the area-averaged static pressure upstream $\overline{p}_1(k)$ of the passage relative to the mass-averaged upstream dynamic pressure $\overline{q}_1(k)$

$$C_p(z, y, k) = \frac{p_2(z, y, k) - \overline{p}_1(k)}{\overline{q}_1(k)}. \tag{12}$$

Note that as suggested by Cumpsty and Horlock [9], area-averaging is used for static pressure and mass averaging is used for total and dynamic pressure.

In comparison to $C_p(z,y,k)$, the total pressure loss coefficient $\zeta^*(z,y,k)$ takes the difference of the total pressure into account

$$\zeta^*(z,y,k) = \frac{\overline{g}_1^*(k) - g_2(z,y,k)}{\overline{q}_1(k)}, \tag{13}$$

with the corrected inlet total pressure of the passage

$$\overline{g}_1^*(k) = \frac{\dot{m}_{psg}(k) \cdot \overline{g}_1(k) + \dot{m}_{act,psg}(k) \cdot p_{act}(k)}{\dot{m}_{psg}(k) + \dot{m}_{act,psg}(k)}, \tag{14}$$

including the contribution of the actuation mass flow as well. Here, $\dot{m}_{act,psg}(k)$ is the measured passage's actuation mass flow and $p_{act}(k)$ is the total pressure of the actuation, defined as the pressure in the actuation stagnation tank.

Since our definition of the actuation amplitude $c_\mu(k)$ (Eq. (1)) does not allow for an evaluation of the overall mass flow effort of a specific passage for pulsed actuation, the last characteristic describes the actuation mass flow of a specific passage $\dot{m}_{act,psg}(k)$ in relation to the passage's inlet mass flow $\dot{m}_{psg}(k) = \dot{m}(k)/n_s$

$$\mu(k) = \frac{\dot{m}_{act,psg}(k)}{\dot{m}_{psg}(k)} \quad . \tag{15}$$

Here, $\dot{m}_{act,psg}(k)$ is approximated with the general jet mass flow $\dot{m}_{jet}(k)$ weighted with the control input $u_{k,psg}$ for the valve of the specific passage

$$\dot{m}_{act,psg}(k) = \dot{m}_{jet}(k) \cdot u_{k,psg} \quad . \tag{16}$$

To compare the effects of different actuation amplitudes c_μ on the passage characteristics, $C_p(z,y,k)$ was area-averaged and $\zeta^*(z,y,k)$ was mass-averaged. Both characteristics were also phase-averaged and divided by the corresponding characteristic value for the case without actuation. Whereas $\mu(k)$ was phase-averaged. The outcome is shown in Fig. 7 over different c_μ values. It can be seen that with increasing c_μ, the positive effect on the static pressure rise of

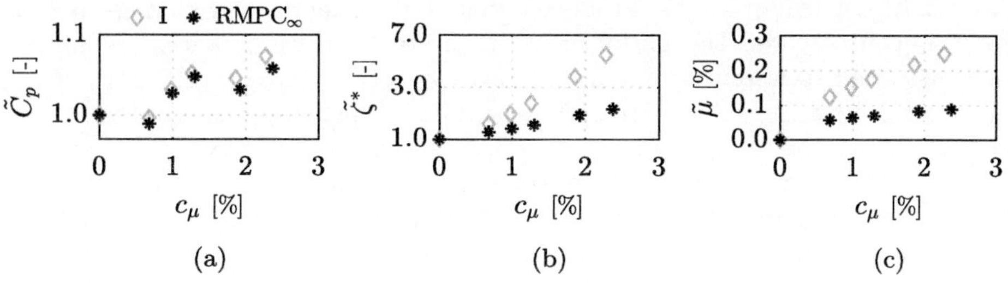

Fig. 7. Normalized mean static pressure rise coefficient (**a**), corrected total pressure loss coefficient (**b**), and actuation mass flow consumption (**c**) for base flow ($c_\mu = 0\%$), steady blowing, and RMPC actuation for different actuation amplitudes.

RMPC-enabled actuation was just moderately lower compared to steady actuation. However, for the latter, the consumption of actuation mass flow was up to three times higher, while the total pressure loss coefficient was up to two times higher than with RMPC actuation. Consequently, using RMPC-enabled AFC, the efficiency drop was significantly lower while producing a comparable rise in static pressure.

5 Conclusion

An annular low-speed compressor stator rig with hub-sided pneumatic AFC was investigated. Downstream of the stator vanes, a rotating throttling disc was installed to mimic the effect of periodic disturbances as they could occur with the usage and integration of pressure gain combustion concepts. Previous investigations of a linear compressor stator rig [11] indicated that although the flow field in a stator row can be manipulated effectively, the overall efficiency of the cascade will decrease with higher actuation amplitudes. Therefore, different advanced closed-loop concepts were applied to test their performance in reducing the impact of the periodic disturbances and decreasing the effort of actuation mass flow compared to steady actuation. To that end, a scalar surrogate control variable was specified for a control blade equipped with high-bandwidth pressure sensors, and an input-output model was identified to predict the influence of the binary actuation on the defined control variable. In addition, two different disturbance models were included in the control algorithms for state estimation and disturbance prediction.

It could be shown that an MPC with the simple disturbance model could not lessen the disturbance impact. In contrast, an MPC with the extended disturbance model could damp the disturbances effectively. The two applied learning controllers, a QILC and an RMPC, performed even better due to the ability to learn from past disturbance cycles. In our case, this resulted in a smooth, converging control trajectory and slightly less actuation effort. Comparing the QILC and the RMPC, the latter converged slightly faster, which is based on the receding horizon principle applied within the RMPC algorithm. With the exception of the MPC with the simple disturbance model, all closed-loop approaches lowered the relative 2-norm of the disturbance impact on the control variable up to 25%. For this control task, the performance with steady actuation was insufficient and not suited to lessening the disturbance's effects. With several RMPC runs, an optimal input trajectory for each actuation amplitude c_μ was obtained and thereafter used for further open-loop wake experiments to obtain more detailed insight into the influence of the considered passage's wake. For comparison, a steady blowing approach was used. The c_μ-values could be estimated much more accurately compared to a previous study due to the data of the conducted actuator slot experiments with an MPT. With the RMPC actuation, a similar static pressure rise of the considered passage was achieved, while the actuation mass flow effort was up to 66% lower depending on the actuation amplitude. This is particularly advantageous because it could be confirmed for the annular test rig that the efficiency of a passage decreases with increasing actuation amplitude. In summary, AFC can improve an operating condition, but should only be used selectively due to actuation costs. For this purpose, in a periodic operation, an optimal closed-loop controller exploiting the periodicity, such as an RMPC, can be a promising option.

Acknowledgements. The authors gratefully acknowledge the support of the Deutsche Forschungsgemeinschaft (DFG) as part of collaborative research center SFB 1029 "Substantial efficiency increase in gas turbines through direct use of coupled unsteady combustion and flow dynamics" on projects B01 and B06.

References

1. Zander, V., Nitsche, W.: Control of secondary flow structures on a highly loaded compressor cascade. Proc. IMechE Part A J. Power Energy **227**(6), 674 (2013)
2. Cattafesta III, L.N., Sheplak, M.: Actuators for active flow control. Ann. Rev. Fluid Mech. **43**(1), 247 (2011)
3. Anthoine, J., et al.: Measurements Techniques in Fluid Dynamics - An Introduction, 3rd revised edn, Von Karman Institute for Fluid Dynamics, Rhode-Saint-Genese (2009)
4. Fietzke, B., King, R., Mihalyovics J., Peitsch, D.: Binary repetitive model predictive active flow control applied to an annular compressor stator cascade with periodic disturbances. In: Proceedings of ASME Turbo Expo 2021: Turbomachinery Technical Conference and Exposition, GT2021-58744, vol. 4, p. V004T05A003 (2021)
5. Schmid, C., Biegler, L.T.: Quadratic programming methods for reduced Hessian SQP. Comput. Chem. Eng. **18**(9), 817 (1994)
6. Morrison, D.R., Jacobson, S.H., Sauppe, J.J., Sewell, E.C.: Branch-and-bound algorithms: a survey of recent advances in searching, branching, and pruning. Discrete Optim. **19**, 79 (2016)
7. Schäpel, J.S., King, R., Yücel, F., Völzke, F., Paschereit, C.O., Klein, R.: Fuel injection control for a valve array in a shockless explosion combustor. In: Proceedings of the ASME Turbo Expo 2018: Turbomachinery Technical Conference and Exposition, GT2018-75295, vol. 6, p. V006T05A007 (2018)
8. Bar-Shalom, Y., Li, X.-R., Kirubarajan, T.: Estimation with Applications to Tracking and Navigation: Theory Algorithms and Software. John Wiley & Sons, Hoboken (2004)
9. Cumpsty, N.A., Horlock, J.H.: Averaging nonuniform flow for a purpose. J. Turbomach. **128**(1), 120 (2006)
10. Steinberg, S.J., King, R.: Closed-loop active flow control of repetitive disturbances in a linear stator cascade. In: 2018 Flow Control Conference AIAA 2018-3689, p. 3689 (2018)
11. Steinberg, S.J., King, R.: Efficiency of active flow control in an unsteady stator vane flow field. In: Radespiel, R., Semaan, R. (eds.) Fundamentals of High Lift for Future Civil Aircraft. NNFMMD, vol. 145, pp. 631–648. Springer, Cham (2021). https://doi.org/10.1007/978-3-030-52429-6_38

Numerical Methodologies for Magnetohydrodynamic Flow Control for Hypersonic Vehicles

Heather Muir(✉), Louisa Michael, and N. Nikiforakis

Department of Physics, Cavendish Laboratory, University of Cambridge, J J Thomson Avenue, Cambridge CB3 0HE, UK
ham38@cam.ac.uk

Abstract. The concept of magnetohydrodynamic (MHD) flow control is of current interest for its applications in spacecraft reentry and aerodynamic control for hypersonic vehicles. This work presents an efficient approach for realistically simulating MHD effects in weakly ionised plasmas produced by hypersonic flows. The governing equations consists of the full Navier-Stokes resistive-MHD system under the low magnetic Reynold's number assumption. The numerical approach employs a Cartesian mesh which facilitates hierarchical adaptive mesh refinement in a highly parallelised framework. Geometry is implemented via a rigid-body Ghost Fluid Method which permits arbitrarily complex embedded boundaries. An advanced 19-species equilibrium air-plasma equation of state (plasma19) has been adopted and extended in this work, for the study of test cases where the assumption of local thermodynamic equilibrium applies. The numerical methodology paired with plasma19 equation of state is shown to effectively capture boundary-layer shock wave interactions in a complex double-cone flow, with imposed magnetic field. The model predicts MHD flow control (augmented shock position) in line with experimental measurements, improving upon previous model predictions.

Keywords: Computational physics · Hypersonics · Magnetohydrodynamics · Flow control

1 Introduction

During hypersonic flight, the conversion of kinetic to thermal energy across the leading shock wave results in severe compression and heating of the gas surrounding the vehicle. Temperatures become so extreme that dissociation and ionisation reactions produce a weakly ionised plasma. This plasma layer has interesting properties which present both challenges and opportunities.

Given that the plasma layer is characterised by heightened electrical conductivity, this property can be innovatively leveraged. An imposed magnetic field generated from within the vehicle is able to exert a Lorentz force on the electrically conductive gas. The Lorentz force acts to increase shock stand-off distance,

R. King and D. Peitsch (Eds.): AFCC 2021, NNFM 152, pp. 336–355, 2022.
https://doi.org/10.1007/978-3-030-90727-3_21

in turn increasing total drag, reducing heat flux to the vehicle and reducing the density of free electrons. This collective mechanism of desirable effects is termed *MHD flow control.*

MHD flow control has a growing number of established and prospective applications. These include active thermal protection systems and the possibility of mitigating radio-black during spacecraft reentry, as well as aerodynamic control including hypersonic inlet control for ramjets and scramjets [1].

Numerous studies have demonstrated the effect of the gas chemistry model on the numerical simulation of weakly ionised plasmas formed from hypersonic flows [2–4]. The choice of gas chemistry model depends both on the ease of computation and on the validity range of its thermochemical assumptions. Additionally, for MHD applications, the computation of electric properties (such as electrical conductivity, and electron number density) are directly linked to the gas chemistry model and of especially high importance.

Studies on MHD control which adopt an ideal gas with supplementary finite conduction model are useful for investigating the flow effects of interest i.e. *relative* influence of Hall effect or magnetic field strength on the flow field, however, the *outright* results are unrealistic for the post-shock temperatures produced in hypersonic flows. Therefore, for MHD control feasibility to be realistically studied, a real gas model is required.

Real gas effects are captured by thermochemistry models which consider the dissociation and ionisation reactions occurring in high temperature air. In the context of reentry vehicles, with consideration of vehicle geometry and altitude-based free stream conditions, a map of valid thermochemical assumptions has been compiled as Fig. 1 from various data sources.

It is recognised that for a large portion of reentry trajectories the gas chemistry is in - or very close to - local thermochemical equilibrium (LTE) [5]. Figure 1 shows that the more ballistic the reentry trajectory the greater the portion of flight occurs under equilibrium conditions. A number of other MHD flow control applications also fall within the LTE valid regime. The major advantage of an equilibrium chemistry model is the use of pre-computed thermodynamic state data to efficiently obtain properties, and whereby the fluid dynamic time step need not be restricted due to the chemistry. A study by Chalot et al. [3] concluded that with optimum implementation, equilibrium chemistry programs may achieve a computing time close to only 20% slower than ideal gas.

As highlighted, the accurate computation of electrical conductivity underpins the realistic study of MHD effects. Bisek [9] also highlights that accurate conductivity modeling in the LTE regime typically involves supplementary models with semi-empirical approximations and remains problematic. As such, MHD effects have scarcely been studied in the equilibrium regime.

D'Angola *et al.* have made significant advances in computing the transport and thermodynamic properties of air plasmas in LTE [10,11]. This theory has formed the basis for the development of a new 19-species equilibrium air-plasma model (plasma19) by Treüble [12] which has been adopted and extended in this work.

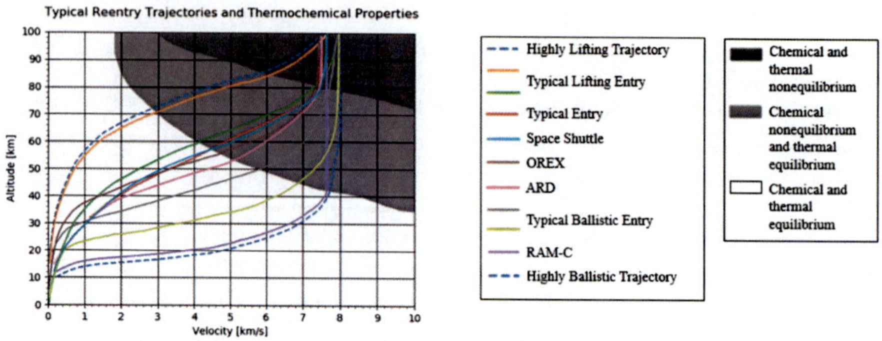

Fig. 1. Plot demonstrating regions of valid thermochemical assumptions, with overlaid reentry trajectories given by velocity-altitude coordinates. The zones of equilibrium states are as classified by Fujino *et al.* [6] for a spherical reentry vehicle of characteristic length O(1). The set of reentry trajectories is adapted from reference trajectory data from [7,8].

This work therefore proposes a computationally efficient approach for the realistic simulation of plasma generation and MHD effects in weakly ionised air-plasma within the equilibrium validity range. A full Navier-Stokes resistive-MHD system is modelled with real plasma gas effects, where electromagnetic properties including electron number density and electrical conductivity are computed directly, without need for supplementary approximation models. The numerical approach of this work combines methodologies which are able to realistically and efficiently capture complex flow interactions produced by non-simple geometries with viscous effects, under the influence of magnetic fields. To demonstrate the efficacy of this approach, this paper presents key validation cases for MHD Lorentz forcing dynamics, and for hypersonic double cone experiments. Model predictions are then compared to experimental results for a hypersonic double cone test case with imposed magnetic effects.

2 Governing Equations

2.1 Resistive MHD System Equations

The magnetohydrodynamic behaviour in this regime of flight is complex, with many interdependent phenomena across disparate scales. For numerical solution to be amenable, assumptions and simplifications are necessary.

In the case of aerospace plasmas - where collision frequency within the plasma is sufficiently high, flow length scales are significantly larger than the Debye length, and time scales are larger than the reciprocal of the plasma resonant frequency - a continuum magnetohydrodynamic (MHD) model governs [13].

Broadly across the literature, hypersonic plasmas are considered in the low magnetic Reynolds number regime [1,6,14] where Re_m is of the order of unity or less. Under this regime the imposed and induced magnetic fields are sufficiently decoupled such that the static imposed magnetic field is unaffected by the flow.

$$Re_m = \frac{\text{Magnetic induction}}{\text{Magnetic diffusion}} = \frac{\sigma_0 V_0 L_0}{\epsilon_0 c^2} \tag{1}$$

The governing system equations are derived by combining the Navier-Stokes equations for fluid dynamics (accounting for fluid viscosity and thermal conductivity) with Maxwell's equations for electromagnetism. The derivation under the low Re_m weakly ionised plasma regime results in resistive MHD equations which account for Lorentz forcing and joule heating.

For studies of axisymmetric vehicle geometries, the axis of symmetry can be exploited, converting the 3D problem into a 2D plane in cylindrical coordinates. And therefore the MHD system equations are given in r-z space as:

$$\frac{\partial \mathbf{U}}{\partial t} + \frac{\partial \mathbf{F}}{\partial r} + \frac{\partial \mathbf{G}}{\partial z} = \frac{\partial \mathbf{F}_v}{\partial r} + \frac{\partial \mathbf{G}_v}{\partial z} + \mathbf{S}^c + \mathbf{S}_v^c + \mathbf{S}_{MHD} \tag{2}$$

U, **F** and **G** denote the state vector, and r-z directional inviscid flux vectors respectively, and $\mathbf{S}^c$ denotes the associated cylindrical source term vector. For the 2D r-z system, we denote the velocity vector in the $(\hat{r}, \hat{z}, \hat{\theta})$ direction to have components (u, v, w), and so the system vectors are defined as:

$$\frac{\partial \mathbf{U}}{\partial t} = \begin{pmatrix} \rho \\ \rho u \\ \rho w \\ \rho v \\ E \end{pmatrix}_t \quad \frac{\partial \mathbf{F}}{\partial r} = \begin{pmatrix} \rho u \\ \rho u^2 + p \\ \rho u w \\ \rho u v \\ u_r(E+p) \end{pmatrix}_r \quad \frac{\partial \mathbf{G}}{\partial z} = \begin{pmatrix} \rho w \\ \rho u v \\ \rho v w \\ \rho v^2 + p \\ v(E+p) \end{pmatrix}_z \quad \mathbf{S}^c = \begin{pmatrix} -\frac{1}{r}\rho u \\ -\frac{1}{r}\rho u^2 \\ -\frac{1}{r}\rho u w \\ -\frac{1}{r}\rho u v \\ -\frac{1}{r}u(E+p) \end{pmatrix} \tag{3}$$

where total energy E is given by: $E = \rho e + \frac{1}{2}\rho \mathbf{u}^2$.

Assuming a Newtonian fluid, the viscous stress tensor is given by:

$$\boldsymbol{\tau} = \mu(\nabla \mathbf{u} + \nabla \mathbf{u}^T) + \lambda(\nabla \cdot \mathbf{u})I \tag{4}$$

where μ is the fluid dynamic viscosity and using stokes hypothesis $\lambda = -\frac{2}{3}\mu$. ζ is the fluid's thermal conductivity.

The flow is assumed to be laminar, and dynamic viscosity is computed via Sutherland's law. Thermal conductivity is computed with constant Prandtl number = 0.71 for the ideal gas case, and ζ has precomputed state data in the plasma19 gas model.

then the viscous flux vectors are derived as follows:

$$\frac{\partial \mathbf{F}_v}{\partial r} = \begin{pmatrix} 0 \\ \tau_{rr} \\ \tau_{rz} \\ \tau_{r\theta} \\ Q_r \end{pmatrix}_r = \begin{pmatrix} 0 \\ \mu(\frac{4}{3}u_r - \frac{2}{3}v_z - \frac{2}{3}\frac{u}{r}) \\ \mu(v_r + u_z) \\ \mu(w_r - \frac{w}{r}) \\ u\tau_{rr} + v\tau_{rz} + w\tau_{r\theta} + \zeta T_r \end{pmatrix}_r \tag{5}$$

$$\frac{\partial \mathbf{G}_v}{\partial z} = \begin{pmatrix} 0 \\ \tau_{zr} \\ \tau_{zz} \\ \tau_{z\theta} \\ Q_z \end{pmatrix}_z = \begin{pmatrix} 0 \\ \mu(u_z + v_r) \\ \mu(\frac{4}{3}v_z - \frac{2}{3}u_r - \frac{2}{3}\frac{u}{r}) \\ \mu w_z \\ u\tau_{zr} + v\tau_{zz} + w\tau_{z\theta} + \zeta T_z \end{pmatrix}_z \tag{6}$$

and the following viscous source term vector emerges from the cylindrical system conversion:

$$\mathbf{S}_v^c = \begin{pmatrix} 0 \\ \tau_{rr} - \tau_{\theta\theta} \\ \tau_{rz} \\ \tau_{r\theta} \\ Q_r \end{pmatrix} = \begin{pmatrix} 0 \\ \frac{2}{r}\mu(u_r - \frac{u}{r}) \\ \frac{1}{r}\mu(u_z + v_r) \\ \frac{2}{r}\mu(w_r - \frac{w}{r}) \\ \frac{1}{r}(u\tau_{rr} + v\tau_{rz} + w\tau_{r\theta} + \zeta T_r) \end{pmatrix} \tag{7}$$

Some additional assumptions and resultant simplifications can be made in defining the $\mathbf{S}_{MHD}$ vector.

The full expansion of Ohm's law includes a Hall current contribution. This additional current drives the generation of an electric field which has a complex interaction with the gas dynamics - termed *Hall effect.* Several authors have shown that for a conductive body, the Hall effect diminishes the MHD flow control [15]. In the case of a perfectly insulated body, however, the overall flow structure is essentially unaffected by the Hall effect [16,17]. Regarding the leading surfaces as an insulated material (via the electrodynamic boundary condition at the body), studies of MHD flow control justifiably neglect Hall effect [18–20], therefore applying the *generalised* Ohm's law as given by Eq. 8:

$$\mathbf{J} = \sigma(\mathbf{E} + \mathbf{u} \times \mathbf{B}) \tag{8}$$

The electric field $\mathbf{E}$ is a conservative field where $\mathbf{E} = -\nabla\phi$ and ϕ is the electric potential. Therefore the governing system is comprised of system of 5 hyperbolic PDE's as defined within Eq. 2, coupled with an additional Poisson-type elliptic PDE maintaining divergence-free current density:

$$\nabla \cdot [\sigma(-\nabla\phi + \mathbf{u} \times \mathbf{B})] = 0 \tag{9}$$

The magnetic field is a static imposed dipole field with dimensional components $\mathbf{B} = (B_r, B_\theta, B_z)$, and defined analytically via:

$$\mathbf{B} = B_0 \left[\frac{3\mathbf{r}(\mathbf{r}\cdot\mathbf{m}) - r^2\mathbf{m}}{r^5}\right] \tag{10}$$

Where $\mathbf{r}$ is the radial vector from the dipole centre, and $\mathbf{m}$ is an orientation vector aligned parallel to the dipole centreline. Within the axisymmetric governing system where $\mathbf{r}$ and $\mathbf{m}$ are defined in the r-z plane, the B_θ magnetic field component is zero.

For electrodynamic boundary conditions which treat the surface as insulated, the solution of equation of 9 can be solved analytically as $\phi = 0$, which means the elliptic current density equation does not require numerical solution and electric current density is given by:

$$\mathbf{J} = -\sigma(u_r B_z - u_z B_R)\hat{\theta} = J_\theta \tag{11}$$

Therefore the Lorentz forcing terms in the MHD source term vector can be defined directly, rendering the computation as vastly more expedient:

$$\mathbf{S}_{MHD} = \begin{pmatrix} 0 \\ (\mathbf{J}\times\mathbf{B})\hat{r} \\ (\mathbf{J}\times\mathbf{B})\hat{\theta} \\ (\mathbf{J}\times\mathbf{B})\hat{z} \\ \mathbf{u}\cdot(\mathbf{J}\times\mathbf{B}) + \sigma^{-1}J^2 \end{pmatrix} = \begin{pmatrix} 0 \\ -\sigma(u_r B_z^2 - u_z B_r B_z) \\ 0 \\ \sigma(u_r B_z B_r - u_z B_r^2) \\ u_r J_\theta Bz - u_z J_\theta B_r + \sigma^{-1}J^2 \end{pmatrix}$$

2.2 Plasma19 Equation of State

This work employs an advanced equation of state (EoS) for an air plasma which considers 19-species reaching chemical and thermal equilibrium. The new *plasma19* EoS leverages a precomputed database recently developed by Träuble in 2021 [12] (utilising theory from D'Angola [10] and Colonna [21]). This database has been extended in this work over a wider density-pressure range necessary for atmospheric reentry conditions, and with an extended implementation for electron number density in line with the extended theory of D'Angola [11].

The 19-species considered in the air-plasma are:

$$\begin{gathered} N_2,\ N_2^+,\ N,\ N^+,\ N^{++},\ N^{+++},\ N^{++++}, \\ O_2,\ O_2^+,\ O_2^-,\ O,\ O^-,\ O^+,\ O^{++},\ O^{+++},\ O^{++++}, \\ NO,\ NO^+,\ e^- \end{gathered}$$

The molar fractions of each chemical species, and bulk thermodynamic properties under equilibrium are precomputed in this database through a Gibbs free energy minimisation method (solved numerically through for e.g. Newton-Raphson iterative scheme). The full set of state properties computed by Plasma19 have been extensively validated against empirical measurements [12].

In the state database previously generated the electron species molar fraction was found to become unreliable below approximately 10,000 K. This was insufficient for the temperature band of interest for hypersonically generated plasmas (approximately 2,000 K–10,000 K). In recognising this issue, D'Angola's proposed a theoretical formulation for electron molar fraction in the extended work [11], which is verified to compute "very accurate" electron molar fraction in air plasmas down to temperatures as low as 50 K. Therefore this theoretical formulation was implemented in this work to generate an extended plasma19 state database, achieving direct and accurate resolution of electron number density over the full property range of interest for equilibrium aerospace plasmas (verified against the results of D'Angola [11]).

For MHD studies, it is critical for the EoS to accurately compute electrical conductivity based on species dissociations and ionisations at the relatively low air-plasma temperature resulting from this regime. The negatively charged O^- and O_2^- species are not captured by an 11-species model but are computed to be present in small concentrations in the air-plasma temperature band of interest.

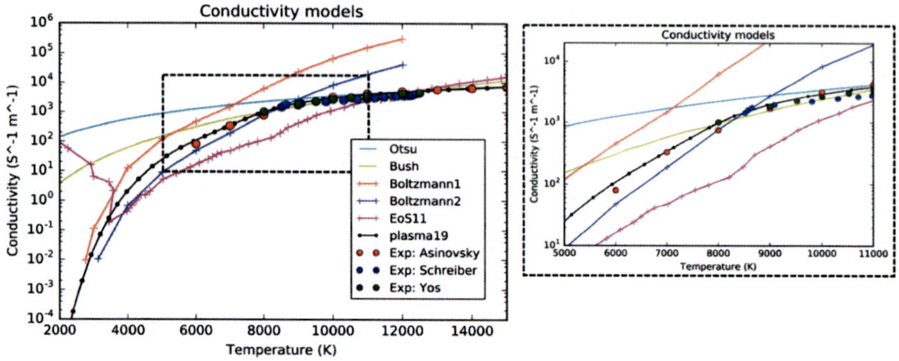

Fig. 2. Electrical conductivity across low air-plasma temperature range at pressure = 1 atm. Comparison of σ computed by plasma19 with other models currently used in the literature and to experimental measurements (refer to text).

Comparison of the computed electrical conductivity across the extended plasma19 validity range shows very good agreement with experimental measurement - critically, in the low temperature range typical of hypersonic flight in the LTE regime. In Fig. 2, three sets of experimental measurements have been obtained from the works of Asinovsky et al. (1973) [22], Schreiber et al. (1971) [23] and Yos (1963) [24] where a low-temperature equilibrium plasma was generated via a stabilised DC arc applied to a column of air maintained at atmospheric pressure. Otsu [16] and Bush [25] use the simple analytic power law approximation defined in next section's Eq. 15 with exponents 2 and 4 respectively. It is clear that the analytic approximations used by Otsu and Bush produce reasonable estimations at temperatures above approximately 9,000 K, but diverge from experimental measurements as temperature decreases. Boltzmann solvers for electrical conductivity involve the complex solution of a large set of coupled ODEs and are computationally expensive. EoS11 is an 11-species equilibrium model as implemented by Villa et al. [26]. As can be seen, the non-physical species molar fractions over the low temperature range, and resultant anomalous electrical conductivity, render EoS11 as problematic for LTE MHD studies. The anomalous conductivity values at low temperatures are not only physically unrealistic, but sharp jumps in properties lead to numerical instability when integrating the MHD forcing source terms.

The efficiency of the overall code depends strongly on the efficiency of the interpolation access routine for the look-up table. Therefore, this work implements an efficient binary search method - finding indices in O(log(n)) operations - with bilinear interpolation between the state data of the tabulated indices.

Plasma19 has the advantage of leveraging the pre-computed σ values without the need for a supplementary conductivity formulation, rendering the model as highly efficient. The smooth and accurate conductivity values computed by plasma19 over the low temperature range enable the direct and realistic simulation of LTE MHD flow control for this regime.

3 Numerical Methods

The model of this work is a new research code developed within the AMReX framework [27]. The underlying computational mesh is regular Cartesian with hierarchical adaptive mesh refinement (AMR), where the domain is subdivided into blocks as represented in Fig. 3. This framework permits massively parallel computation of the domain sub-blocks, resulting in efficient and highly scalable computation.

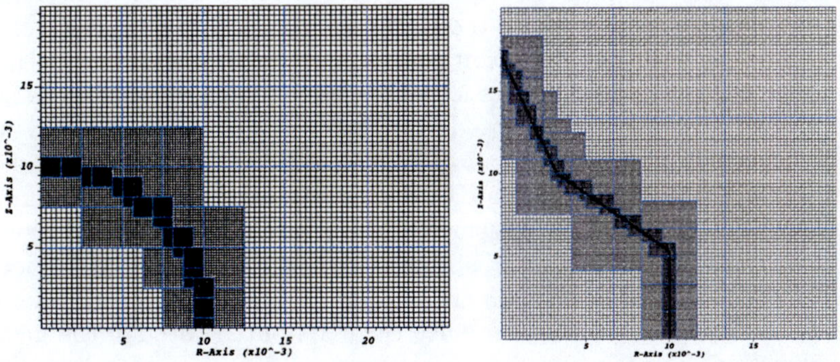

Fig. 3. Cartesian mesh configuration with two levels of hierarchical AMR. Blue borders show subdivision of the domain into 'blocks' which can be computed in parallel.

Geometry is implemented via a rigid body Ghost Fluid Method (GFM), such as that implemented by Bennett et al. [28], based on the approach of Sambasivan and UdayKarmar [29]. By implementing embedded boundary geometry via a rigid body GFM, the mesh is not subject to distortions and skewing which sometimes occur in the case of body fitted meshes, especially as geometries become more complex, and the adaptive refinement approach ensures high resolution is achieved automatically and locally in the vicinity of high density gradients. This combination of methods therefore permits the accurate, robust and efficient modeling of complex geometries in the presence of strong shocks, with multiple levels of hierarchical AMR applied to achieve high effective resolution and solved efficiently via MPI-based computation with high scalability across many nodes.

The system is solved via an operator split strategy whereby the hyperbolic and diffusive fluxes are computed successively with the result of each timewise integration input as the initial condition for the subsequent operation.

Fluxes are also dimensionally split in this manner. Cylindrical and MHD source terms are integrated via simple Euler time step integrations.

The stable time-step is restricted as the minimum of the hyperbolic and diffusive stable time steps, given by:

$$\Delta t = \min[\Delta t_{hyp}, \Delta t_{diff}] \tag{12}$$

$$= \min[\frac{\mathrm{CFL} \cdot \Delta x_{d,i}}{S_{max,i}}, \frac{(\Delta x_{d,i})^2}{2 \cdot \max[\frac{\mu_i}{\rho_i}, \frac{\zeta_i}{(\rho c_p)_i}]}] \tag{13}$$

where $\Delta x_{d,i}$ is each dimensional spatial step size. Under the operator split solution strategy the explicit time-marching evolution is stable in 2-dimensions and including all source terms for CFL = 0.9.

The hyperbolic system fluxes are evaluated numerically via a complete wave approximate Riemann solver (HLLC) with high order extension (MUSCL) which is second order accurate in space and first order accurate in time. Complete wave Riemann solvers are susceptible to the Carbuncle instability for test cases in the hypersonic regime. A shock detection with transverse flux stabilisation method is implemented for effective suppression of Carbuncle instability. The transverse flux stabilisation employs a weighted combination flux (complete and incomplete wave solver fluxes) in the vicinity of strong, normal, grid-aligned shocks - an approach which is similar to the HLLE++ method described by Tramel [30], but treated here as a directly augemented flux rather than system eigenvalue modification, and meliorated in this case to a Cartesian mesh with hierarchical AMR. Viscous derivative terms are evaluated using the central differencing method at cell interfaces with divergence then computed through the finite volume construction. The transient solver can be used for time accurate simulation of transient test cases, and also reaches a stable and convergent steady state (converges to equivalent result above sufficient resolution) for steady test cases.

4 Validation

4.1 MHD System Validation Under Ideal Gas Law

For the purpose of validating the underlying resistive MHD model, the model is first reduced to its inviscid Euler equivalent, closed with an ideal gas equation of state for comparison with the validation test cases in the literature. The most prevalent validation test is the case originally studied by Poggie & Gaitonde [18], and replicated subsequently by Damevin & Hoffmann [31] and most recently by Fujino et al. [32].

The free-stream conditions for the flow over a 10 mm radius sphere are:

Free-stream conditions:

M = 5, Re = 80,000, $T_\infty = 100$ K,

$V_\infty = 1002.25$ m/s, $\rho_\infty = 0.0798\, kg/m^3$, $p_\infty = 2290.85$ Pa

The MHD effect is shown to increase shock stand-off distance with increased magnetic interaction parameter (Q_{MHD}) defined as:

$$Q_{MHD} = \frac{\text{Lorentz force}}{\text{Fluid inertial force}} = \frac{\sigma_0 B_0^2 R_0}{\rho_\infty V_\infty} \tag{14}$$

where σ_0 is taken to be the maximum post-shock conductivity (in the stagnation region), B_0 is the maximum absolute magnetic field strength outside of the vehicle, and the characteristic length R_0 is the radius of the spherical body.

Since the gas conductivity is not captured by the ideal gas EoS, a supplementary analytic equation for conductivity is required. Models from various studies [14,16,18] assume an approximate conductivity model of the following form:

$$\sigma = \sigma_0 \left(\frac{T}{T_{ref}} \right)^n \tag{15}$$

where T_{ref} is taken to be the maximum temperature in the stagnation region.

Poggie & Gaitonde investigate different valued exponents (n = 0–4), and conclude that the exponent, and conductivity model more generally, has a notable impact on the MHD control effects. For these initial set of validation tests they assume a constant conductivity of $\sigma_0 = 300(\Omega m)^{-1}$, i.e. n = 0, in the post-shock region. Magnetic field strength is varied corresponding to Q_{MHD} values of 0–6.

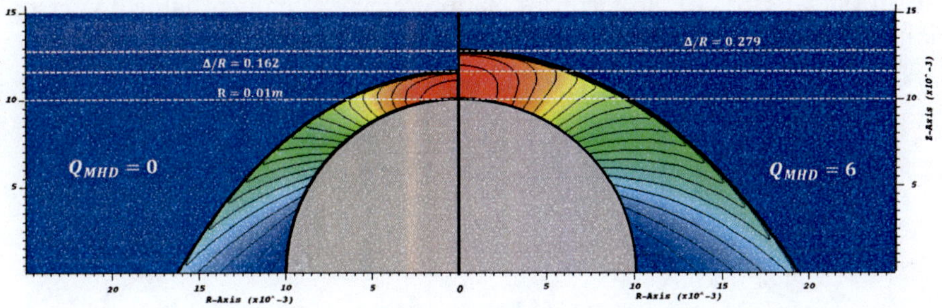

Fig. 4. Pressure profiles with contours as computed in this work for $Q_{MHD} = 0$ and $Q_{MHD} = 6$.

The full pressure profile for the $Q_{MHD} = 0$ and $Q_{MHD} = 6$ cases are shown in Fig. 4 and are in very good agreement with the literature. Examining the result quantitatively, Fig. 5 shows that the shock stand-off distances computed under 6 different magnetic field strengths exhibit excellent agreement with literature results and theory.

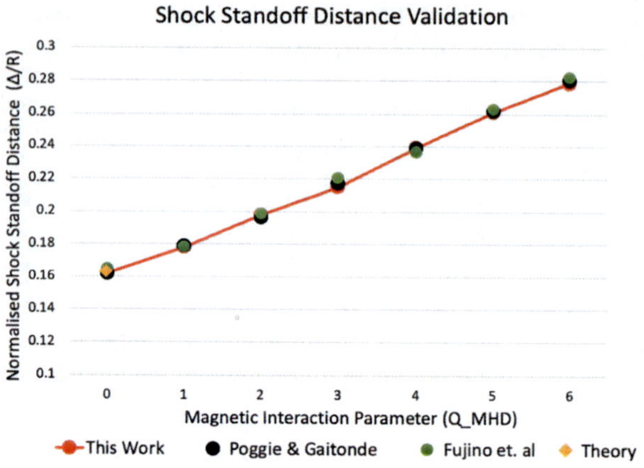

Fig. 5. Quantitative comparison of shock stand-off distance for this work vs previous studies for $Q_{MHD} = 0–6$.

4.2 Hypersonic Double Cone Tests

The hypersonic double cone tests are famously used for the validation of Navier-Stokes codes to model non-simple geometries and capture boundary layer shock wave interactions. Experimental data has been obtained from a large number of tests in the CUBRC LENS facilities [33] producing a high enthalpy, hypersonic, laminar flow with shock wave boundary layer interactions.

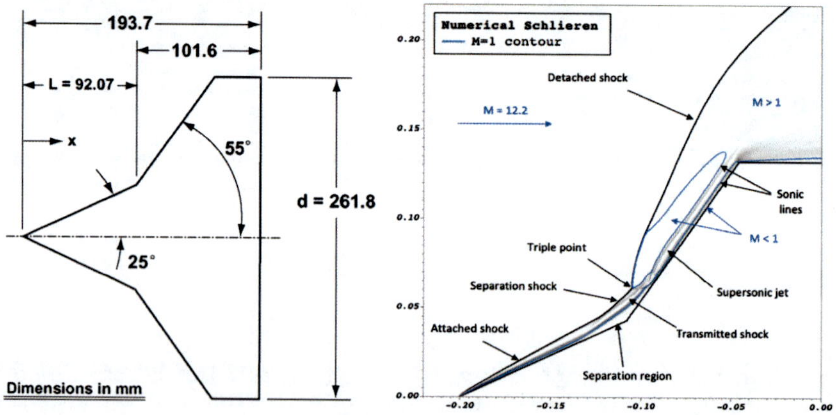

Fig. 6. Left: geometry specifications of the CUBRC 25/55° double cone model, right: numerical Schlieren result of this work from test conditions of the CUBRC run 1 flow.

Figure 6 shows the model dimensions for the CUBRC experiments, and a numerical Schlieren result of the run 1 conditions, demonstrating the ability of the Navier-Stokes rigid body GFM model to capture the complete feature formation at steady state.

To validate the system quantitatively a pressure trace is taken along the surface of the vehicle and compared with the results of the experiments and other simulations. The first case considers the test run 7 conditions for non-reactive nitrogen (N_2) gas.

<u>CUBRC run 7 conditions:</u>

$$\text{M} = 15.6,\ V_\infty = 2073 \text{ m/s},\ T_\infty = 42.6 \text{ K},\ T_w = 300 \text{ K},$$
$$\rho_\infty = 1.57 \times 10^{-4}\ kg/m^3,\ p_\infty = 2.23 \text{ Pa, Gas} = N_2$$

For other cases including air test runs, where real reactive gas effects are at play, perfect quantitative agreement between experiment and simulation remains illusive and has been the subject of many studies: [33–35] among others. Though all models struggle to exactly match experiments in terms of heat flux and pressure distribution, generally good agreement is found between numerical models. Zuo recently (2021) [36] showed that for runs with air as the test gas, similar results are obtained between perfect gas, thermochemical equilibrium and non-equilibrium models in terms of qualitative flow structure and surface properties. Another recent (2020) study by Holloway [37] showed that equilibrium and non-equilibrium models demonstrated reasonably good agreement in terms of overall shock structure and property fields for the air-gas composition test case.

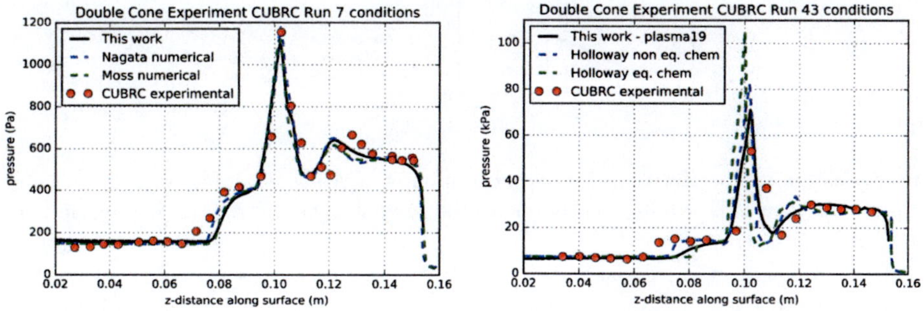

Fig. 7. Pressure trace along surface of double cone: left: CUBRC run 7 compared with experimental result and simulations of Nagata et al. [38] and Moss et al. [39], right: CUBRC run 43 compared with experimental result and simulations of Holloway [37] under different thermochemistry assumptions.

The air test case of CUBRC run 43 is studied by Holloway [37] and compares results for 11 species equilibrium and 11 species non-equilibrium (two-temperature) models. The quantitative results of these models are compared to the result of this work and the experimental data, as shown in Fig. 7.

<u>CUBRC run 43 conditions:</u>

$$\text{M} = 8.87,\ V_\infty = 4267 \text{ m/s},\ T_\infty = 576 \text{ K},\ T_w = 296.2 \text{ K},$$
$$\rho_\infty = 2.134 \times 10^{-3}\ kg/m^3,\ p_\infty = 352.77 \text{ Pa, Gas} = \text{Air}$$

Whilst none of the models perfectly capture the position of the transmitted shock upstream of the separation region, the result of this work shows very good agreement with prevailing models which account for real gas effects. In fact the peak pressure and post-peak dip in pressure computed from this work are seen to match more closely with the experimental data. Properties along the secondary surface of the double done are more critical for the ensuing MHD double cone test, since magnetic interaction dominates downstream of the kink.

5 Results

5.1 Hypersonic Double Cone with Electrodynamic Effect

Wasai et al. [40] conducted an experiment on magnetohydrodynamic control of shock interactions over a 25°/55° double cone geometry (Fig. 8) in hypersonic flow where results were obtained to measure the MHD flow control effect. Simulations of the experimental conditions have been conducted by Nagata et al. [38,41].

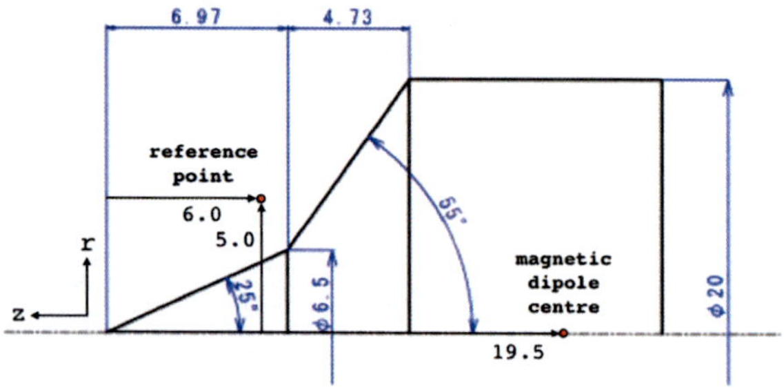

Fig. 8. Geometry specifications for the experiments of Wasai et al. [40] with dimensions in mm. The case of B = 0.36 T is the measured magnetic field strength at the reference point.

Results of their numerical simulations showed that for a magnetic field strength matched with the experimental condition the MHD effect was unobservable in the computed steady state flow field solution. However, the experiment does in fact reveal a small but observable increase in shock stand-off distance due to MHD flow control. By initialising the magnetic field strength in the numerical simulation to be much higher than that of the experiment, Nagata et al. were able to demonstrate the qualitative effect an imposed magnetic field has on the flow, which agrees with the qualitative effect observed in the experiment.

The experimental inflow test gas was pre-heated, and Wasai [40] and Nagata [41] consider it to have reached a state of thermochemical equilibrium, determining the initial (single) temperature and species composition as such. Experimental results for the test run of flow velocity of 11.6 km/s are presented by Wasai [40] and therefore used as the test flow condition for this work.

Test run conditions:

$$\mathrm{M} = 5.6,\ V_\infty = 11.6\ \mathrm{km/s},\ T_\infty = 6110\ \mathrm{K},\ T_w = 300.0\ \mathrm{K},$$
$$\rho_\infty = 2.52 \times 10^{-3}\ kg/m^3,\ p_\infty = 7.22\ \mathrm{kPa},\ \mathrm{Gas} = \mathrm{Air}$$

Importantly the flow is characterised with the following parameters:

$$\text{Reynolds number} = 1.77 \times 10^3$$
$$\text{Magnetic Reynolds number} = 0.2$$
$$\text{Hall parameter} = 0.61$$

Additionally, the model surface is treated as an insulated surface. Therefore it can be assessed that the governing model assumptions of low magnetic Reynolds number, and simplified Ohm's law neglecting Hall effect are valid for this test case.

The increase in shock stand-off position is observed via a brightness intensity rendering of the flow field with and without the imposed magnetic field, and also, by taking a trace through lines parallel to the z-axis of symmetry at different radial distances (r = 4 mm, 8 mm and 10 mm) along the model to determine shock position more precisely. These results are presented for the 11.6 km/s test case in reference [40], and the increase in leading shock position due to MHD effect of the imposed magnetic field is measured from these graphs, presented with uncertainty margins in the table of Fig. 9.

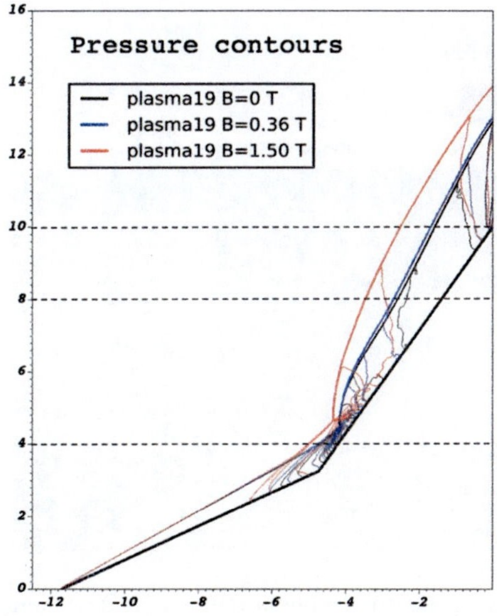

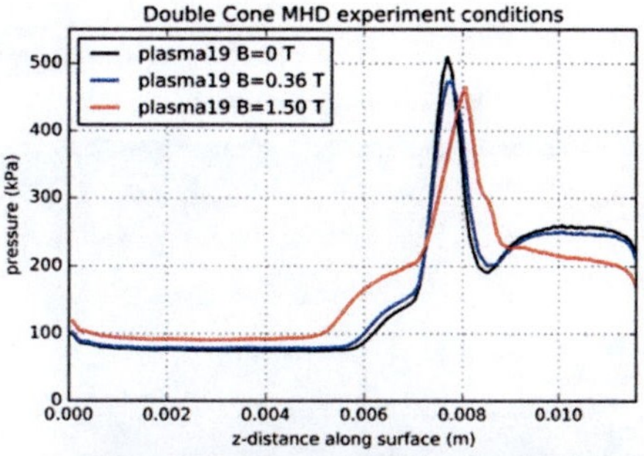

Radial distance	MHD shock stand-off enhancement – Experiments (mm)	MHD shock stand-off enhancement – This work (mm)
R = 4 mm	0.046 - 0.264	0.052
R = 8 mm	0.081 - 0.366	0.117
R = 10 mm	0.079 - 0.373	0.115

Fig. 9. Left: contours of pressure for the case without magnetic field, magnetic field initialised to experimental value B = 0.36 T, and a stronger magnetic field of B = 1.5 T intensity, with radial lines of 4 mm, 8 mm, 10 mm marked. Upper right: comparison of pressure trace along surface. Lower right: increase in shock stand-off distance from the series of experiments by Wasai et al. [40]- measured range of values from experimental graphs, compared with measured increase in leading shock stand-off distance computed from this work with experimental value of magnetic field strength B = 0.36 T.

In the table of Fig. 9, the measured increase in leading shock stand-off from the simulations of this work are compared to the measured increase from the trace taken along the radial lines in the results of Wasai et al. [40]. Results show an increase in shock stand-off distance in line with the range of results measured from the experiments. The shock stand-off enhancement effect can be visualised through the pressure contour plot of Fig. 9.

The MHD effect as depicted by the overlaid pressure contours for different magnetic field strength (Fig. 9) agrees very well in terms of qualitative phenomenology of the simulations Nagata et al. [38,41]. The change in pressure trace property along the vehicle surface also shows good qualitative agreement with traces computed by Nagata et al. Further, the predictions of this work are quantitatively in-line with experimental measurements for matched Magnetic field strength. Therefore, the model of this work appears highly suitable to capture realistic MHD flow control effects in hypersonic flows with complex geometries.

The model permits us to investigate many other properties of interest in the weakly ionised plasma layer formed around the vehicle, both in the absence of, and under the influence of, an applied magnetic field. A key property of interest for many applications is electron number density (N_e). Figure 10 shows the computed electron distribution for the case with and without magnetic field activated.

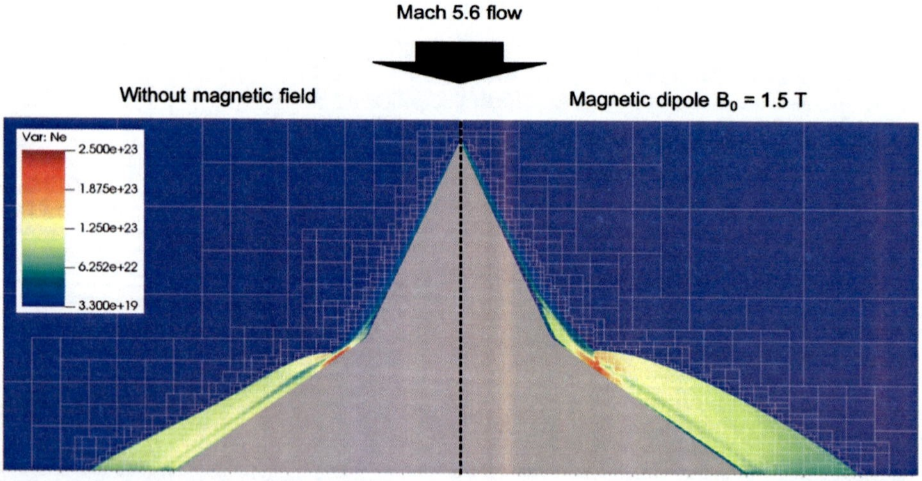

Fig. 10. Comparison of steady state solution for electron number density (N_e) without magnetic field (left plane) and for the case of applied magnetic field of strength B = 1.5 T. Regions of mesh refinement are shown by white block borders.

There are a number of additional electromagnetic properties which result directly from the activation of the magnetic field. Figure 11 shows (for the B = 1.5 T case) the induced electric current in the circumferential directions, revealing there exists an anticlockwise current flow in the supersonic shock regions of the flow structure and a clockwise current flow in the separation and subsonic regions. The Lorentz force magnitude with directional vector plot shows peak magnetic

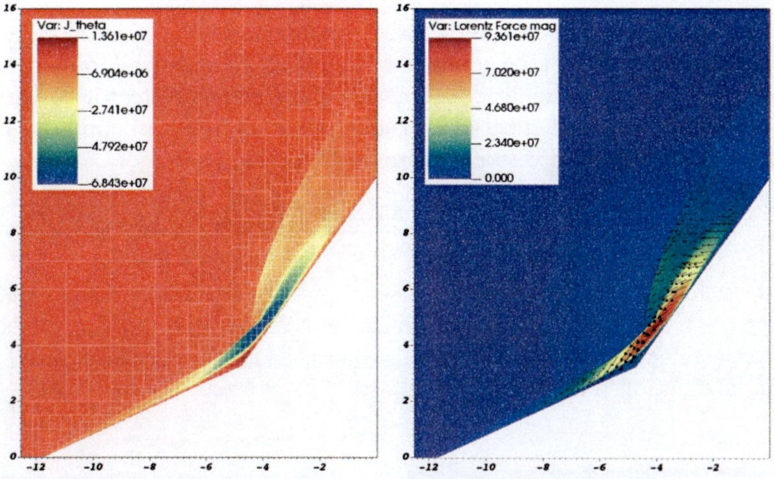

Fig. 11. Key electromagnetic properties- left: circumferential component of electric current density, and right: Lorentz forcing magnitude and directional vectors.

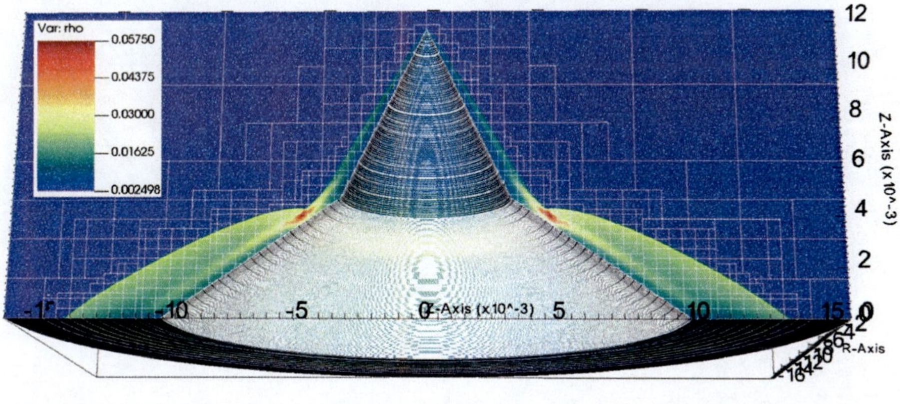

Fig. 12. Full 3D rendering of computed density field for case of imposed magnetic field strength $B = 1.5\,T$.

forcing in the region of peak anticlockwise electric current flow. The directional vectors indicate why a notable movement in separation shock position was previously observed in the pressure contour plot. Resolving these parameters reveal useful insights for understanding the progression of shock structure phenomenology enacted by electrodynamic influence (Fig. 12). Since the magnetic interaction parameter is scaled by flow momentum, we see from this result that low velocity flow regions are determined as having high magnetic interaction.

Assessment of LTE Validity for MHD Effects

LTE validity is primarily determined by the comparison of the rate of chemical reaction to the rate of fluid transport. Specifically, in the case of quantifying MHD effects, the relevant consideration is that the system tends toward equilibrium specifically where the magnetic interaction is dominant. Therefore to assess the appropriateness of the LTE model for the MHD flow control double cone test case, the magnitude of magnetic interaction parameter (indicator of where the flow is electromagnetically affected by the imposed **B**-field) is compared with an indicative measure of fluid residence time (τ_r), defined as:

$$\tau_r = \frac{\text{vehicle length scale}}{\text{total velocity}} = \frac{L_0}{|\mathbf{v}|} \tag{16}$$

Where the velocity approaches a very low magnitude (such as subsonic and stagnation regions), and therefore the residence time is high, the flow can be assumed to be approaching equilibrium. Figure 13 shows the comparison of defined τ_r and magnetic interaction parameter Q:

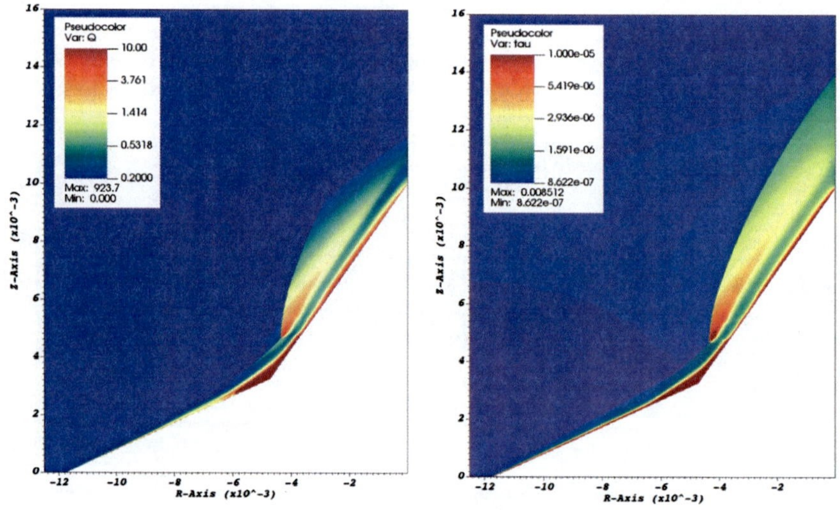

Fig. 13. Left) magnetic interaction parameter: $Q = (\sigma |B|^2 R_0)/(\rho |\mathbf{u}|)$, compared with right) residence time indicator of regions approaching equilibrium: $\tau_r = \frac{L_0}{|V|}$

As can be seen from the comparison in Fig. 13, the regions with dominant magnetic interaction, predominantly producing the measured MHD effects, map very well with the regions where the flow is expected to approach equilibrium based on the residence time indicator τ_r. This is an important relationship to draw: regions which tend to zero velocity, also tend towards equilibrium in terms of the chemical state, and correlate with the key regions of the flow structure where the Lorentz forcing magnitude vs flow momentum is high such that electromagnetic forcing affects the flow.

6 Conclusions

This work reports on a new research code with a methodology for the realistic and efficient simulation of weakly ionised air plasmas in the LTE regime, and associated MHD phenomena. The Navier-Stokes resistive-MHD system (low Re_m assumption) is solved for simple and complex geometries with imposed magnetic fields. MHD flow control dynamics were validated in line with literature results and complex double cone flows with shock wave boundary layer interaction and real gas chemistry were effectively captured as compared with available experimental data.

The model employs a highly efficient implementation whereby plasma19 EoS leverages pre-computed state data, and geometry is modelled via an embedded boundary Ghost Fluid Method, this facilitates an underlying mesh consisting of hierarchical AMR within the AMReX framework which offers a very high degree of parallelism and scalability. On equivalent meshes, the model run time is close (within 24%) to the efficiency of the simple ideal gas model often employed in previous studies for its ease of computation. Unlike equivalent ideal gas models, this model is able to directly compute electron number density and electrical conductivity, which are important for the prediction of plasma properties and MHD effects.

Simulations were produced of the experimental test conditions producing a non-simple shock structure around a double-cone geometry under electrodynamic influence. Numerical results were produced in line with experiments for shock structure and increase in stand-off distance, therefore improving upon previous numerical predictions which could not resolve an MHD effect for the experimental value of magnetic field intensity.

There are many other MHD applications of interest under this regime, with high relevance to emerging flight control and reusable space technologies.

Acknowledgements. I (H.A. Muir) would like to acknowledge the General Sir John Monash Foundation and the Laboratory for Scientific computing, by whom the work has been funded. I would also like to acknowledge and thank Leoni Pugh for working with us to develop Fig. 1 of this paper.

References

1. Bityurin, V.A., Bocharov, A.N.: MHD flow control in hypersonic flight. In: AIAA/CIRA 13th International Space Planes and Hypersonics Systems and Technology. Institute for High Temperatures of Russian Academy of Sciences, Moscow, Russia (2005)
2. Edwards, T.A.: Fluid/chemistry modeling for hypersonic flight analysis. Comput. Math. Appl. **24**(5–6), 25–36 (1992)
3. Chalot, F., Hughes, T.J.: A consistent equilibrium chemistry algorithm for hypersonic flows. Comput. Methods Appl. Mech. Eng. **112**(1–4), 25–40 (1994)
4. Sawada, K., Dendou, E.: Validation of hypersonic chemical equilibrium flow calculations using ballistic-range data. Shock Waves **11**(1), 43–51 (2001)

5. Lani, A., Molnar, J., Abeele, D.V., Rini, P., Magin, T., érard Degrez, G.: Numerical study of elemental demixing in atmospheric entry flow regimes near local thermodynamic equilibrium. In: European Conference on Computational Fluid Dynamics, Von Karman Institute for Fluid Dynamics (VKI), Aeronautics and Aerospace Department (2006)
6. Fujino, T., Ishikawa, M.: Numerical simulation of control of plasma flow with magnetic field for thermal protection in earth reentry flight. IEEE Trans. Plasma Sci. **34**(2) (2006)
7. Anderson, J.D.: Hypersonic and High Temperature Gas Dynamics. McGraw-Hill, New York (1989)
8. Savino, R., Paterna, D., De Stefano Fumo, M., D'Elia, M.: Plasma-radiofrequency interactions around atmospheric re-entry vehicles: modelling and arc-jet simulation. Open Aerosp. Eng. J. **3**, 76–85 (2010)
9. Bisek, N.J.: Numerical study of plasma-assisted aerodynamic control for hypersonic vehicles, Ph.D. thesis, Department of Aerospace Engineering, The University of Michigan (2010)
10. D'Angola, A., Colonna, G., Gorse, C., Capitelli, M.: Thermodynamic and transport properties in equilibrium air plasmas in a wide pressure and temperature range. Eur. Phys. J. D **46**, 129–150 (2008)
11. D'Angola, A., Colonna, G., Gorse, C., Capitelli, M.: Thermodynamic properties of high temperature air in local thermodynamic equilibrium: II accurate analytical expression for electron molar fractions. Eur. Phys. J. D **65**, 453–457 (2011)
12. Träuble, F., Millmore, S., Nikiforakis, N.: An improved equation of state for air plasma simulations. Phys. Fluids **33**(3), 036112 (2021)
13. Goedbloed, J.P., Poedts, S.: Principles of Magnetohydrodynamics: With Applications to Laboratory and Astrophysical Plasmas. Cambridge University Press, Cambridge (2004)
14. Dias, F., Xisto, C.: Numerical computations of MHD flow on hypersonic and reentry vehicles. In: Proceedings of the ASME 2018 IMECE (2018)
15. Fujino, T., Matsumoto, Y., Kasahara, J., Ishikawa, M.: Numerical studies of magnetohydrodynamic flow control considering real wall electrical conductivity. J. Spacecraft Rockets **44**(3), 625–632 (2007)
16. Otsu, H., Konigorski, D., Abe, T.: Influence of hall effect on electrodynamic heat shield system for reentry vehicles. AIAA J. **48** (10)
17. Matsuda, A., Otsu, H., Kawamura, M., Konigorski, D., Takizawa, Y., Abe, T.: Model surface conductivity effect for the electromagnetic heat shield in re-entry flight. Phys. Fluids 20 (2008)
18. Poggie, J., Gaitonde, D.V.: Magnetic control of flow past a blunt body: numerical validation and exploration. Phys. Fluids 1**14**, 127103 (2008)
19. Khan, O.U., Hoffmann, K.A., Dietiker, J.: Computational aspects of high-speed flows with applied magnetic field. IEEE Trans. Magn. **42**(3), 389–397 (2006)
20. Yoshino, T., Fujino, T., Ishikawa, M.: Possibility of thermal protection in earth re-entry flight by MHD flow control with air-core circular magnet. Trans. Electr. Electron. Eng. (4), 510–517 (2009)
21. Colonna, G.: Improvements of hierarchical algorithm for equilibrium calculation. Comput. Phys. Commun. **177**, 493–499 (2007)
22. Asinovsky, E., Kirillin, A., Pakhomov, E., Shabashov, V.: Experimental investigation of transport properties of low-temperature plasma by means of electric arc. Proc. IEEE **59**(4), 592–601 (1971)
23. Schreiber, P., Hunter, A., Benedetto, K.: Electrical conductivity and total emission coefficient of air plasma. AIAA J. **11**(6), 815–821 (1973)

24. Yos, J.: Transport properties of nitrogen, hydrogen, oxygen, and air. AVCO/RAD-TM **63**(7) (1963)
25. Bush, W.B.: Magnetohydrodynamic-hypersonic flow past a blunt body. J. Aerosp. Sci. **25**(11), 685–690 (1958)
26. Villa, A., Malgesini, R., Barbieri, L.: A multiscale technique for the validation of a numerical code for predicting the pressure field induced by a high-power spark. J. Phys. D Appl. Phys. **44** (165201)
27. Almgren, A., Bell, J. (primary developers): AMReX - a software framework for massively parallel, block-structured adaptive mesh refinement (AMR) applications, opensource (2020). https://amrex-codes.github.io/amrex/
28. Bennett, W.P., Michael, L., Nikiforakis, N.: Cartesian cut-cell and GFM approaches to free-surface and moving boundary interaction. In: 54th AIAA Aerospace Sciences Meeting 0602 (2016)
29. Sambasivan, S.K., UdayKumar, H.S.: Ghost fluid method for strong shock interactions part 2: immersed solid boundaries. AIAA **47**(12), 2923–293 (2009)
30. Tramel, R., Nichols, R., Buning, P.: Addition of improved shock-capturing schemes to overflow 2.1, p. 3988 (2009)
31. Damevin, H.-M., Hoffmann, K.A.: Numerical simulations of magnetic flow control in hypersonic chemically reacting flows. J. Thermophys. Heat Transf. **16**(4), 498–507 (2002)
32. Fujino, T., Matsumoto, Y., Kasahara, J., Ishikawa, M.: Numerical analysis of reentry trajectory coupled with magnetohydrodynamics flow control. J. Spacecraft Rockets **45**(5), 911–920 (2008)
33. Holden, M., Wadhams, T., Boyd, I., Wang, W.: CFD validation for hyperconic flight: hypersonic double-cone flow simulations. In: 40TH AIAA Aerospace Sciences Meeting and Exhibit, Reno, NV, pp. 14–17 (2002)
34. Hao, J., Wang, J., Lee, C.: Numerical simulation of high-enthalpy double-cone flows. AIAA J. **55**(7), 2471–2475 (2017)
35. Hao, J., Wen, C.-Y.: Numerical investigation of oxygen thermochemical nonequilibrium on high-enthalpy double-cone flows. Int. J. Heat Mass Transf. **127**, 892–902 (2018)
36. Zuo, F.-Y., Hu, S.-L.: Thermochemical non-equilibrium effects on aerothermodynamic prediction of laminar double-cone flow. Acta Astronautica (2021)
37. Holloway, M.E., Hanquist, K.M., Boyd, I.D.: Assessment of thermochemistry modeling for hypersonic flow over a double cone. J. Thermophys. Heat Transf. **34**(3), 538–547 (2020)
38. Nagata, Y., Yamada, K., Abe, T.: Numerical investigation of the electrodynamic flow control around the double-cone configuration. In: 49th AIAA Aerospace Sciences Meeting including the New Horizons Forum and Aerospace Exposition (2011)
39. Moss, J.N., Bird, G.A.: Direct simulation Monte Carlo simulations of hypersonic flows with shock interactions. AIAA J. **43**(12), 2565–2573 (2005)
40. Wasai, K., Makino, H., Nagata, Y., Hiraoka, K., Yamada, K., Abe, T.: Electrodynamic control of shock interactions in a 25°/55° double cone model in hypersonic flow. In: 48th AIAA Aerospace Sciences Meeting Including the New Horizons Forum and Aerospace Exposition (2010)
41. Nagata, Y., Yamada, K., Abe, T.: Hypersonic double-cone flow with applied magnetic field. J. Spacecraft Rockets **50**(5), 981–991 (2013)

Author Index

R. King and D. Peitsch (Eds.): AFCC 2021, NNFM 152, pp. 357–358, 2022.
https://doi.org/10.1007/978-3-030-90727-3